MAKING THE UNSEEN VISIBLE

ABOUT THE COVER ART

Fat Man by Yukiyo Kawano is part of her *Suspended Moment* project (2016–) and is reproduced courtesy of the artist. The shape of the Fat Man bomb was created from a kimono from Hiroshima owned by Kawano's grandmother and sewn together with the artist's own hair, melding the DNA of generations of atomic bomb survivors.

Fat Man (after Matsuo Bashō)

The sculpture is covered in excerpts from *The Narrow Road to The Deep North* (奥の細道). The celebratory journal written by Haiku master Bashō in 1689 is full of tasks for locating difficult-to-recognize beauty, and graceful acceptance of restraint, inconvenience, and uncertainty, through his trip to the northern part of Japan that includes modern day Fukushima, the site of nuclear disaster since the 2011 Daiichi nuclear plant explosions.

Nuclear accidents forever change the land, divide the community, and suppress the already vulnerable. The artwork was made for an exhibition in Richland, WA, in 2016, part of Kawano's storytelling about people who are suffering from radiation exposure to connect stories of the hibakusha with the downwinders of atomic bomb production and the downwinders of the nuclear disaster in Fukushima, Japan.

With a close look, viewers witness the threads of crude hair that was used to put the pieces together which suggests the DNA that I inherited from my mother, and from my grandfather who was in Hiroshima, under the mushroom cloud, on that critical day.

Yukiyo Kawano
http://yukiyokawano.com

Cover photograph by Stephen Miller

Making the Unseen Visible

SCIENCE AND THE CONTESTED HISTORIES OF RADIATION EXPOSURE

Edited by

Jacob Darwin Hamblin and Linda Marie Richards

Oregon State University Press Corvallis

Library of Congress Cataloging-in-Publication Data is available from the Library of Congress.

∞ This paper meets the requirements of ANSI/NISO Z39.48-1992 (Permanence of Paper).

First published in 2023 by Oregon State University Press
Printed in the United States of America

Oregon State University Press
121 The Valley Library
Corvallis OR 97331-4501
541-737-3166 • fax 541-737-3170
www.osupress.oregonstate.edu

Oregon State University Press in Corvallis, Oregon, is located within the traditional homelands of the Mary's River or Ampinefu Band of Kalapuya. Following the Willamette Valley Treaty of 1855, Kalapuya people were forcibly removed to reservations in Western Oregon. Today, living descendants of these people are a part of the Confederated Tribes of Grand Ronde Community of Oregon (grandronde.org) and the Confederated Tribes of the Siletz Indians (ctsi.nsn.us).

Contents

PART II: INTERNATIONAL DISCOURSE ON HARM

PART III: REMEMBERING AND FORGETTING

MAKING THE UNSEEN VISIBLE

Introduction

JACOB DARWIN HAMBLIN AND LINDA MARIE RICHARDS

Invisibility is one of the defining features of radiation. It is what made the discovery of x-rays in 1895 so remarkable—here was a physical phenomenon that could not be seen or heard, yet evidence of it could be detected indirectly, as when Wilhelm Röntgen first produced a photographic plate of his wife Anna's skeletal hand. The invisibility of radiation from x-rays, radioactive materials, and from the process of nuclear fission also has captured our interest because of the harm to human and animal bodies.[1] After the atomic bombings of Hiroshima and Nagasaki, the vague term "radiation sickness" emerged as a condition afflicting Japanese survivors of the blast itself, and anyone connected with the construction of atomic bombs knew that there were special harms from processes they might never see. In 1946, during one laboratory test in the New Mexico desert, Canadian physicist Louis Slotin accidentally killed himself simply by letting a screwdriver slip, losing control of an experiment with plutonium. He seemed uninjured at first, but vast amounts of radiation had touched him. He soon began to vomit, and his skin later blistered—he was dead nine days later from invisible harm.[2]

The links between radiation and health effects have been contested for many decades, especially in situations where large numbers of people have gotten sick, either from radiation sickness in the short term or cancers in the long term. Among the most dismissive of such harms was Alvin C. Graves, the scientific director of a 1951 series of United States nuclear tests at Enewetak, in the Marshall Islands. He had been in the same room as Louis Slotin, but had not died. After receiving the highest dose of any of the survivors, he had the same disturbing side effects as some of the Japanese survivors of the atomic bombings: vomiting, fever, chronic fatigue, and loss of hair. His sperm count went to zero, his skin was red, and he developed cataracts. He convalesced for about six months before recovering his strength and returning to work.[3] By the time he went to participate in the Enewetak experiments, he believed he had completely recovered. Indeed he became an outspoken skeptic about others' claims of illness from radiation exposure, and in the early 1950s he met with civic leaders and university faculty members in Nevada and Utah to reassure them that the dangers from radioactive fallout from nearby atomic tests were exaggerated.[4] Perhaps because he had survived one of the most egregious cases of accidental exposure, he felt he had earned the right to be dismissive of others' concerns.

For Graves, time would tell the tale. In 1955, nine years after the Los Alamos accident, he experienced a myocardial infarction (heart attack). He seemed to recover, but never felt normal afterward, gained weight, and had elevated cholesterol. His doctors diagnosed this as myxedema, an affliction of the thyroid gland. Through treatment with a thyroid hormone, he felt better; but in 1964 he was surprised to see that a medical x-ray revealed his heart to exceed the upper limits of normal size by some 17%. He suffered a fatal heart attack in 1966, at the age of 54.[5] An autopsy revealed his heart to be greatly enlarged. His coronary arteries had severe arteriosclerosis, commonly thought of as the "hardening" of the arteries. His testes had degenerated considerably. As for the thyroid gland, it "was so atrophied that it was difficult to identify." Because Graves's father also had suffered a heart attack, however, the scientists were unwilling to attribute his death to radiation exposure. "Since the thyroid gland is readily destroyed by radiation, one might suspect that the myxedema of Case 4 [Graves], presumably radiation induced, may have promoted his coronary disease by elevating the blood cholesterol. This could well have precipitated the first heart attack. . . . Our statistician friends, however, assure us that it is dangerous to draw definite conclusions from such small numbers of cases in such limited populations, as there may be etiologic factors other than radiation exposure." Thus, as Graves himself had done to others, his colleagues now stopped short of acknowledging the link between radiation exposure and the state of his body.[6]

We mention the death of Graves because his story shows us how easily the ties between radiation exposure and health harm can be contested or even cast aside. Invisibility does not just apply to radiation itself, but to the narratives we create about the lives of people connected to it. In this edited collection, what we mean by invisibility is in part about science, with evidence of harm so blurred by other comorbidities that it proved immensely frustrating for those with thyroid cancer or other ailments to prove in court that their conditions were due to living near a nuclear facility, participating in a bomb test, or working in a uranium mine.

But we also mean invisibility in how we tell stories about the past. Whom do we see, and what voices do we hear? We know now, for example, that atomic veterans received treatments for numerous health ailments that were not officially attributed to atomic tests. For soldiers who also served in Vietnam, it was easier to get compensation for exposure to Agent Orange than to bother trying to fight the uphill battle of proving their physical problems stemmed from nuclear tests. Will that harm be invisible in our historical narratives, even as it is invisible to science?[7] We also know that dosimetry badges, so prevalent among atomic workers and soldiers in nuclear facilities and weapons tests, were not commonly distributed to miners. The uranium mining on Navajo land—or indeed in locations as wide-ranging as Madagascar, Czechoslovakia, and India—was invisible in discussions of safety, just as the miners themselves rarely feature prominently in narratives about nuclear weapons or nuclear power.[8]

What we now know about the second half of the twentieth century compels us to tell the story of radiation effects in ways that acknowledge invisibility. The end of the Cold War stimulated numerous efforts to achieve transparency about past human exposure to harmful, cancer-causing radiation. In the 1990s, operators of the Hanford nuclear site in Washington—where plutonium was produced for the US arsenal—launched an environmental dose reconstruction project and asked scientists to determine whether the site's releases of radioactivity into the air and nearby river were responsible for spikes in thyroid cancer in the Pacific Northwest.[9] Around the same time, scientists and government officials in Russia pointed out case after case of Soviet-era secrecy: harmful effluent in rivers near villages, widespread harm from the 1986 Chernobyl accident, and even the dumping of entire reactors at sea.[10] In the United States, investigative journalists exposed so many examples of Cold War–era involuntary human experimentation with radiation that the Clinton administration launched a massive project of its own to shine a light on it. It made documents available to the public and conducted oral histories of scientists from the early years of the Atomic Energy Commission.[11]

Despite such actions in the name of transparency, clashes soon arose about how to frame and interpret histories of radiation exposure. For example, the Hanford studies were meant to lay to rest public outcries by enrolling scientists, but their findings and authoritative statements were met with bitter criticism from different scientists and from other stakeholders—residents, activists, and nearby Native American tribal councils. Thousands sued the Hanford operators, embroiling the region in legal battles for more than two decades.[12] In Russia (and Belarus, near the Chernobyl site), the environmental politics of the 1990s later yielded to the authoritarian politics of downplaying the harm of reactor accidents and highlighting instead the importance of nuclear power as a form of national heritage.[13] And even as the Clinton investigations turned up cases of hidden harms, such as to the Marshall Islanders who lived adjacent to nuclear weapons tests in the South Pacific, they found many instances not of remorse but defensiveness.[14] Longtime Atomic Energy Commission official

Merril Eisenbud, impatient at the repeated calls for compensation to victims, observed: "Oh, I've got a great deal of sympathy for the Marshall natives. I think they were treated very badly. But enough is enough. I mean, they've been given a half a billion dollars, or something like that, and you know, they were victims in the backwash of a war that killed 55 million people. Those were terrible times."[15]

Such varied positions about radiation exposure convey conflicts in values, attitudes, and political goals, and these have only begun to manifest in historical narratives. Although the layperson might be tempted to imagine that scientific knowledge can put to rest any discrepancies in interpretation about radiation effects, that rarely has been the case. Scientific research about radiation, historically, has been integrated into a range of priorities by governments, industry, activists, and victims.[16] For example, in the 1950s, amid a global controversy about the effects of fallout from nuclear weapons tests, the National Academy of Sciences attempted the first definitive summary of the state of the science. The scientists found that oceanographers, geneticists, and health physicists were all asking different kinds of questions and arriving at different conclusions about the harms from the tests. And throughout, they tempered their statements about biological harm by acknowledging the need to protect the US's ability to conduct weapons tests and to encourage the nuclear power industry to grow.[17] Scholars of the nuclear sector have routinely pointed out that the science that emerged from it rarely came across as disinterested, detached work that could be trusted. Indeed, trust has been a key challenge to the nuclear industry over several decades. Even when legal challenges have favored nuclear sites, resentments lingered about the ways the science was framed to ignore voices of victims.[18]

Behind the legal battles over victim compensation are broader questions about the uses of science and the pursuit of justice, and these will remain long after the courts have had their say. Many disputes about radiation exposure already have been addressed in judicial decisions, have disappeared through legislation, or have literally died along with litigants as they have aged and passed away.[19] In our view, while legal cases may be closed, the historical research is only just beginning. Histories of radiation exposure reflect some of the most controversial and complex legal and political struggles of the second half of the twentieth century. They raise scores of environmental, scientific, human rights, health, and legal questions, and they connect to broader histories of science, activism, social movements, public policy, and diplomacy.

Connecting to environmental justice, we can point to the long-standing practice of mining uranium on Indigenous lands—on Navajo lands in the Southwest United States or aboriginal land in Australia, for example—and the rise in cancers among miners all over the world.[20] Touching on public health, we may look to the siting of nuclear reactors and explore the consequences of major accidents such as Three Mile Island (1979), Chernobyl (1986), and Fukushima (2011).[21] Scientists have argued for decades about the genetic consequences of exposure to radiation from the fallout of nuclear tests or to proximity to radioactive waste sites.[22] Diplomats have debated radiation exposure to negotiate key treaties such as the 1963 Limited Test Ban Treaty and the 1972 London Convention (banning ocean dumping of radioactive waste).[23] Environmental dimensions of radiation exposure are wide-ranging, raising questions about consequences for wildlife and humans living near nuclear power reactors, weapons production facilities, and waste disposal sites—including the health effects of those living downriver and downwind. Many of the victim advocacy groups describe themselves as "Downwinders," explicitly self-identifying their own bodies as transformed by the nuclear age.[24]

Our Intervention

The essays in this book explore nuclear history with such goals in mind, to make the unseen visible. They focus not just on invisible radiation, but also on overlooked people and unheard voices. Yes, radiation is famous for being undetectable to the human senses, and we rely on Geiger counters to tick loudly in its presence. In addition, understanding and documenting the myriad forms of radiation sickness—leukemia, various cancers, sterility, spontaneous abortions, thyroid and heart problems, early deaths, autoimmune disorders, and birth defects of many kinds—rendered radiation invisible because it was virtually impossible to show that any particular manifestation of illness could be linked to a radiation source.[25] The political and legal battles have rendered voices unheard and people unseen. Often these have been Indigenous peoples mining uranium, working-class people tasked with cleanup activities, military service people put into harm's way, or people scattered across large areas with no unified political voice. Even as we examine a phenomenon—radiation—that is itself not detectable without technology, we are also striving to draw attention to narratives that move human experiences from the unseen to the visible.[26]

The style of our intervention is to address a need in the literature for a hybrid treatment of the history of radiation exposure that captures new scholarly work along with conversations from key voices outside academia, across continents and historical contexts. The existing literature on radiation exposure is rich, yet it typically explores site-specific or individual stories of harm. There is excellent work on disasters such as Chernobyl and Three Mile Island, and even dramatic retellings in other media that capture important issues, such as HBO's 2019 mini-series *Chernobyl*.[27] We see fewer synthetic treatments of radiation exposure that attempt to draw together strands that are common across contexts. One excellent effort to do so was Arjun Makhijani and his colleagues' 1995 *Nuclear Wastelands*, written from a critical scientific standpoint.[28]

Unlike Makhijani and colleagues' laudable work, our aim is to focus on humanities scholarship, particularly history, to highlight conceptual tensions that emerge as we tell the multifaceted stories of radiation exposure. We aim to identify key narratives and voices, to make the volume useful as an introduction to the topic, and we also will make a scholarly intervention about themes in nuclear history that connect disparate times and geographic spaces. We are inspired by the work of Kate Brown, whose 2013 book *Plutopia* was able to draw out themes uniting two communities near sites of plutonium production (Hanford in the United States; Mayak in the Soviet Union) despite dramatic differences in political ideology, economic system, and numerous other contrasts.[29] We hope to highlight themes that empower general readers to think about the issues while also informing future historiography in academic scholarship.[30] Here, by combining several site- and subject-specific essays in a collected volume, along with reflective essays, we offer a geographically diverse collection that draws together themes about the contested histories of radiation exposure.

Most of the contributors to the volume have participated in one of our three annual workshops connected to the Downwinders Project at Oregon State University. This was a multi-year study (2017–2021) funded by the National Science Foundation to collect oral histories, support archival development, and to generate scholarship about problems related to radiation effects. The participants in our workshops were scholars, activists, scientists, and others who shared lived experiences of radiation exposure. After the first workshop, we published some of the scholarship in a special issue of the *Journal of the History of Biology* on "telling the stories" of radiation exposure.[31] We were proud of this special issue, but we also felt that it lacked the inclusive spirit of our workshops because it did not draw on any of the voices outside conventional academia. Participants in our second and third workshops, drawn from scholars in several different countries, were keen to work together on a book that also included reflective essays, with the work of historians and the voices of participants in this lived history complementing one another.

The essays in this volume are informed by insights from environmental history, especially place-based accounts of illness. We are drawing on a robust literature on the pervasiveness of exposure in the decades after World War II. Some of these emphasize the global nature of nuclear trade networks, the flow of radionuclides in ecosystems, or the widespread and unnoticed practices in fields such as medicine.[32] We expect an important audience to be those focused on environmental justice.[33] Naturally, we see our work as an addition to what we think of as the "Downwinders" literature.[34]

Yet our work speaks to issues beyond radiation effects, touching on fundamental issues about human rights and relations with environmental contaminants. Pesticides and other carcinogens were introduced to the natural environment routinely in the twentieth century and found their ways to permeable and interconnected human bodies. Existing work touches on subjects such as Agent Orange in Vietnam, cigarette smoke, industrial toxins, and even climate change. Scholars point out that exposure became, in the twentieth century, impossible to avoid, and seemingly accepted as a necessary trade-off. Scholars have shown us how pesticides and pharmaceutical drug contaminants in water and food sources posed inescapable risks in ordinary daily living for everyone, including future generations.[35] Numerous studies acknowledge how scientific work reflected social and political values, as when carcinogens such as insecticides and mercury were perceived as the price of "modernity" in Japan.[36] Even when attempting to ban substances such as DDT, markets were flooded with other dangers that escaped regulation.[37] As with the history of radiation exposure, these involved trade-offs between human health and some perceived gain, whether it was economic prosperity, agricultural production, industrial growth, or national defense.

Suffering and deaths can be as invisible as the radiation itself. We often speak of the *hibakusha*—those Japanese survivors of the bombs that destroyed Hiroshima and Nagasaki. Their lives and those of their descendants have been sources of inspiration and controversy for nearly eight decades.[38] Some scholars have broadened the term to include all of those in the world exposed to harm from radiation.[39] Accounting for such harm remains difficult. French diplomat Marc Finaud marked August 6, 2020, with an opinion piece, "75 Years Later, Nuclear Weapons Still Kill," and tallied up the numbers of casualties from a few of the studies that attempted to account for nuclear weapon pollution. A 1991 study by the International Physicians for the Prevention of Nuclear War estimated 2.4 million preventable and unnecessary deaths due to global nuclear weapons testing and fallout alone.[40] We may never fully see these casualties. But we need to try. Downwinders and other affected communities have comorbidities, enhancing their risks to illnesses. But they also have knowledge to provide, about what is it like to live in that pain, in a space where maladies have no name.

We have organized the collection according to three broad themes, and in keeping with our aim of integrating academic and community voices, it begins with the community. The poems and essays in Part I develop ideas about communities and trust. Scientific claims have extraordinary social and legal power because they have the appearance of objectivity, yet they mask decisions that represent cultural, political, and social values. Our book begins with a pairing of two voices—Kathleen Flenniken and Patricia Hoover. Flenniken, the former Poet Laureate of the US State of Washington, graciously let us reprint three of her poems about living downwind of the Hanford nuclear site. We have paired two of these with Hoover's reflections about starting a conversation about radiation among those victimized by it. Hoover is herself a Downwinder and has spent years organizing victims of thyroid diseases, including actively contesting the scientific results that emerged from the Hanford Environmental Dose Reconstruction Project with a survey of Hanford Downwinders that collected self-reported health outcomes.[41] By starting our collection with Flenniken and Hoover, we hope to create a space for the kinds of objections about contamination that have emerged in communities exposed to radiation. It also shows how history in the hands of survivors can become living art to reveal and heal. We also have included an essay from Adrian Monty, who relates her experiences encountering these communities for the first time as a "fledgling scholar" working on our project.

Following these opening reflections, we proceed to several essays about the emerging science of radiation and historical efforts to understand the effects on exposed peoples. What we find are numerous stories of people trying to make sense of science that comes from others who are not part of their communities. These examples often go beyond failing to earn trust, and instead they show us ways in which such trust would not have been justified in any event. Joshua McGuffie writes about early accounts of radiation sickness, and American attempts to study the victims of Hiroshima and Nagasaki. Prerna Gupta is the first in our collection to highlight the invisibility of workers, in this case uranium miners in India, and reflects on the power of the concept of risk in obscuring claims of those exposed. Oliver George Tapaha continues the discussion with a powerful reflection about a Diné uranium miner in his own family, honoring one unique life while making the intergenerational harm visible. Such radiological harm is not conveyed by statistical estimates of risk. Sasha Stiles and Edward Granados, in discussing contamination at the Rocky Flats nuclear site in Colorado, then give us a sense of the ways that scientific arguments can be constructed in alternative ways to show harms that are often overlooked. Magdalena Edyta Stawkowski reveals to us the "malignant infrastructures" that continued to afflict people in post-Soviet Kazakhstan.

Another set of essays, in Part II, "International Discourse on Harm," highlights the role of scientists in shaping international conversations about harms from radiation. Austin R. Cooper's discussion of "how to hide a nuclear explosion" shows us the connections between French nuclear tests, secrecy, and the contested territories of North Africa during an era of decolonization. Matthew

Adamson reveals the collaboration between scientists and the state—and ultimately the diplomatic compromises—in Morocco, neighboring the French nuclear testing region of the 1960s. Joshua McMullan highlights public understanding of harm from Chernobyl in British discourse, specifically looking at contested knowledge of contaminated sheep in Wales. William Knoblauch reminds us of how much the dangers of radiation exposure became vehicles for diplomatic negotiation in the years after the Three Mile Island controversy, and especially after the disaster of Chernobyl. What can we make of the evolution of public discourse about radiation effects? Jaroslav Krasny weighs in with an essay about violations of international law inherent in "unnecessary suffering" due to radiation exposure. Helen Jaccard, a Veterans For Peace (VFP) activist, shows us the ways in which her work aboard the VFP-restored ship *The Golden Rule* led her to see radiation-harmed communities in far too many places.

A thematic area in Part III, "Remembering and Forgetting," is related to public history. The authors of our final essays all reveal that the historical understanding of past harms is deeply tied to heritage, pride, and public memory.[42] In this way science itself has been drawn into contested narratives of other kinds. Desmond Narain Doulatram provides a profound discussion of the ways that Marshallese history continues to be shaped by events during the era when the United States tested nuclear weapons in the Pacific region. He connects the role of public understanding to the contemporary climate crisis faced by Pacific Islanders. Jeffrey C. Sanders provides a historical look at the B Reactor at the Hanford nuclear site and the public representations of its history. Sarah Fox shows us the multiple ways of knowing the past in Richland, Washington, a small-town community where commemoration and forgetting go hand in hand. We end our collection with artists. First is Yukiyo Kawano, whose art speaks to being "unseen." She is a third generation *hibakusha,* a grandchild of survivors of the 1945 bombings, here reflecting back on Hiroshima and Nagasaki after three quarters of a century. Final words come from the poetry of Kathleen Flenniken.

Many of the questions posed by the authors are about sovereignty and agency on multiple levels. They consider contamination and political sovereignty, as with the invisible radioactive fallout that respected no borders. Yet they also ask that readers consider more intimate details, such as contaminated people's bones, urine, hair, and nails. They ask us to imagine the power and authority of science and the implications for bodily sovereignty. Our authors highlight scientific uncertainty and other blind spots in scientific work, such as gender and racial biases, or inattention to radiation effects in children. From the academic to the very personal, we see nuclear abuse and pain in different contexts and places. These diverse contributions concern national, political, legal, organizational, human, and animal bodies. The stories here demand radiation contamination be seen, to be better understood in the present and future.

We imagine our audience to be the scholarly community, students, and what some might call "stakeholders" in radiation exposure issues—such as scientists who work on radiation effects, activists who continue to seek redress, and people whose loved ones have been affected in some way. The historical legacy of radiation exposure worldwide reaches far beyond academia. In the case of Hanford alone, the Downwinders lawsuits consumed a quarter-century and touched the lives of some five thousand individuals and fourteen legal offices. The scholarship here includes discussion of American controversies, but also has new work on Chernobyl, mining controversies in India, and French nuclear testing in the context of pan-Africanism, to name a few topics. The reflective essays speak to living abandoned in a world of emotional, mental, and physical atomic violence against bodily sovereignty. We wish to draw connections among multiple stories—of Hiroshima survivors, Marshall Islands evacuations and exposures, sickness among Navajo (Diné) uranium miners, and living downwind of nuclear sites—to provide an introduction to issues in the history of radiation exposure. It is an invitation to listen, and to see. We hope that our work will empower all readers to explore the nexus where science, history, and public activism for health and justice meet.

Notes

1 On public perceptions about radiation effects, see Matthew Lavine, *The First Atomic Age: Scientists, Radiations, and the American Public, 1895–1945* (New York: Springer, 2013).

2 For a compelling version of the Slotin exposure, see Alex Wellerstein, "The Demon Core and the Strange Death of Louis Slotin," *New Yorker*, May 16, 2016, http://www.newyorker.com/tech/elements/demon-core-the-strange-death-of-louis-slotin (accessed November 8, 2016).

3 Philip L. Fradkin, *Fallout: An American Nuclear Tragedy* (Boulder, CO: Johnson Books, 2004), 90-91. On the exposure discrepancy, see 264n33.

4 Fradkin, 114.

5 Louis Henry Hempelman, Clarence C. Lushbaugh, and George L. Voelz, "What Has Happened to the Survivors of the Early Los Alamos Nuclear Accidents?" October 2, 1979, paper submitted to Conference for Radiation Accident Preparedness, Oak Ridge, TN, October 19–20, 1979, LA-UR-79-2802, http://www.orau.org/ptp/pdf/accidentsurvivorslanl.pdf (accessed September 26, 2016). See pages 10–12.

6 Ibid., 15–16.

7 On veterans' perspectives, see F. Lincoln Grahlfs, *Voices from Ground Zero: Recollections and Feelings of Nuclear Test Veterans* (Lanham, MD: University Press of America, 1996).

8 On uranium mining and the long-term effects on workers and public health in a variety of contexts, see Gabrielle Hecht, *Being Nuclear: Africans and the Global Uranium Trade* (Cambridge, MA: MIT Press, 2012); Doug Brugge, Timothy Benally, and Esther Yazzie-Lewis, eds., *The Navajo People and Uranium Mining* (Albuquerque: University of

New Mexico Press, 2006); Michael A. Amundson, *Yellowcake Towns: Uranium Mining Communities in the American West* (Boulder: University Press of Colorado, 2004); Laurel Sefton MacDowell, "The Elliot Lake Uranium Miners' Battle to Gain Occupational Health and Safety Improvements, 1950–1980," *Labour/ Le Travail* 69 (2012), 91–118; Norman Naimark, *The Russians in Germany* (Cambridge,, MA: Harvard University Press, 1996), 238–250.

9 A discussion of the Hanford project can be found in D. B. Shipler, B. A. Napier, W. T. Farris, and M. D. Freshley, "Hanford Environmental Dose Reconstruction Project: An Overview," *Health Physics* 71, no. 4 (1996): 532–544.

10 Numerous Soviet-era activities were made public by the Russian Federation in 1993 and are discussed in A. V. Yablokov, "Radioactive Waste Disposal in Seas Adjacent to the Territory of the Russian Federation," *Marine Pollution Bulletin* 43, no. 1 (2001): 8–18.

11 *Final Report of the Advisory Committee on Human Radiation Experiments* (Washington, DC: Government Printing Office, 1995); Eileen Welsome, *The Plutonium Files: America's Secret Medical Experiments in the Cold War* (New York: Dial Press, 1999); Alex Wellerstein, *Restricted Data: The History of Nuclear Secrecy in the United States* (Chicago: University of Chicago Press, 2021).

12 Trisha T. Pritikin, *The Hanford Plaintiffs: Voices from the Fight for Atomic Justice* (Lawrence: University Press of Kansas, 2020).

13 See Tatiana Kasperski, "Nuclear Dreams and Realities in Contemporary Russia and Ukraine," *History and Technology* 31, no. 1 (2015): 55–80; Eglė Rindzevičiūtė, "Nuclear Power as Cultural Heritage in Russia," *Slavic Review* 80, no. 4 (2021).

14 On the first tests in the Marshall Islands, see Jonathan M. Weisgall, *Operation Crossroads: The Atomic Tests at Bikini Atoll* (Annapolis, MD: Naval Institute Press, 1994). On the exposures and lives of Marshallese since then, see Arjun Makhijani, "Never-Ending Story: Nuclear Fallout in the Marshall Islands," *Nonproliferation Review* 17, no. 1 (2010): 197–204; Holly M. Barker, *Bravo for the Marshallese: Regaining Control in a Post-Nuclear, Post-Colonial World* (Belmont, CA: Wadsworth, 2013); Martha Smith-Norris, *Domination and Resistance: The United States and the Marshall Islands during the Cold War* (Honolulu: University of Hawai'i Press, 2016).

15 *Human Radiation Studies: Remembering the Early Years, Oral History of Merril Eisenbud, DOE/EH-0456* (Washington, DC: United States Department of Energy, Office of Human Radiation Experiments, 1995), 12.

16 The science of radiation effects is linked closely with the politics of nuclear testing and radioactive fallout. See Toshihiro Higuchi, *Political Fallout: Nuclear Weapons Testing and the Making of a Global Environmental Crisis* (Stanford, CA: Stanford University Press, 2020); M. Susan Lindee, "Human Genetics after the Bomb: Archives, Clinics, Proving Grounds and Board Rooms," *Studies in History and Philosophy of Science Part C: Studies in History and Philosophy of Biological and Biomedical Sciences* 55 (2016): 45–53; Alison Kraft, "Dissenting Scientists in Early Cold War Britain: The 'Fallout' Controversy and the Origins of Pugwash, 1954–1957," *Journal of Cold War Studies* 20, no. 1 (2018): 58–100.

17 Jacob Darwin Hamblin, "'A Dispassionate and Objective Effort': Negotiating the First Study on the Biological Effects of Atomic Radiation." *Journal of the History of Biology* 40, no. 1 (2007): 147–77.

18 On the role of trust in the evolution of radiation protection recommendations and regulations, see J. Samuel Walker, *Permissible Dose: A History of Radiation Protection in the Twentieth Century* (Berkeley: University of California Press, 2000); Soraya Boudia, "Global Regulation: Controlling and Accepting Radioactivity Risks," *History and Technology* 23, no. 4 (2007): 389–406; Néstor Herran, "'Unscare' and Conceal: The United Nations Scientific Committee on the Effects of Atomic Radiation and the Origin of International Radiation Monitoring," in *The Surveillance Imperative: Geosciences during the Cold War and Beyond*, ed. Simone Turchetti and Peder Roberts (London: Palgrave, 2016), 69–84.

19 On the legal struggles of Downwinders in the US West, for example, see Pritikin, *The Hanford Plaintiffs.*

20 Stephanie A. Malin, *The Price of Nuclear Power: Uranium Communities and Environmental Justice* (New Brunswick, NJ: Rutgers University Press, 2015).

21 Recent work on the consequences and continued social meaning of these accidents include M. X. Mitchell, "Suffering, Science, and Biological Witness after Three Mile Island," *Journal of the History of Biology* 54 (2021): 7–29; Natasha Zaretsky, *Radiation Nation: Three Mile Island and the Political Transformation of the 1970s* (New York: Columbia University Press, 2018); and Kate Brown, *Manual for Survival: An Environmental History of the Chernobyl Disaster* (New York: Norton, 2020).

22 The links between radiation exposure and disease have been contested for decades. For the perspective of a key administrator of scientific work for the Atomic Energy Commission, see Merril Eisenbud, *An Environmental Odyssey: People, Pollution, and Politics in the Life of a Practical Scientist* (Seattle: University of Washington Press, 1990). Excellent scholarly works on the contested relationship between radiation exposure and disease include Gerald Kutcher, *Contested Medicine: Cancer Research and the Military* (Chicago: University of Chicago Press, 2009); Angela N. H. Creager, "Radiation, Cancer, and Mutation in the Atomic Age," *Historical Studies in the Natural Sciences* 45, no. 1 (2014): 14–48; Donna M. Goldstein and Magdalena E. Stawkowski, "James V. Neel and Yuri E. Dubrova: Cold War Debates and the Genetic Effects of Low-Dose Radiation," *Journal of the History of Biology* 48, no. 1 (2015): 67–98.

23 For a discussion of radiation effects and diplomacy, see Maria Rentetzi, "Determining Nuclear Fingerprints: Glove Boxes, Radiation Protection, and the International Atomic Energy Agency," *Endeavour* 41, no. 2 (2017): 39–50. See also Jacob Darwin Hamblin, *Poison in the Well: Radioactive Waste in the Oceans at the Dawn of the Nuclear Age* (New Brunswick, NJ: Rutgers University Press, 2008).

24 See Valerie L. Kuletz, *The Tainted Desert: Environmental and Social Ruin in the American West* (New York: Routledge, 1998); Sarah Alisabeth Fox, *Downwind: A People's History of the Nuclear West* (Lincoln: University of Nebraska Press, 2014); Chip Ward, *Canaries on the Rim: Living Downwind in the West* (New York: Verso, 1999).

25 On scientific work that emerged partly in response to Downwinders activism, see Dianne Quigley, Amy Lowman, and Steven Wing, eds., *Tortured Science: Health Studies, Ethics and Nuclear Weapons in the United States* (New York: Routledge, 2012).

26 On the scientific and social dimensions of the invisible nature of radiation in the first half of the twentieth century, see Luis A. Campos, *Radium and the Secret of Life* (Chicago: University of Chicago Press, 2015). Some scholars, such as Spencer Weart, have maintained that such invisibility sometimes fed an irrational fear of nuclear energy. See, for example, Spencer R. Weart, *The Rise of Nuclear Fear* (Cambridge, MA: Harvard University Press, 2012).

27 On post-Chernobyl lives, see Adriana Petryna, *Life Exposed: Biological Citizens after Chernobyl* (Princeton, NJ: Princeton University Press, 2013). For a discussion of the Chernobyl HBO program as a public history narrative, see Eglė Rindzevičiūtė, "*Chernobyl* as Technoscience," *Technology and Culture* 61, no. 4 (2020): 1178–1187.

28 Arjun Makhijani, Howard Hu, Katherine Yih, eds., *Nuclear Wastelands: A Global Guide to Nuclear Weapons Production and Its Health and Environmental Effects* (Cambridge, MA: MIT Press, 1995).

29 Kate Brown, *Plutopia: Nuclear Families, Atomic Cities, and the Great Soviet and American Plutonium Disasters* (New York: Oxford University Press, 2013).

30 One of us (Hamblin) did this, for example, in a 2012 essay in *Environmental History* called "Fukushima and the Motifs of Nuclear History," drawing together some recurring themes that manifested after the 2011 crisis in Japan. Another of us (Richards) has argued that the assumptions of radiation health safety science make up one thread that unites seemingly disparate Cold War narratives of harm.

31 Jacob Hamblin and Linda M. Richards, "Connecting to the Living History of Radiation Exposure," *Journal of the History of Biology* 54 (2021): 1–6.

32 Similar approaches include Angela Creager's *Life Atomic: A History of Radioisotopes in Science and Medicine* (Chicago: University of Chicago Press, 2013) and Tom Zoellner's *Uranium: War, Energy, and the Rock that Shaped the World* (New York: Penguin, 2009).

33 Douglas Brugge et al., *The Navajo People and Uranium Mining* (Albuquerque: University of New Mexico Press, 2006); Valerie Kuletz, *The Tainted Desert* (New York: Routledge, 1998); Gabrielle Hecht, *Being Nuclear: Africans and the Global Uranium Trade* (Cambridge, MA: MIT Press, 2012).

34 In addition to historical works such as those by Kuletz, Pritikin, Fox, and Ward, the Downwinders literature includes a range of genres including memoir and poetry. Personal reflections about living near nuclear sites include Teri Hein, *Atomic Farmgirl: Growing Up Right in the Wrong Place* (New York: Houghton Mifflin, 2003), and Kristen Iversen, *Full Body Burden: Growing Up in the Nuclear Shadow of Rocky Flats* (New York: Broadway Books, 2013). Kathleen Flenniken's poetry is excerpted in several places in this volume. See Kathleen Flenniken, *Plume* (Seattle: University of Washington Press, 2012).

35 See Linda Lorraine Nash, *Inescapable Ecologies: A History of Environment, Disease, and Knowledge* (Berkeley: University of California Press, 2007); Sandra Steingraber, *Living Downstream: An Ecologist looks at Cancer and the Environment* (New York: Da Capo, 2010); and Nancy Langston, *Toxic Bodies: Hormone Disruptors and the Legacy of DES* (New Haven, CT: Yale University Press, 2011).

36 Brett Walker, *Toxic Archipelago: A History of Industrial Disease in Japan* (Seattle: University of Washington Press, 2011).

37 Frederick Davis, *Banned: A History of Pesticides and the Science of Toxicology* (New Haven, CT: Yale University Press, 2014).

38 The classic study of the Atomic Bomb Casualty Commission's work in Japan after World War II is M. Susan Lindee, *Suffering Made Real: American Science and the Survivors at Hiroshima* (Chicago: University of Chicago Press, 2008). On radiation exposure in Japan after the era of thermonuclear tests, see Lisa Onaga, "Measuring the Particular: The Meanings of Low-Dose Radiation Experiments in Post-1954 Japan," *Positions Asia Critique* 26, no. 2 (2018): 265–304. The politics of radiation exposure in Japan saw a resurgence after the 2011 Fukushima disaster. See Maxime Polleri, "Post-political Uncertainties: Governing Nuclear Controversies in Post-Fukushima Japan," *Social Studies of Science* 50, no. 4 (2020): 567–588.

39 Robert A. Jacobs, *Nuclear Bodies: The Global Hibakusha* (New Haven, CT: Yale University Press, 2022).

40 Marc Finaud, "Opinion: 75 Years Later, Nuclear Weapons Still Kill," *Geneva Center for Security Policy*, August 6, 2020, https://www.gcsp.ch/global-insights/75-years-later-nuclear-weapons-still-kill (accessed December 20, 2021).

41 Rudi H. Nussbaum, Patricia P. Hoover, Charles M. Grossman, and Fred D. Nussbaum, "Community-Based Participatory Health Survey of Hanford, WA, Downwinders: A Model for Citizen Empowerment," *Society and Natural Resources* 17, no. 6 (2004): 547–559.

42 On the special link between sites of past contamination and public memory, see Anna Storm, *Post-Industrial Landscape Scars* (London: Palgrave, 2014). On nuclear issues specifically, see Susanne Bauer, Karena Kalmbach, and Tatiana Kasperski, "From Pripyat to Paris, from Grassroots Memories to Globalized Knowledge Production: The Politics of Chernobyl Fallout," in *Nuclear Portraits: Communities, the Environment, and Public Policy*, ed. Laurel Sefton MacDowell (Toronto: University of Toronto Press, 2018), 149–189.

PART I

Communities and Trust

Richland Dock, 1956

KATHLEEN FLENNIKEN

Someone launched a boat into the current,

caught and delivered fish to the lab
and someone tested for beta and P-32.
Someone with flasks and test tubes tested
And re-tested to double check the rising values.

And someone drove to the public dock
with a clipboard and tallied species and weight.
Chatting with his neighbors, *Which fish*
are you keeping? How many do you eat?

And someone with a slide rule in a pool of light
figured and refigured the radionuclide
dose. Too high. Experimented frying up
hot whitefish. No. No. Then someone decided

all the numbers were wrong. Someone
from our town. Is that why we
were never told? While someone fishing—
that little boy; the teacher on Cedar Street—

caught his limit and never knew.

Whole-Body Counter, Marcus Whitman Elementary

KATHLEEN FLENNIKEN

> "The mobility of this new laboratory provides versatile capabilities for measuring internally deposited gamma-ray-emitting radionuclides in human beings."
>
> —*Health Physics*, November 1965

We were told to close our eyes.
Everyone was school age now, our
kindergarten teacher reminded us,

old enough to follow directions
and do a little for our country.
My turn came and the scientists

strapped me in and a steady voice
prompted, The counter won't hurt,
lie perfectly still, and mostly I did

and imagined what children
pretend America is, parks
bordered by feathery evergreens,

lawns so green and lush
they soothe the eyes, and pupils
open like love—

a whole country of lawns
like that. Just once I peeked
and the machine had taken me in

like a spaceship and I moved
slow as the sun through the chamber's
smooth steel sky.

I shut my eyes again and pledged
to be still; so proud to be
a girl America could count on.

Reprinted with permission of University of Washington Press, *Plume* (University of Washington Press, 2012), 16.

How to Start a Conversation

PATRICIA HOOVER

As I look back, my notable milestones are clear. By far the most significant event to shape my life's trajectory occurred on August 19, 1990. This is the day my life, as I thought I knew it, was irreversibly shaken to its core. I was forty-three years old. From birth I had dealt with an extensive array of inexplicable medical issues. From a PBS NOVA program titled "The Bomb's Lethal Legacy," I learned why. I was exposed to hundreds of thousands of toxic radionuclides growing up in the 1940s, '50s, and '60s downwind and downriver from the Hanford plutonium production facility near Richland, Washington. I was a Hanford DOWNWINDER!!

Like many residents of the Inland Northwest, it took my entire life to that point in 1990 to learn what had long been suspected as the cause of many Downwinder's health problems: stillbirths, thyroid diseases, birth defects, infertility, elevated cancer rates, and more.

My transition from ignorance to knowing was an abrupt shock on that August day. My identity was forever changed.

An additional glance back reveals another watershed occurrence in 2012. A fellow activist and dear Seattle friend heard a local poet named Kathleen Flenniken read from her recently published book titled *Plume.* The poems chronicled the author's girlhood growing up in Richland, where her father worked at Hanford. I bought the book and was immediately moved and emotionally captivated by Kathleen's story. Her poems are part memoir and documentary. With my lived experience as a Hanford Downwinder, *Plume* was a touchstone. We were plutonium pals without even knowing each other. I identified in a parallel yet disparate way with Kathleen's life as described in her reflective *Plume* poetry.

In my world in northeast Oregon, I had *no idea* of the dangers posed by Hanford. In Kathleen's world in the "Atomic City" of Richland, just a few dozen miles north of my childhood home, the government *knew* it was putting people in danger, but constantly reassured townspeople that they were safe, protected, and had nothing to worry about.

I met Kathleen Flenniken in 2012, and we have spent time growing a friendship over the years. In 2017, we combined her artful poems and my lived experiences in a presentation we call "A Hanford Conversation." She reads one of her evocative poems from *Plume* and I extemporaneously express how the sentiments of the poem trigger memories for me. This literary interplay debuted in White Salmon, Washington, and was reprised in Eugene, Oregon.

We were invited to present our "Conversation" at the Oregon State University National Science Foundation grant workshop on June 20, 2019. Two of the six poems we discussed that day are included in this essay.

For example, in the poem "Richland Dock, 1956" the absolute prevarication of the government operating Hanford and simultaneously protecting citizens from harm is clearly divulged. It is a chilling poem with the final line a double meaning with Kathleen's creative use of the word "limit."

This poem evokes many images for me:

- For years before 1956, the local Native American fishermen noticed and talked about severely deformed and mutated fish in their catches.

- The vacillating safety standards, to protect nuclear workers and local residents, were "adjusted" to fit constantly fluctuating radiation releases into the Columbia River.

- To me, worrying about the "hot" radioactive fish caught at a dock in Richland, just a few miles downriver from Hanford seemed quixotic, at best. Hundreds of miles downriver from the nuclear plant, where the Columbia River empties into the Pacific Ocean, radioactive shellfish were gathered in bays on both Oregon and Washington coastlines. Is there really any safe dose?

- The government knew it needed to close the river to fishing and recreating, but chose not to alarm the public with such action. (This Downwinder grew up swimming, water skiing, fishing, and camping on the same polluted toxic waterway.)

- In 1956, the Hanford facility had been operating for over twelve years, yet the plant managers continued to recalculate supposedly "safe" doses. Accurate science and reality bounce back and forth in this revealing and provocative poem.

The poem "Whole Body Counter, Marcus Whitman Elementary" is absolutely heart wrenching. I identified completely with the innocent five-year-old kindergartners. How frightening for a young child to be strapped in and loaded into a dark, metal cylinder and told to be good and not move. Kathleen's description of coping is inventive, naive, and poignant.

Here is part of my response in our "Conversation":

- I, too, as a child in junior high school had men in white lab coats (doctors?) visit my health class, where students had to stand in line while these unidentified strangers walked behind us and felt our necks (presumably looking for enlarged thyroid glands).

- Girls with severely ridged and cracked fingernails were given bottles of nail "polish" supplied to our physical education teacher in junior high school, from an unknown source (Hanford labs?).

- This poem reminds me that I didn't need a ride in a body-counter or any stranger's hands on my neck because at age eleven my thyroid gland completely stopped functioning, and I became sick. Years later a large tumor and the thyroid gland had to be surgically removed. That is also how I acquired my permanent jewelry, my Hanford "necklace" . . . the half-moon scar around my neck that so many of us in the region wore. The Feds called it our Hanford "smile," as if thyroid disease is anything to be pleased about.

- Hanford's poisonous radiation releases were not just of plutonium. Hundreds of dangerous radionuclides are released in the production of plutonium. Kathleen's "spaceship ride" was looking for many different types of radiation. Strontium-90 mimics the element calcium and the body does not differentiate between the two substances. Strontium causes a "honeycombing" and thus a weakening of bones. I am reminded by the verse of all the excessive broken bones my classmates suffered when we were young. Plaster casts on arms, hands, legs, and feet were the norm, not an exception in my childhood.

- Kathleen again used an imaginative double meaning with the word "count" in the concluding line of this haunting poem.

The bottom line for this literary "duet" Kathleen and I present is the indisputable fact that the government *knew* it was harming people. Period. Kathleen's personal story in the *Plume* poems and my experiential Downwinder narrative depict how the two of us have become keen and perceptive witnesses to Cold War America. The juxtaposition of two innocent young girls' childhood circumstances growing up in the shadow of a nuclear production facility deeply connects us. That bond is apparent in our very meaningful and powerful "Hanford Conversation" . . . so will *we* start a conversation?

A Fledgling Scholar's Encounter with the Downwinder Project

ADRIAN MONTY

The June 2018 Oregon State University Downwinders Project workshop, "Telling the Stories of Radiation Exposure," was my first official encounter with the invisible nature of stories of radiation exposure. I entered the conference room, out of breath from my hot Oregon summer bike ride to campus, nervous about the days to come. As a fledgling scholar in the Environmental Arts and Humanities graduate program at OSU, I was exhilarated to be asked to take notes as part of the project. I had no way of knowing how much my views of the world would be substantially questioned and altered by this experience.

The multifaceted nature of the topics I heard were stories of nuclear wreckage: the displacement of Marshallese Islanders, the invisibility of the radiation-exposed, the continued devaluation of anecdotal evidence, and the nature of uncertainty in science. Watching these subjects becoming visible in my mind caused the shaping of an invisible but all-too-real ball in my throat. Prior to hearing numerous scholarly accounts of the global history of nuclear exposure, the only information I knew about nuclear history stemmed from an undergraduate ethics class in which the Last Resort debate concerning the US decision to drop nuclear bombs on Japan was discussed ad nauseam. I thought about all of the history and science classes I had taken and not once hearing about these other important issues. The suppression of this knowledge runs deep. By the end of the workshop, I knew I needed to be part of the solution to make these stories more visible.

Included at the first workshop was Patricia Hoover whose story became an integral part of my thesis project. Later, when writing my piece on the Hanford Nuclear Reservation, I constantly thought of Patricia and her story, how much she inspired me with her narrative and her strength. She is a Downwinder, a journalist, citizen-scientist, and an activist. Patricia sought answers about the illnesses that plagued her. She along with others, including Dr. Rudi Nussbaum, created the Northwest Radiation Health Alliance in 1993. They compiled a citizen science health survey in the 1990s. Their results were published in *Society and Natural Resources* in 2004 along with seven other peer-reviewed published papers.[1] Through this effort, Patricia and the NWRHA furthered the efforts of Downwinders to survey more extensively the afflictions put on them, introduced the term Downwinder into a more popular light, and made contributions to the field of community epidemiology.

A major component of the Downwinders Project is collecting oral histories. After the workshop, my role grew with the project, and I began to arrange, participate in, and transcribe oral history interviews. This aspect of the project was what really got me invested; this unique opportunity to hear in person from the people who lived this history. It is one thing to read about the stories and react to the writing, yet it is another thing altogether to hear about their personal hardships and stories from their own mouths. Having conversations with people like Patricia is so vital—reading about numbers is one thing, but talking to a Downwinder is something else entirely—transformative, visceral, visible.

Requesting someone to record an oral history asks them to relive their trauma of being part of the nuclear narrative. Participants put themselves in their past mind-sets and emotions so we can have their history officially recorded. It is often difficult to listen to these stories, in the sense that they are emotionally complex and taxing. Listening to each participant and figuring out how to capture their unique voice and historical perspective on the page as I transcribed became an important part of my process.

The first interview I transcribed was between Dr. Hamblin, Dr. Richards, and John Till. Till, a radiobiologist and nuclear engineer, was the chair of the Technical Steering Panel formed by the Department of Energy (DOE) to respond to public outcry when documents from Hanford were declassified in 1986. I had little to no context about the project and did not realize who John Till really was until after I typed out his side of the story.

It was interesting to hear the same story from multiple vantage points, something that continued to deepen my interest in the project and help shape my storytelling endeavor.

The second interview I transcribed was with Jim Thomas, a member of the Hanford Education Action League (HEAL), who worked tirelessly to help file the Freedom of Information Act requests that led to the release of some of the secret history at Hanford. He read all of the documents (roughly 20,000 pages) released by DOE at the time. Jim also corresponded with John Till on numerous occasions in an attempt to make sure the TSP was covering all of their bases when it came to sharing information with the public.

The metaphorical pieces of red yarn on my imaginary corkboard started to reveal the bigger picture once Jim Thomas mentioned John Till in his interview.

These overlapping strings started to create a unified picture when I had the opportunity to speak with Larry Shook, a veteran, HEAL member, and journalist who I sought out for an oral history interview. I had read about Larry Shook while reading a vital text concerning Hanford and its Downwinders, *Atomic Harvest* (1993) by Michael D'Antonio. Being part of this interview with Shook later in my writing process boosted the passion I had for my own project. Hearing Larry's story about Hanford and the work he did with HEAL firsthand was an important experience for me. Being able to actively piece together the moving parts of the history as he talked made me realize how much information I have about this fraught and complicated history that so few seem to know about. Being part of this project ultimately tied me to the communities most affected by these invisible strings.

I wanted to make what I had heard into something all people could understand and discuss. The stories of those harmed by radiation slowly became the heart of my thesis, their words stitched and pieced together through my own research. Deciphering my array of notes, mind maps, and imaginary red-yarn-adorned corkboard became my ultimate challenge. I needed to interweave the already recorded history and previously diminished anecdotes to make a cohesive narrative.

With the oral histories of Patricia Hoover, Jim Thomas, John Till, and Larry Shook in my literary artillery, I wrote a piece titled "The Poisonous Plumes of Hanford." This piece mainly concerns the intentional 1947 release of radioactive chemicals known as The Green Run. Along with the official narrative, I weave in Patricia Hoover's childhood experiences of falling ill and being unknowingly observed and tested by Hanford officials, Jim Thomas helping with the FOIA to release confidential documents concerning the Green Run and other experiments, John Till's involvement on the Technical Steering Panel for the DOE, and my own accounts of Richland, Washington, and the Hanford grounds.

During my research and my work with the Downwinders Project, I also transcribed two interviews involving Atomic Veterans, Keith Kiefer and Fred Walden. These I threaded into a piece titled "Secrecy Strings: Stitching Together the Gaps of the Atomic Veteran Narrative Since 1945." Keith Kiefer, now an Atomic Veteran and the Commander of the National Association of Atomic Veterans (NAAV), was assigned to the clean-up mission on the Enewetak Atoll of the Marshall Islands in 1977 as a ground radio repairman. He worked on the contaminated land for over six months. "At forty years old a doctor told me that I had the bone structure of a ninety-year-old." Since Keith was part of the Radiological Clean-up Project on the Enewetak Atoll, he was exposed to a myriad of different chemical elements. One of these elements, strontium-90, shares some properties with calcium. When absorbed by the human body, strontium-90 acts like calcium and goes straight to the bones. Keith Kiefer continued in this vein by saying, "What's interesting is, the documents show that the government didn't go into it blind. In many cases, they knew what they were doing, and they were using the soldiers as guinea pigs, test subjects." Kiefer still fights for compensation for second and third generation atomic soldiers.[2]

Fred Walden is an original Atomic Veteran who served with the US military for over thirty years. The first time Fred shared his story with us we had technical mishaps. The entire conversation we had with him was incredibly moving, but we found out later, not digitally recorded. This error was devastating to me at the time because his story was so striking and vital. However, we were able to speak with him a second time about much of the same content, with a feeling of déjà vu. His oral history is in our archives and woven into my article.

The inception of his military service was in 1957 when he was deployed at the Nevada Test Site. Fred Walden recalled how he witnessed the 74-kiloton nuclear detonation test known as Shot Hood as part of Operation Plumbbob on July 5, 1957: "We actually went up to Ground Zero and back out while it was still smoking." He could see each individual pebble and crevice in the sand on the ground beneath him through a stark black and white lens. He tells us that he fumbled to get his helmet on and could see his skeleton through his hands like an x-ray. Although Fred and his fellow soldiers were exposed to a bomb almost five times the size of the Hiroshima bomb in witnessing the tests in Nevada, he and the others were told they "weren't close enough for anything to happen" to them. But Fred witnessed these explosions and was there for the aftermath. There were test dummies set up at the blast sites, now called Survival Town, like you might see in an old episode of *The Twilight Zone*. Fred told us that the first few rows of dummies were still on fire when they got there. "Then the last three or four had been vaporized, there was nothing." In discussing Fred's medical issues since his time in the military, his wife Anne chimed in from the background of the phone call to confirm that Fred had over thirty surgeries to remove his government-sponsored cancerous nodules.[3]

Each of the stories I transcribed, whether I was present for the interview or not, struck me to my core. Stories shape our lives. In hearing them, we carry them, and pass them on. These oral histories are the stories of those who were told they did not matter or were wrong based on the official narrative posed by those in power. They were rendered invisible just like the radiation they were exposed to. The workshops mattered not only to those whose stories were finally being told and recognized, but to the way history is recorded and shaped. Who controls the narrative? While transcribing interviews, I absorbed the stories I was listening to, grew attached to them, and felt the draw to make them more visible.

I listened to and transcribed the stories of Downwinders and Atomic Veterans and learned about nuclear history with headphones behind a computer screen for about a year. A month before finishing my graduate project I had the opportunity to visit Hanford and take tours of the Hanford Site Clean-up and B-Reactor. With this trip, I was able to personally witness the visible narrative upheld in 2019.

Surrounded by brush desert, agriculture, a french fry production facility, and the occasional manicured façade, stands the 586-square-mile campus of the Hanford Nuclear Reservation. Since 1943 the management prided itself on being innovative, efficient, and safe. This attitude was later essential to the goals of the government to "Keep the Cold War safe." Hanford management has extended this pride to industrial development, conservational efforts, environmental consciousness, and progress. All of it seems a bit counterintuitive if you look across their massive span of land marked mainly by warning signs of radiation.

The site's number one goal now is the completion of the Waste Treatment & Immobilization Plant (WTP). On the tour we were told that this is where the innovative vitrification process of turning radioactive waste into glass will one day occur. In the meantime, they are frantically transferring high level dangerous waste that has been stored in single-shell tanks to double-shells in the time it takes for the WTP to be built. Then they will transfer the waste from the double-shell tanks to the WTP for vitrification, then to big metal milk-bottle shaped canister storage. Sounds to me like a very risky game of hot potato.

The most visceral part of the tour was of Hanford's B-Reactor or what is now called the Hanford Unit portion of the Manhattan Project National Historical Park. As part of a tour, I wandered through the factory where thousands of people worked various steps to create what seemingly ancient signs denoted as "The Metal" and "The Product." Unbeknown to them was the danger residing in the materials they were working with and around for hours a day. If they did know more about their work, friendly reminders of "Security" resided in their break rooms and other areas within the B-Reactor museum. These seem frozen in time in 1944 so tourists can continue to be amazed by what this building did for our country and the war effort while the radiation and bombs remain unseen.[4] One of these reminders, posted on a "Security" board reads: "Protection for all *Don't talk* Silence Means Security."

A letter from the Secretary of War dated July 10, 1944, is also posted in this room where workers may have once eaten their lunches and learned about the news of their labor unions. Addressed "To the Men and Women at The Hanford Engineer Works," this letter states that the work they were doing was "urgent and essential," and "The very fact that the War Department cannot give you any information about this project is the surest guarantee of its great value." These secrets were kept not only from our enemies, but the very people who were part of the effort.

There are parts of this story that still stay invisible and contained just like some of the waste at Hanford and elsewhere around the world. This project helps to share the stories that are often left untold along with the context of what people experienced within the government's narrative of the Atomic Age; more specifically, how they experienced being an unwitting part of that nuclear narrative. By telling these stories and integrating them with the official narrative, we can take a peek through the cracks of that recorded history and allow the rays of truth to shine through. We can continue to dosimetrically piece together the narrative by reconstructing history along with those who were only in the background before. Making the unseen visible.

Notes

1. Patricia Hoover shared three oral histories and two public presentations as part of the NSF Downwinders grant from 2017 to 2021; Rudi H. Nussbaum, Patricia P. Hoover, Charles M. Grossman, and Fred D. Nussbaum, "Community-based Participatory Health Survey of Hanford, WA, Downwinders: A Model for Citizen Empowerment," *Society and Natural Resources* 17, no. 6 (2004): 547–559.
2. Quotations are from an unpublished interview with Keith Kiefer by Linda M. Richards, September 24, 2018. The transcription will become part of Oregon State University's Special Collections and Archives Research Center.
3. Quotations are from an unpublished interview with Fred Walden by Linda M. Richards and Jacob Darwin Hamblin, November 30, 2018. The transcription will become part of Oregon State University's Special Collections and Archives Research Center.
4. Linda Marie Richards, "Review: The B Reactor National Historic Landmark. The Hanford Site Manhattan Project National Historical Park," *Public Historian* 38, no. 4 (2016): 305–317, https://doi.org/10.1525/tph.2016.38.4.305.

The First Accounts of Radiation Sickness

JOSHUA MCGUFFIE

Familiar Tools for a Novel Disease

"Suddenly," recalled Dr. Michihiko Hachiya, "a strong flash of light startled me."[1] Hachiya directed the Communication Bureau's hospital in central Hiroshima. Immediately after the flash on August 6, 1945, the blast bowled his house over, injuring him and his wife, Yaeko-san. They lived just over a mile north and west of the blast site. The force tore open his cheek and lodged a massive shard of glass in his neck. Yaeko-san suffered burns and bleeding wounds. The couple struggled to his hospital that morning, though most of the building burned in the aftermath of the blast. "The sky became bright as flames from the hospital mounted."[2] They survived their wounds, largely with help from two doctor colleagues who also made the trek to the ruined facility. During the day's travails, Hachiya remained convinced that the US had dropped a 500-ton conventional bomb on the city. What else could have caused such destruction? Despite his wounds and uncertainty about the blast, Hachiya had to figure out what to do next, how to treat the host of patients pouring into the ruins of his hospital and exhibiting a suite of symptoms far more complex than burns and broken limbs.

The invisible but deadly radiation produced by the bomb that destroyed Hiroshima, and by the bomb that destroyed Nagasaki three days later, precipitated both a public health crisis and a knowledge problem for the doctors who responded to the catastrophes. Local doctors in the two destroyed cities had to treat patients whose symptoms defied conventional wisdom about infectious disease. They also had to grapple with the trickle of news that gradually let them know that their patients suffered from radiation exposure. What was an atomic bomb? How did imperceptible radiation cause such horrendously palpable symptoms? What did it mean for the palliative care they should offer? While doctors in Hiroshima and Nagasaki groped in the dark, another group of doctors who belonged to the various branches of both the Japanese and American medical corps worked to understand the sickness as well. These doctors had a different knowledge problem. They had no practical experience with the biological effects of the free neutrons and gamma rays produced by the new fission. But many of these outsiders brought their experience with laboratory x-ray experiments on animals. Despite this background, they faced uncertainty in August 1945. Would fission behave like x-rays? Would human beings respond like dogs, rabbits, or salmon?

To confront these questions, both local clinicians and military outsiders fell back on familiar practices and ways of knowing. On the ground, the variety of local Japanese doctors forced into clinical work embraced basic practices for a public health disaster. Before they knew about the radiation danger, they quarantined patients to prevent the spread of disease. Doctors thought they had cases of dysentery in Hiroshima and cases of cholera in Nagasaki.[3] They collected case histories, did rudimentary blood work, and performed autopsies in order to learn about the mechanisms that drove their patients' symptoms. Meanwhile, the cadre of American and Japanese military experimentalists turned back to over two decades of knowledge created by animal x-ray experiments. They chose to assume that fission would behave like x-rays and that animal reactions could inform the kind of data they should collect from suffering human beings. Though the experiences of local doctors and the outsider experimentalists had little in common, they shared a turn to tried-and-true practices that allowed them to make some sense of the new sickness amid difficult circumstances. Comparing and contrasting their responses shows how the first accounts of radiation sickness grew up from prewar practices, from older ways of knowing molded to the contingencies of atomic fission.

Personal journals and early institutional reporting offer pathways to trace these ways of knowing as they developed in the chaotic weeks and months after the bombings. For the initial local and clinical responses, the translated journals of Michihiko Hachiya, doctor and director of the Communication Ministry's hospital in Hiroshima, and Raisuke Shirabe, doctor and professor of surgery at Nagasaki Medical University, constitute the main sources.[4] The travel diaries of Stafford Warren, doctor and chief medical officer of the Manhattan Engineer District (MED), and Ashley Oughterson, doctor and member of the US army's medical corps, provide a foundation for the experimentalists' story. Placed in conversation with journal publications from the animal x-ray tradition, these travel journals cast light on how the experimentalists used interspecies knowledge in the wake of the bombings. Finally, reports submitted in November 1945 from the MED and the Army Medical College in Tokyo show the earliest examples of how the experimentalists' interspecies knowledge led to conflicting national formulations of the sickness. Masao Tsuzuki, an experimentalist who led the Japanese National Research Council's medical section, and his lieutenant, the hematologist Hitoshi Motohashi, play important roles throughout this story.

This essay examines the two earliest medical attempts to confront the crisis of novelty presented by the biological effects of the atomic bombs. These unfolded unevenly, eliciting distinct responses. The bombs dropped on Hiroshima and Nagasaki produced invisible radiations that in turn caused a suite of symptoms, both visible and devastating, in those unfortunate people irradiated by them. Doctors, confronted with the symptoms, did not immediately set out to describe a single new disease. Instead, Japanese and American doctors relied on pre-bomb practices to mediate their experiences with the symptoms and challenges they encountered. Local clinicians used traditional responses to a public health crisis in ways that allowed them to treat patients and exert some control over the ruins of their hospitals. Outsider experimentalists cast human data in dyes left over from animal experiments in the 1910s, 1920s, and 1930s.

These first accounts, which grew up in the weeks and months after the bombings, mattered. Japanese clinicians and experimentalists argued that the new fission induced a novel disease. Despite seeing symptoms with their own eyes, American experimentalists from the Manhattan Engineer District (MED) claimed it did not. The inability to quantify or concretely describe the mechanisms by which invisible radiations caused so much illness in Hiroshima and Nagasaki allowed the Americans to dismiss any notions that tied fission to a distinct disease. This would not be the last time that US atomic officials dismissed ties between radiation and ill health since the MED's account of the bombs' effects went on to become the normative viewpoint of the future US Atomic Energy Commission's atmospheric testing program. Radiation's subtlety caused many Downwinders in the US to suffer in the twilight between conspicuous symptoms and uncertain diagnosis, as a number of authors in this volume show. This contested and uneven medicalization of exposure to invisible radiation went on to become the stuff of diplomatic wrangling, long-term genetics research, and cultural introspection in the Cold War.[5]

Making New Knowledge Locally

As sick and injured patients streamed into his hospital the day after the bombing, Michihiko Hachiya still assumed that he needed to treat conventional maladies created by a conventional weapon. Since vomiting and diarrhea began the day after the bomb was used, Hachiya ordered an isolation ward organized to manage what he thought was an outbreak of dysentery. Over the next few days diarrhea turned to bloody diarrhea. The hospital ruins became more crowded and Hachiya struggled to isolate all the patients he thought suffered from the outbreak. Dr. Hanaoka, who ran the Communication hospital's outpatient clinic, began systematizing patients' symptoms on August 9. He reported three groups of patients to Hachiya:

1. Those with nausea, vomiting, and diarrhea who were improving.
2. Those with nausea, vomiting, and diarrhea who were remaining stationary.
3. Those with nausea, vomiting, and diarrhea who were developing hemorrhage under the skin or elsewhere.[6]

Hanaoka described no conventional disease, but perhaps created the first systematic description of radiation sickness.

On the day Hanaoka reported in Hiroshima, the third atomic bomb ever built destroyed Nagasaki. Raisuke Shirabe recalled "a bright blue flash shone in my eyes."[7] Destruction engulfed Nagasaki Medical University. "We saw that the hospital, basic science classrooms, and all the wooden buildings had collapsed and were burning."[8] He and the surviving faculty, nurses, and students fled up the hillside behind the campus, where they camped on the evening of August 9. The next day they began to treat patients in the school's air raid shelters. Shirabe traveled a few miles north, to the less damaged Nameshi neighborhood. There he arranged to use a social club as a relief station for patients from the Medical School's ruins. When patients began to arrive at the temporary clinic, bloody diarrhea had taken hold. "Suspecting cholera," Shirabe and his colleague Dr. Kido "moved these patients to a corner of the room to isolate them."[9]

In the first week after the bombings, the fact that the bombs were atomic meant little to Hachiya, Shirabe, and the doctors and nurses scrambling to treat sickening patients. The military medical establishment in Tokyo did fear a problem with radiation if indeed the bombs were atomic, per President Truman's radio message of August 6. They acted quickly to confirm the news. On August 10 a joint army-navy survey team pinpointed where the bomb had exploded over Hiroshima. They also found photographic film exposed even though it had been safely stored away from any light source.[10] News of the radiation traveled out of the city, but not within the city. The bombs were already creating new geographies of knowledge, in which national interests outweighed local ones. Clinicians in the two cities did not know that the patients who seemed well one day and sick the next were dying from the inside out because of radiation.

Facing this bewildering public health crises, Hachiya and Shirabe provided what palliative care they could. Shirabe moved patients to relative safety in Nameshi to encourage his patients' recovery. "It would not be possible to give good patient care in the ruins . . . on the bare earth or concrete floors in the ruins."[11] In Hiroshima, Hachiya worked to improve his facilities in place. His old pharmacy storeroom became a dining room. There patients shared what little food was available. Faced with a shortage of nurses, he directed that ambulatory patients be taught how to dress their own wounds. He decided "to place a crock of Remaon's solution [a mild germicidal solution] near the entrance of the hospital, and notices posted instructing patients to soak their dressing in the solution before covering their wounds."[12] The system worked. The crock became a spot for socializing. Despite these improvements, doctors and nurses in both cities spent much of the first week after the bombings piling up the dead for cremation.

The clinicians longed to know about the mechanisms behind their patients' illness. On August 13, Hachiya remarked that "the most popular explanation was still that some poison gas had been liberated and was still rising from the ruins."[13] Frustrated with his lack of understanding, Hachiya ordered two of his doctors to take a thorough case history of each patient in the hospital on August 16. By the 20th, the new wave of death began to crest. At that point Hachiya suspected a "suppression of the white blood cells" because previously healthy patients bled so much and suffered petechiae, pinpoint-sized hemorrhages, all over their bodies.[14] He therefore rejoiced when a microscope arrived from the main Communications Bureau Hospital in the capital on August 20.[15] With it, his pathology staff could make rudimentary investigations into patients' blood picture.

Hachiya's suspicion about low white blood cell counts proved prescient. Dr. Hanaoka shared data from the microscope after dinner on the 22nd: The white blood count in persons exposed in the Ushita area, between two and three kilometers from the hypocenter, ranged from 3,000 to 4,000. Patients nearer the hypocenter, although fewer in number, had counts around 1,000. Severely ill patients had counts lower than 1,000, and the nearer the hypocenter the patient had been, the lower the white count.[16] Counts like these fell far below what the doctors would have expected even in patients sick with a bacterial infection. Hanaoka's speedy processing of 50 blood samples in two days was impressive, working with no electric light. Pairing his data with patient location from verbal case histories gave it radiological significance. The veil began to fall. "Our preliminary blood findings filled us with excitement and the feeling that for the first time we were coming to grips with this unknown enemy."[17] Now they could explain their patients' weakness and susceptibility to infection.

The doctors wanted a macroscale view of their ailing patients' bodies, so they began to prioritize autopsies. On August 26, Hachiya and a colleague performed their first since the bombing. They found the dead woman's body cavity filled with blood that had not coagulated. The blood had hemorrhaged from countless points on her internal organs. To Hachiya, the autopsy proved an epiphany. "If we had begun to do autopsies sooner, perhaps we would not have been so in doubt about our patients' signs and symptoms."[18] He had been working on a text for an informative broadside designed to educate the local population about their symptoms. After the autopsy, Hachiya said, "I tore up what I had done and started over."[19] The next day, his staff posted the revised text around the hospital and Communications ministry.[20] The broadside pointed to the need for control over the continuing public health crisis. Hachiya encouraged those who felt well to continue working. He assured survivors who were losing their hair that they would likely live and ought not swamp his hospital based on that symptom alone. The broadside described a conventional and manageable disease. Hachiya only mentioned radiation and uranium in passing at the end of the text.

Combining information from blood sample analysis and visible symptoms uncovered during autopsies allowed the Hiroshima clinicians to zero in on one of the new radiation's most devastating effects, lowered platelet levels. Continued blood work showed an almost universal dearth of platelets in the sickest patients. Hachiya marveled at this news. He ran to the autopsy shed to share it with Chuta Tamagawa, the professor of pathology from Hiroshima's Medical School who had fled the city but then returned. "Is that so!" he exclaimed. "Well! That explains everything . . . that's why blood hasn't clotted even after seven hours!"[21] Low platelet counts provided a mechanism for patients' inexplicable deaths by hemorrhaging. "We had interpreted," Hachiya wrote, "the low white count as characteristic of the disease, but it became obvious that this was only one feature of a disease that involved platelets as well. . . . We had overlooked the platelets because they are more difficult to evaluate than white blood cells."[22]

The Nagasaki doctors never started any significant lab work in their makeshift relief stations due to the extent of the destruction in that city. By September, they considered their situation untenable. They arranged with Rear Admiral Kodo Yasuyama, a graduate of Nagasaki's medical school, to move the faculty and their patients at Nameshi to his Naval Hospital in the nearby port city of Omura. Located twenty-two miles from Nagasaki's city center, Omura suffered only minor damage from the bomb, and the hospital had sent medical staff to Nagasaki to operate relief stations.[23] The hospital also received medical evacuees who arrived by train to its well-maintained grounds and 1,700-bed hospital. By bringing the medical faculty to Omura, Yasuyama added to the number of doctors and nurses he had on his staff. Shirabe made the trip to Omura on September 26, among the first of the faculty to arrive.[24] Moving out of Nagasaki, the medical faculty positioned itself to care for and study patients in a controlled, familiar medical context. Though the Hiroshima clinicians performed medical investigations first, the work of Nagasaki doctors at Omura became key for knowledge about the new radiation sickness.

In the hands of local clinicians, radiation sickness scarcely had anything to do with radiation in the first weeks after the bombings. Cut off from information from the capital and without even basic laboratory equipment, they treated patients' symptoms as best they could. They cleaned wounds. They tried to help patients keep water down. They distributed preciously scarce food even as patients vomited it up or passed it in bloody diarrhea. News about radiation from the new bombs trickled in to their hospitals and relief stations even as they did this. The news made little difference initially, and in neither city did surviving radiologists lead the effort to treat the novel sickness. In Hiroshima, blood work offered insights into the strange symptoms killing patients. In Nagasaki, only the hope of escape offered the city's doctors the possibility of understanding the disease. On the ground, fission remained largely absent from radiation sickness as August 1945 rolled into September.

Outsiders and X-Rays

The symptoms that the *hibakusha* experienced systematically became the stuff of radiation when doctors from outside the cities, led by Masao Tsuzuki and Stafford Warren, arrived to collect biological data about the new fission. They did so armed with a host of data about animals exposed to x-rays, a knowledge base they felt comfortable using as a guide for understanding the effects of fission in human beings. Their experience with this interspecies knowledge guided them as they chose what data to collect from patients exposed to fission's invisible insults. Warren and Tsuzuki prioritized learning about the survivors' blood picture. They also wanted tissue samples from the blood forming organs for analysis under the microscope—a practice called histology. They assumed that radiation from fission would injure blood just like x-rays did and that irradiated human beings would respond to those insults like animals. Armed with these convictions, they ventured from their labs to the devastated cities.

Though wartime adversaries, Masao Tsuzuki and Stafford Warren shared much in common professionally. Warren took his MD at the University of California's Medical School in San Francisco. Warren's mentor there, George Whipple, first irradiated dogs with x-rays from that school's new and powerful Coolidge Tube in 1919.[25] He observed the dogs, noting an inexplicable lag between exposure and the onset of clinical symptoms. In the early 1920s, Warren joined Whipple to continue the dog experiments. Together, they published six articles in 1922 and 1923 in the *Journal of Experimental Medicine*.[26]

Methodologically, they relied on autopsy, urine analysis, and histology. Tsuzuki took his MD at the University of Pennsylvania. In 1926, he ran his own x-ray rabbit experiment. He worked to establish effects of x-rays on particular organs in healthy rabbits, especially using histological practice.[27] Warren knew the piece and called it "a very rational attempt to arrive at the sensitivity of the various organs."[28] Both experimentalists grew up with animal x-ray research. In August 1945, both found themselves leaders in their respective military's medical corps.

How did the two doctors deploy their animal knowledge in the wake of the bombings? Most importantly, they made the assumption that animal data mapped onto human data. In 1923, Warren and Whipple argued that the data from their recently completed dog experiments could describe human reactions to x-rays. In the *Journal of the American Medical Association,* they wrote, "Evidence from animal experiments and scattered clinical observations is convincing that the human intestinal mucosa is peculiarly sensitive to the hard and short wave length roentgen rays."[29] In the 1920s they worried about the lining of the intestine, but by 1945, they wanted to look for damage to the blood. Tsuzuki studied the bone marrow of the rabbits he irradiated in 1926. By the 1930s, Warren could outline the timing of insults to the blood. "A profound leucopenia appears after 5 to 6 days. . . . The platelets suddenly disappear from the blood smears the day before death."[30] When the experimentalists arrived in Hiroshima and Nagasaki, they looked for insults to the blood, treating human survivors of the bomb as pieces in an animal puzzle they had been trying to solve for decades.

Tsuzuki got the jump on studying what the bombs had done because he could get to Hiroshima sooner than Warren. A rear admiral in the Imperial Navy, he used his administrative clout to initiate an epidemiological survey in Hiroshima in mid-August.[31] He managed to have volunteers distribute over 100,000 survey forms in the city. Hachiya never noted them, an indication of how chaotic things were. Hachiya did receive Tsuzuki's invitation to a symposium in Hiroshima on September 3. The director eagerly anticipated the meeting. "Since Professor Tsuzuki was going to speak on radiation sickness this afternoon, I went to the wards after breakfast and spent most of the morning reviewing our records, questioning patients, and making notes so I might be prepared to comment if the occasion arose."[32] He brought local clinical data, exactly the kind Tsuzuki did not have, to the symposium. At the meeting, Tsuzuki gave a talk on the biological effects of radiation, and his colleague, Dr. Miyake, discussed autopsy data. Questions, answers, and discussion of the data followed the formal talks.

Tsuzuki immediately took the information he learned at the symposium back to Tokyo, which had become the node for both Japanese and US experimentalists by early September. Hachiya complained that the "outside investigators only stayed a short while and thus could never acquire the intimate knowledge of the situation permitted to those who were here all the time."[33] But Tsuzuki did not come in search of the "intimate knowledge" the clinicians had gained from sitting by bedsides and speaking with patients. He wanted quantifiable data that he could share with the military medical establishment in the capital. He also knew the Americans would demand data. The US Navy, the Army, and the MED sent medical teams to Japan. Warren, head of the MED's team, arrived on September 6 after a harrowing journey across the Pacific. The next day he met up with Ashley Oughterson from the Army. The two men interviewed a Colonel M. Hiraga that morning on medical data from Hiroshima. "Warren took" everything Hiraga said "down in long hand."[34] Later that afternoon, the they met Tsuzuki and his assistant, Hitoshi Motohashi from the Army Medical College. For the next month-and-a-half, Tsuzuki and Warren existed in a strange symbiosis. Warren needed Tsuzuki for access to data in Hiroshima. Tsuzuki found Warren helpful as he tried to navigate the occupation government's new power structures. Each doctor found the other useful.

Warren benefited from this relationship with Tsuzuki once he and his team arrived in Hiroshima because they were unprepared to navigate cultural differences or to work in the utter devastation that befell the city. Once there, they haphazardly collected any Japanese data they could. On the 9th of September, Tsuzuki arranged for Warren, Oughterson, and Shinohara, a translator from Tokyo, to make the rounds of the devastated city. They spent the morning at the Red Cross hospital, which sat less than a mile from ground zero.[35] Next, they interviewed the chief of Hiroshima's provincial medical department during lunch at the central police station. On the 10th, they went to Ujima Hospital. "Spent morning at Military hospital seeing patients, hematology and pathology."[36] In the afternoon, they shared tea with Colonel Subayashi at the Ono Military Hospital. After engaging in this important social exchange, the Americans listened to medical reports presented by a team of doctors from the imperial university in Kyoto who had been studying patients at Ono. None of these interactions could have occurred in such a timely and at least nominally organized manner had Tsuzuki not smoothed the way for the Americans.

Warren longed for a situation in which his men could control a systematic effort to collect data. This he found not in either of the two destroyed cities but at the Naval Hospital in Omura. Warren learned of Omura as he and Tsuzuki flew into its airport on their way to Nagasaki on September 17.[37] He sent a team of four doctors to the Naval hospital on the 27th, a day after Shirabe arrived from Nameshi. Captain George Whipple, the son of Warren's mentor and coauthor of the dog experiments, led the team that included Lieutenant Joseph Howland.

The lieutenant's credentials commended him for the expedition. He had a PhD in zoology and an MD. But more importantly, he injected Ebb Cade, an African American man who had been in a car accident on his way to work at the secret uranium facilities at Oak Ridge, with plutonium back in March 1945.[38] Cade had no idea that Howland injected him with the dangerous alpha emitter. But Howland valued his orders from Warren and the MED leadership more than he valued Cade's health. The lieutenant's willingness to extract data from human beings paid dividends in Japan because Admiral Yasuyama, director of the hospital in Omura, allowed the American access to patients. MED doctor Brichard Brundage noted, "Examinations and complete histories (with interpreter assistance) were made on all of the patients by our medical officers."[39]

By the end of September, the US occupation had advanced to the point that Warren's doctors could collect data in Omura with near impunity. Shirabe, still working at the hospital after his effort to evacuate patients from Nagasaki, noted the American's activities. On September 28, he wrote, "An American was selecting specimens and wanted to bring home some interesting cases. He was the one who gave me cigarettes yesterday. With a smile, he said he got a backache from bending to get specimens."[40] These specimens largely went back to the University of Rochester Medical School.[41] The day after the doctor, likely Howland, complained about a bad back from all the lifting, Warren arrived at the naval hospital to meet with Yasuyama. The admiral wined and dined him. Americans were confiscating Japanese medical facilities for use by the occupation government across the country. Warren's good graces went a long way toward his keeping control in Omura. Shirabe described the visit as jocular. "In the conference room we could see that a group of visiting foreigners, the superintendent [Yasuyama], and Professor Tsuzuki were enjoying dinner with drinks."[42] The Nagasaki professor joined the affair later that evening and met Warren. The MED colonel noted meeting Shirabe and described the visit with joy.[43] He had reason to be happy. The overwhelming bulk of the data that Warren's Medical Section would use for its reporting on the biological effects of radiation from the bombs came from Omura.

Laboratory experience with animal experiments gave Warren and Tsuzuki a road map for very quickly delving into human radiological research in September and October 1945. Like the local clinicians, they fell back on familiar practices to understand the new fission. They trusted their x-ray research as a guide, hoping that fission behaved similarly to the rays their laboratory machines created. Leaning on experience with x-ray data, they took case histories and blood samples from the living. From the dead they took tissue samples. These human samples they treated as commensurable with data from animal tissue samples. While Hachiya relied on his pathology lab to find the platelet problem in humans in early September 1945, Warren and Tsuzuki knew about it from animal experiments conducted over a decade before the bombs fell. In their hands x-rays insinuated themselves into the novel sickness caused by fission.

The Experimentalists' First Reports

Despite their shared scientific foundation, the Japanese and American doctors betrayed competing political commitments as they began to publish using their data from the *hibakusha* in late autumn 1945. Warren's Medical Section forwarded a short but synoptic report to General Leslie Groves, officer in charge of the Manhattan Engineer District, on November 27.[44] The faculty of the Army Medical College in Tokyo released a significantly more in-depth report three days later.[45] Motohashi, the college's professor of hematology and Tsuzuki's assistant, co-authored the section on what they called radiation disease. The two reports systematized and visualized the onset of symptoms in comparable ways. They also shared a concern with blood formation and blood picture. Epistemologically, they came right out of the x-ray animal tradition. But these commonalities in form yielded no common conclusion. The Japanese experimentalists discerned a discrete atomic bomb disease while the MED dismissed the idea of any discrete and discernible disease at all.

In the report sent to General Groves, a single table showing the onset of symptoms encapsulated the MED's argument about how the biological effects of radiation had unfolded in Hiroshima and Nagasaki.[46] In this view, the timing of the onset of symptoms determined life or death. The leftmost column counted days after the blast. The three columns to its right showed the onset of index symptoms, such as vomiting and bloody diarrhea, in three sets of survivors grouped by exposure: most severe, moderately severe, and mild. The table shared striking similarities with one created by George Whipple to describe the final days of one of his x-ray dogs in 1919.[47] Again, the leftmost column highlights time after the moment of irradiation. The "remarks" column offers a condensed prototype of the three right-hand columns in the MED report. The key index symptoms appear. Next to each other, the two graphs present a unified experience of radiation, erasing most of the distinctions between x-rays and fission. Only the existence of quantified data in dog 18-45's table indicated difference in context. In the lab, the doctors could take measurements. In the ashen remains of Hiroshima and Nagasaki, such figures proved impossible to come by in the first days after the blast.

While the MED report adhered to norms from the animal x-ray tradition, it also benefited from contemporary animal research. Just a month before the

Preliminary Report came out, Art Welander, a graduate student at the University of Washington's Applied Fisheries Laboratory, published a report on his long-term x-ray experiments with chinook salmon in 1943 and '44. Warren organized the lab back in 1943 and had initiated its researchers into the animal x-ray tradition.[48] So in November 1945 Welander reported that "during the course of the experiment it became clearly apparent that the hemopoietic [blood-forming] tissue . . . was injured as much, if not more, than any other tissue."[49] The report to Groves followed this lead: "The important laboratory findings related primarily to disturbances in the hematopoietic function. . . . The most striking findings at autopsy were signs of destruction of the bone marrow."[50] Of course, fish lack bone marrow—they make blood in their kidneys—but both reports singled out blood-forming tissue as a key histological interest. The MED's active x-ray animal research program allowed them to argue that they found fission symptoms "which would have been predicted from animal experiments."[51]

Motohashi's reporting mirrored the MED's, a sign of the international norms established within the experimentalists' community. He argued that the radiation disease moved through the *hibakusha* in three stages. These temporal units roughly fit with the three groups in the MED report. He agreed with the MED doctors that "those who happened to be near the centre of bombing and have received a good deal of gamma-rays and neutrons died in a few days or at least in about 10 days."[52] His timings for symptoms in stages 2 and 3 also lined up with the American data. He created a bar graph that showed the timing of symptoms in groups who died in stage 1, stage 2, and who survived through stage 3.[53] The image accomplished somewhat more elegantly what the MED's table attempted to. Having shown his readers the graph, he then quantified the occurrence of symptoms among those sample populations from Hiroshima. For example, "nausea and vomiting were pronounced symptoms on the day of the bombing, appearing in proportion of 99/287 for survivors and 38/228 for the dead."[54] Across the board, Motohashi presented more fine-grained data than the MED did.

Motohashi's attention to detail perhaps anticipated differences that would develop with the US experimentalists. While the question of blood picture pushed the Americans toward contemporary fish data, Motohashi zeroed in on individual human beings in his report.[55] In a series of four graphs, he described blood sedimentation, a measurement of inflammation based on the behavior of red blood cells. In each graph he described an individual whom he named. Certainly Motohashi benefited from access to patient data from army hospitals in Hiroshima in his native tongue. But graphing the lives and deaths of individuals pulled his atomic practice away from the Americans. Though portrayed according to norms from the x-ray animal tradition, in these graphs

Motohashi presented a properly atomic symptom. Data from fission stood on its own and could be interpreted on its own.

As the winter of 1945 spilled into '46, the Japanese and American experimentalists increasingly diverged as they interpreted the data collected from the first three months after the bombings. In the suite of symptoms they quantified, the Japanese doctors identified a discrete disease with a clear genesis. Motohashi and the Army Medical College called it radiation disease in their November report. Here he closely followed Hachiya's practice on the August 27 broadside on radiation sickness in Hiroshima. In February 1946, Tsuzuki argued that "we would like to call such a pathologic condition as a whole an 'Atomic Bomb Disease.'"[56] In contrast, the MED never saw any wholeness that constituted a disease. Their November report referred to "the biological effects of radiation," a term directly plucked from the animal x-ray tradition.[57] In a comprehensive report published in June 1946 they advocated for x-ray injury. Warren and Henry Barnett, who had been at Omura, concluded, "Radiation injury has the advantage of custom, since it is generally understood in medicine to refer to x-ray effect."[58] In Warren's view, fission remained tethered to x-rays.

The experimentalists parted ways at the end of 1945 not because of differences in practice but because they disagreed about how to see the biological action of the new fission. The Japanese doctors saw a unity, a single disease easily traceable to the moment of fission. They perceived the sickness as a continuation of the bombs' total effect. Invisible radiations created burns, broken bones, and insults to the blood and organs. The Americans never saw fission in that way. Its blast effects and heat effects were essentially like those from conventional bombs. Its radiation behaved essentially like x-rays. In the lab, they had never described exposure to x-rays as a disease. Why should they with fission? Instead, they saw a suite of symptoms, unevenly distributed among those exposed to the bombs' free neutrons and gamma rays. Charles Rosenberg has argued that "the existence of a disease as *specific* entity is a fundamental aspect of its intellectual and moral legitimacy."[59] Based on laboratory experiments with animals, the MED never afforded that legitimacy to the *hibakusha*.

Conclusion: The American Failure to See a New Disease

Biologically, what did it mean that two atomic bombs razed Hiroshima and Nagasaki in early August 1945? For a host of unsuspecting men, women, and children, it meant instant death. For those who survived the blasts, the answers to this question were complex and unstable, like the radioisotopes that fueled the two bombs. Michael Gordin has argued that the bombs themselves experienced an instability of meaning even as they were used. Initially, just very powerful firebombs, they underwent an "apotheosis" after the emperor announced Japan's speedy surrender in mid-August.[60] In Gordin's estimation, their destructive power mated with their political power to give "atomic" a new and singularly special definition. In their own contextual ways, the local doctors and the military experimentalists worked out what it meant that the visible symptoms ailing and killing the *hibakusha* were atomic, resulting from fission's invisible offspring. Not content to simply catalogue symptoms, they worked to make visible the imperceptible mechanisms unleashed by fission's new radiations by studying blood and tissues under the microscope. This essay has argued that in the first three months after the bombings, radiation sickness existed both without reference to radiation and without the distinction of actually being a sickness. No straightforward path led the suite of biological symptoms toward reification as a discrete disease, let alone one easily tied to fission, in the autumn of 1945.

Instead, the biological effects of the bombs became tied to the new fission in fits and starts. The novel disease took root amid prewar practices and knowledge traditions. Local doctors forced into clinical service in Hiroshima and Nagasaki reached backwards to classic responses in cases of public health emergencies. They treated infections. They created separate wards and worked to enhance sanitary conditions for patients whom they believed suffered from cholera and dysentery. In many ways their medical responses in the first few days after the bombings looked much like those in cities that had been firebombed by conventional ordinance. But as patients inexplicably died, they turned to traditional hospital practices like taking case histories and, when they got the equipment, taking white blood cell and platelet counts. They created knowledge, new to them, even as information about the bombs' radiation trickled in to their destroyed cities. At the same time, outside doctors from the Japanese and US militaries reached into the past to deploy knowledge they had created in the laboratory using x-rays. They tied these two distinct radiations, one novel and one known, together. From the start, fission's agency remained in the background.

When accounts of the bombs' effects began to stabilize in late 1945 and early 1946, they treated fission's work distinctly. Japanese accounts pointed to a clearly atomic disease, born of the new technology that ravaged Hiroshima and Nagasaki. Local clinicians made this argument by appealing to their personal experience. They suffered from the same disease they struggled to understand and treat. Japanese experimentalists privileged the action of the new fission by assigning to it a distinct disease. In many ways, the American-led Atomic Bomb Casualty Commission (ABCC), whose researchers studied the bombs' genetic fallout, tacitly acknowledged these conclusions in the decades after the bombings. "*What happened* to the survivors—the slow and invisible internal pathologies of their bodies over the decades," says Susan Lindee, "was gradually made visible and real by the science of the ABCC."[61] But such a disease, a coherent group of pathologies, never became clear to the Manhattan Engineer District, the progenitors of the bombs. For Warren and his cadre of doctors and scientists, x-rays and fission existed as practically comparable forms of radiation until the end of atmospheric testing in 1963. So would the tradition that data from animals and humans could interchangeably describe radiation's biological effects. Based on this research tradition, the Medical Section saw data, neatly tabulated, when they envisioned the biological effects of the new fission. Doing so, they turned a blind eye to the suffering and very real sickness that the *hibakusha* suffered because of the atomic bombs.

Notes

1 Michihiko Hachiya, *Hiroshima Diary: The Journal of a Japanese Physician August 6–September 30*, trans. Warner Wells (Chapel Hill: University of North Carolina Press, 1955), 1.
2 Ibid., 6.
3 Ibid., 21, and Raisuke Shirabe, *A Physician's Diary of the Atomic Bombing and Its Aftermath*, trans. Aloysius Kuo (Nagasaki: Nagasaki Association for Hibakusha's Medical Care, 2002), 31, http://www.nashim.org/e_pdf/phy/a_physicans_diary.pdf.
4 See notes 1 and 3.
5 For the disease and diplomacy, see John Beatty, "Scientific Collaboration, Internationalism, and Diplomacy: The Case of the Atomic Bomb Casualty Commission," *Journal of the History of Biology* 26, no. 2 (Summer 1993): 205–231, and M. Susan Lindee, "The Repatriation of Atomic Bomb Victim Body Parts to Japan: Natural Objects and Diplomacy," *Osiris* 13 (1998): 376–409. For the bomb and genetics research, see M. Susan Lindee, *Suffering Made Real: American Science and the Survivors at Hiroshima* (Chicago: University of Chicago Press, 1994). For a study of survivors' responses to the sickness, see Naoko Wake, "Atomic Bomb Survivors, Medical Experts, and the Endless Radiation Illness," in *Inevitably Toxic: Historical Perspectives on Contamination, Exposure, and Expertise*, ed. Brinda Sarathy, Vivien Hamilton, and Janet Brodie (Pittsburgh: University of Pittsburgh Press, 2018.
6 Hachiya, *Hiroshima Diary*, 36.
7 Shirabe, *A Physician's Diary*, 3.
8 Ibid., 5.
9 Ibid., 31.
10 The Committee for the Compilation of Material on Damage Caused by the Atomic Bombs in Hiroshima and Nagasaki, *Hiroshima and Nagasaki: The Physical, Medical, and Social Effects of the Atomic Bombings*, trans. Eisei Ishiwaka and David Swain (New York: Basic Books, 1981), 504.
11 Shirabe, *A Physician's Diary*, 14–15.
12 Hachiya, *Hiroshima Diary*, 74.
13 Ibid., 69.
14 Ibid., 99.
15 Ibid.

16 Ibid., 108–109.
17 Ibid., 109.
18 Ibid., 124.
19 Ibid.
20 Ibid., 125.
21 Ibid., 147.
22 Ibid., 148.
23 Nobuko Margaret Kosuge, "Prompt and Utter Destruction: The Nagasaki Disaster and the Initial Medical Relief," *International Review of the Red Cross* 89, no. 866 (June 2007): 290.
24 Shirabe, *A Physician's Diary*, 42. Shirabe began journaling again when he arrived at Omura.
25 C. C. Hall and G. H. Whipple, "Roentgen-Ray Intoxication: Disturbances in Metabolism Produced by Deep Massive Doses of the Hard Roentgen Rays," *American Journal of the Medical Sciences* 157 (1919): 453–482.
26 See *Journal of Experimental Medicine* 35, no. 2 (January 31, 1922), and 38, no. 6 (November 30, 1923).
27 Masao Tsuzuki, "Experimental Studies on Action of Hard Roentgen Rays," *American Journal of Roentgenology and Radium Therapy* 16 (1926): 134–148.
28 Stafford Warren, "Organ and Body Systems," in *Biological Effects of Radiation: Mechanism and Measurement of Radiation, Applications in Biology, Photochemical Reactions, Effects of Radiant Energy on Organisms and Organic Products*, ed. Benjamin Duggar (New York: McGraw-Hill, 1936), 475.
29 S. L. Warren and G. H. Whipple, "Roentgen-Ray Intoxication: Roentgenology in Man in the Light of Experiments Showing Sensitivity of Intestinal Epithelium," *Journal of the American Medical Association* 81, no. 20 (November 17, 1923): 1673–1675.
30 Samuel Shouse, Stafford Warren, and George Whipple, "Aplasia of Marrow and Fatal Intoxication in Dogs Produced by Roentgen Radiation of All Bones," *Journal of Experimental Medicine* 53, no. 3 (1931): 421–435.
31 Lindee, *Suffering Made Real*, 26.
32 Hachiya, *Hiroshima Diary*, 158.
33 Ibid., 160.
34 *Oughterson Daily Journal*, September 7, 1945, Box 292, Folder "Naval Mission to Japan," Stafford Leak Warren papers (Collection 987). UCLA Library Special Collections, Charles E. Young Research Library, University of California, Los Angeles (hereafter cited as MSS Warren).
35 *Oughterson Daily Journal*, September 9, 1945, MSS Warren.
36 Ibid.
37 Joseph Howland to Robert Buettner, November 5, 1947, Box 65, Folder 3, Reel 13.1, MSS Warren.
38 See Eileen Welsome, *The Plutonium Files: America's Secret Medical Experiments in the Cold War* (New York: Dial Press, 1999), 83–86. Welsome provides an exhaustive history of the plutonium injections at Oak Ridge, Rochester, and elsewhere.
39 Brichard Brundage to Stafford Warren, November 12, 1947, Box 65, Reel 13.1, MSS Warren.
40 Shirabe, *A Physician's Diary*, 46.
41 Howland to Buettner, October 28, 1947, Box 65, Folder 3, MSS Warren.
42 Shirabe, *A Physician's Diary*, 49.
43 *Warren Daily Journal*, September 29, 1945, MSS Warren.
44 Preliminary Report of Findings of Atomic Bomb Investigating Groups at Hiroshima and Nagasaki, November 27, 1945, Box 298, MSS Warren.

45 Army Medical College–The First Tokyo Army Hospital, Medical Report of the Atomic Bombing in Hiroshima, November 30, 1945, 16, Box 61, Reel 11.1, MSS Warren.
46 Table I, a, Preliminary Report, November 27, 1945, MSS Warren.
47 Hall and Whipple, "Disturbances in Metabolism," 1919.
48 See chapter 1 in Neal Hines, *Proving Ground: An Account of Radiobiological Studies in the Pacific, 1946–1961* (Seattle: University of Washington Press, 1962).
49 UWFL-2, "Studies of the Effect of Roentgen Rays on the Growth and Development of the Embryos and Larvae of the Chinook Salmon (*Oncorhynchus tschawytscha*)," October 1945, 54, Box 9, Volume 1, Accession No. 00-065, University of Washington, Laboratory of Radiation Biology records. Special Collections Division, University of Washington Libraries, Seattle, Washington.
50 Preliminary Report, November 27, 1945, MSS Warren.
51 Ibid.
52 Medical Report, November 30, 1945, MSS Warren.
53 Ibid.
54 Ibid.
55 Ibid.
56 Masao Tsuzuki, "Report on the Medical Studies of the Effect of the Atomic Bomb," February 28, 1946. Translation in *General Report: Atomic Bomb Casualty Commission January 1947*, 74, US National Research Council, http://www.nasonline.org/about-nas/history/archives/collections/organized-collections/atomic-bomb-casualty-commission-series/abccrpt_pt3app9ch1.pdf.
57 The US National Research Council sponsored the preparation of the 1936 book *Biological Effects of Radiation*, edited by Benjamin Duggar of the University of Wisconsin. This volume contained the article in which Warren praised Tsuzuki's rabbit experiments. See note 31.
58 Manhattan Engineer District, "The Atomic Bombings of Hiroshima and Nagasaki," June 29, 1946, 67–68.
59 Charles Rosenberg, "Framing Disease: Illness, Society, and History," in *Framing Disease: Studies in Cultural History*, ed. Charles Rosenberg and Janet Golden (New Brunswick, NJ: Rutgers University Press, 1992), xvi.
60 Michael Gordin, *Five Days in August: How World War II Became a Nuclear War* (Chicago: University of Chicago Press, 2007), 14.
61 Lindee, *Suffering Made Real*, 256.

Reason and Risk

Challenging the Expert and Public Divide in the Risk Debates on Uranium Mining in India

PRERNA GUPTA

For the last two decades, roughly from the period following the nuclear weapon tests conducted by the country in May 1998, India's oldest uranium mines have been mired in controversy. These mines located in the Adivasi (or Indigenous) land are suspected of having caused adverse health effects on mine workers and the inhabitants of the villages nearby. The controversy has involved contentious claims about the veracity of these health effects, the causes and the linkages with radiation, with different positions staked out by local anti-nuclear activists, NGOs, physicians, physicists, and officials from the state-owned Uranium Corporation of India Limited (UCIL), who have all used different methodologies to analyze the situation and seek attribution for the ill health.

Perception of risks from radiation has come to acquire a central space in this debate. Some people, mine workers and activists around the mines, are convinced that the radiation from the mines and the tailing ponds are the source of the ill health they have observed in the area over generations. The nuclear establishment officials (UCIL and other affiliated government officials) often assert their scientific-technical authority in this controversy to dismiss these claims as "myths" rather than "facts."

This essay deconstructs these claims by employing cultural and sociological studies on risk. Mainstream discourse and many scholars of risk (especially some psychometric studies) make a distinction between objective assessment of risk by experts and subjective—value or emotion-based perception of risk by the public.[1] This essay adds to the literature in risk studies that questions this binary by extending three arguments connected to this case study. First, risk definitions and assessment methodologies by experts are not as unanimous as often perceived, as experts argue and dispute each other's definitions. Second, people's perception of risks is not merely based on values and emotions, devoid of any cognitive assessments. They make judgments based on their past and present social experience of institutions that are involved in making and managing risks. Third, (technical) experts do not exist in contexts free of politics and are not immune to the politicization of risk controversies.

History of the Debate

To dive deeper into the politics of this controversy we first need to understand the historical context of the debate. The newly independent Indian state under the leadership of Jawaharlal Nehru was focused on pushing forward development and modernity through the miracles of science and technology.[2] Nuclear technology was fast acquiring the status of the epitome of modern science in the context of the bombing of Hiroshima and Nagasaki and the arms race of the cold war.[3]

Uranium was important to fuel what the Indian state likes to call the indigenous nuclear program.[4] India's Atomic Energy Commission (AEC) was established in 1948 soon after the independence of India. Efforts were launched soon after to look for uranium, and in 1967 the first mine in Jadugoda became operational. Activists and locals recount that their rice fields were taken away to make room for the construction of the first two tailing ponds, that hold processed waste from the mine.[5]

In 1974, while progress in nuclear energy for electricity generation remained slow, India tested its first nuclear explosive.[6] This shocked the supplier countries who formed the Nuclear Suppliers Group (NSG) in order to control the export of materials, equipment, and technology that can be used to manufacture nuclear weapons.[7] They banned India from trade in nuclear technology, including importing uranium. This ban, along with the continued promotion of aggressive nuclear energy targets, only put more pressure on India's demands for uranium mining.

This growing demand meant more displacement and dispossession for the Adivasis of the region. By the mid-1980s, there were already two valley-dam types of tailing ponds at Jadugoda used for storing the waste from the uranium mine.[8] However, both were almost full, and UCIL wanted to construct a third tailing dam. In 1985, it served the notice to the inhabitants of Chatikocha, a tola or hamlet of Bhatin, that their land was needed. Chatikocha was prosperous, with good crops, forest products, and the seat of famous theater groups.[9] Naturally, the villagers didn't want to be displaced. For nine years nothing happened. In 1994, "the villagers were directed to appear at the UCIL offices to collect their compensation because their land had been acquired. Most families were deeply offended by the pitiful compensation offered by UCIL and refused the offer, delivering instead a set of demands which were ignored."[10]

On January 27, 1996, the UCIL management, assisted by the Central Reserve Police Force, the central industrial security force, the Bihar police, a set of heavy bulldozers and a high-powered shovel, entered Chatikocha and started razing down the houses without any warning. The houses of over thirty families were leveled together with their fields and crops, their sacred groves, and graveyards.[11] Soon, the locals, mostly Adivasis belonging to the Santhal, Ho, and Munda communities, had mobilized a large number of people from nearby villages in support of the people of Chatikocha, and UCIL suddenly found it had an uprising on its hands. It apologized, and the demolition was temporarily suspended, but it did not rehabilitate the people.[12] Student leader Ghanshyam Biruli along with other leaders of the Jharkhand movement took up the issue with the aim of demanding rehabilitation and compensation for the displaced families.

They formed a local organization called the Jharkhand Adivasi Visthapit Berozgar Sangh (JAVB) on June 5, 1995, to this end.[13] However, during this time they started receiving literature on uranium mining and radiation after the historic first World Uranium Hearing in 1992, where Indigenous speakers and scientists from all continents testified to the health and environmental problems of uranium mining and processing, nuclear power, nuclear weapons, nuclear tests, and radioactive waste disposal.[14] As they came to learn more about the hazards of uranium mining, the activists were compelled to ask for a complete stop to uranium mining in the region.

Ionizing radiation was a new and frightening concept. It quickly became central to their campaign against uranium mining. Biruli now linked his father's death by lung cancer in 1984 to his work in the uranium mines. Biruli's father, like many others in the region, worked as contract labor who extracted ore and shoveled yellowcake into drums without adequate safety equipment.[15]

In 1998, Ghanshyam Biruli along with other local activists established Jharkhandi Organisation Against Radiation (JOAR) with the aim of demanding a complete stop to all uranium-related mining in the region.[16] When activists started talking to people in the region, they found that people were experiencing inexplicable impacts on their health and the environment. These were impacts that their elders had never seen and were commonly believed by villagers to be the doing of evil spirits.

In 1999, filmmaker and activist Shri Prakash made a documentary film *Buddha Weeps in Jadugoda* in collaboration with JOAR. Turning the code name "Smiling Buddha" for India's bomb test of 1974 on its head, the film recorded testimonies of people experiencing health effects in their daily life. The film shows several cases of congenital defects in newborn babies and sterility in young women. Jairam Murmu, a young man in his twenties, says that their elders hadn't ever seen diseases like they witness these days, for instance, the teeth of the cattle rotting and falling out. Kundu Melgandi, a local farmer, shows that he has been finding more deformed seeds in a local fruit called "Tiril." Hari Podo Murmu talks about a common belief held in the community that evil spirits residing near the mines and tailings were leading to miscarriages, which he now attributes to radiation.[17]

The film also documents negligent work practices of UCIL, including footage of leaking drums carrying yellowcake being handled by barehanded workers, barefoot teenagers playing with these cylindrical drums, dust blowing toward the villages from open tailing ponds, mine workers testifying that they are not provided gloves or proper masks, and so on. Viewers of the film cannot but connect these practices with the health impacts that are also documented. As opposition to uranium mining gathered momentum in the late 1990s, reports about the situation in Jadugoda were being widely published.[18]

People of Jadugoda were observing an increase in health issues. For example, a midwife who had helped deliver babies for around forty years told the activists that there were more abnormal babies born and miscarriages now than when she was young. She believed that evil spirits were eating up babies. She had observed children being born with the crown of their skull not fully formed, and one out of ten babies she delivered in a single year had this condition.[19] Mine workers observed that when they were cut by the rock, their wounds would not heal quickly.[20] Dopan Majhi (sixty-five), Pradhan of Tilaitand, said: "When UCIL started its operations, a lot of my contemporaries were employed there. They are all dead now."[21] Another elder of the community, Manghal Majhi said he had been a sick person ever since he returned from Jadugoda. Under the company treatment, he was told he has tuberculosis and was treated for that, but when he consulted a private doctor, he was told that he doesn't have tuberculosis. He was then convinced that he was "sick of drinking uranium-contaminated water." Explaining his claim, he said he remembers a tree next to a river that slowly died along with a lot of animals and fish. That is where he used to collect water for himself, and he thinks the water had made him sick.[22]

As the news spread, UCIL stepped in to dismiss these claims as myths that are not based on science. Even in 2019, UCIL's website dismissed media reports as "stories of human interest invariably spiced with melancholy and drama using telling pictures of human sufferings to condition the viewer."[23] It emphasized that an independent Health Physics Unit had been regularly monitoring the radiation levels in the environment and had found no significant impact on background radiation or ground and surface water bodies due to mining. UCIL and its affiliates also claim that in their health surveys they have not observed any radiation-related illness in the region. However, these findings have been contested by independent experts. These debates among experts are explored in detail in the next section.

1. Experts Debate: Challenges to Objective Risk Assessment

The gap between objective risk assessment by experts and the public's subjective perception of risk continues to dominate the popular discourse on risk. Even after much social science analysis rejects the notion that risk is objective, the dichotomy persists in media, policymaking, and even among some scholars of risk. Some researchers, for instance, have even argued that the public's perception of risk is less cognitive than has previously been believed and that such factors as attitudes and moral values play a more crucial role.[24] This is the dichotomy that UCIL and their affiliates emphasize in their narrative of the risks from uranium mining. A section of their website called "Myth of UCIL" claims that people's apprehensions about uranium mining in Jadugoda are based on "a false understanding of the facts" and goes on to elaborate on "facts" based on the studies carried out by their experts.[25] However, the conclusions from these studies have been contested by independent experts, especially the methods used to calculate radiation doses and health impacts.[26]

1.1 Experts Debate How to Measure Radiation

When their claims were dismissed as irrational by UCIL, Ghanshyam Biruli and other local activists decided to do an independent study in the region to gather evidence in support of their claims. It sought out the Gandhian peace organization Sampoorna Kranti Vidyalaya (SKV), run by physician Sanghamitra Gadekar and her husband, physicist Surendra Gadekar. The couple had been actively advocating against nuclear weapons and nuclear energy and had been publishing a magazine called *Anumukti* or *Liberation from the Atom* since 1987. The Gadekars had experience in conducting health surveys around nuclear power plants and explosives' testing sites. They reached Jadugoda to conduct an independent inquiry into the situation in September 2000.[27]

Bhabha Atomic Research Center (BARC)[28] had taken readings from the area around the tailing pond and 2 km around the pond from 1993 to 1997. The survey conducted by Dr. Jha, a scientific officer of the Health Physics Unit of BARC, concluded that the radiation levels were well within the standard limits of 1 milliSievert per year.[29] The BARC claimed that their readings also demonstrated that there was no risk from the tailings pond (which had become the center of media reports and people's apprehensions) and that there had been no significant radiation impact because of UCIL's operations in the preceding thirty-three years. According to BARC, the radiation level from the pond fell back to natural background levels 20 meters from the pond. This was demonstrated by another scientific officer at BARC, A. H. Khan. In 1998 in the presence of journalists Khan found that the radiation readings, though they were as high as 0.76 microGray an hour (or mSv) inside the tailing pond, fell to the permissible limit of 0.11 microGray outside the pond.[30]

Although Surendra Gadekar agreed that the radiation levels fell to normal right outside the ponds, he raised doubts about the averages calculated by BARC for the nearby villages. In an article in *Anumukti,* he recounts an interesting incident out in the field. "One day while we were doing data collection during the survey, Dr. Jha himself arrived and complained that we had been publicizing falsehood with regard to radiation readings in the neighbourhood. . . . The upshot of the discussion was that I agreed to accompany him and take joint readings. . . . Just outside his office, was a pile of rocks. The readings there were three times higher than the ones we had been taking. I asked him why he did not take readings near this pile or many other like piles spread all over Jadugoda. His answer was that this was mine overburden rock and readings near such piles were bound to be higher." [31]

This, Surendra believed, was the crux of the controversy, observing that since radiation levels vary greatly within short distances and times, it was easy to get widely varying numbers and different averages. Surendra argued what matters to people's health is that averages should reflect where people spend most of their time. Talking to the villagers, Surendra further found that UCIL had been selling the overburden rocks to construction companies for very low prices to get rid of them easily. As a result, these rocks had been used to build people's houses and even children's schools. Readings taken by the SKV team in these houses range from 0.25 to 0.35 μSv/hr, double the permissible limit of 0.11 μSv/hr. Along with that, the main intersection in the village where people gather often was covered by dust from the trucks carrying the overburden rocks from the mines. Here, where people are often seen sitting together on the ground, the readings ranged from 0.38 to 0.44 μSv/hr. Hence, Surendra argues that the annual average radiation dose for an individual living in such a house would be 1.5–2 mSv/ year, which exceeds the annual allowed dosage by at least 50 percent.[32]

This gap between UCIL experts' readings and lived reality of local communities in Jadugoda was reminiscent of another case about radiation exposure in the United Kingdom, as recounted in Brian Wynne's remarkable ethnographic study of herding communities living near the Sellafield nuclear plant in the UK. When high levels of cesium were found in the sheep near the Sellafield processing unit in the 1980s, the experts repeatedly assured the people that the levels would fall back to normal very soon. These experts assumed that the high readings were the result of the Chernobyl accident, while sheep farmers had suspected for a long time that the Sellafield plant itself was the source of the contamination in their herds. To their own embarrassment, experts' predictions failed repeatedly, and yet they continued to ignore people's crucial "knowledge of their local environments, hill-sheep characteristics and hill-farming management realities such as the impossibility of grazing flocks all on cleaner valley grass, and the difficulties of gathering sheep from open fells for tests."[33]

Strikingly similar observations were made by Surendra and another independent physicist Koide from Kyoto University four years later. They found that the cattle belonging to farmers in Jadugoda would wander off to exhausts from the Bhatin mines that exist at ground level. These exhausts let out a cool wind and were a source of relief for the cattle in the heat of scorching summer. But, in fact, these exhausts had been installed in mines to continuously vent out radon gas. Thus, the cattle were inhaling a stream of radon gas by standing there. Not only the cattle but farmers themselves used to come to take a whiff of the cool wind freely in the lack of any security or barricade to keep the public out. Koide collected radon samples from the vent and found the concentration to be 2400 Bq/m^3.[34] To put this number into perspective average outdoor radon level varies from 5 to 15 Bq/m^3.[35]

Ignoring such hazardous concentrations of radon, the studies done by the nuclear establishment in this case also focus only on the averages calculated from the samples collected at twenty to thirty locations around the mines. A study done in 2011 by Bhabha Atomic Research Centre (BARC), even as it notes some variation, found the *average* indoor radon to be well within the limit of 100 Bq/m^3 prescribed by the WHO: "The atmospheric outdoor radon levels around the villages vary from 9 to 36 Bq/m^3 with an average of 18 Bq/m^3, which is slightly higher than that of the global average of 10 Bq m^3.[36] The indoor radon levels in the region vary from 45 to 135 Bq/m^3, with an average value of 70 Bq/m^3."[37] There is no reflection in their papers on the choice of certain locations over others. Even while remaining skeptical of the results from the independent studies, Surendra and Koide's epistemic approach raises important questions: What are the radiation levels where people are spending most of their time? Where are the higher radiation readings located? How much are people exposed to these places? Simply calculating averages from randomly sampled areas disregards the lived realities of people and are hence of little consequence to them. And if one goes by

Surendra's account of interaction with Dr. Jha of BARC, then the samples are not so randomly collected either.

1.2 Experts Debate How to Measure the Health Impact of the Mines

The health reports from UCIL and their affiliates again conflict with independent researchers and the observations of local communities. The conflict here too partly arises from the use of different methodologies. UCIL and their affiliates like BARC have used population surveys in the Jadugoda region with clinical confirmation. Most of these surveys are not available publicly. The surveys discussed here have been accessed via the documents submitted to the court, which are merely the summaries, and some that have been reported in news. It is unclear why they chose to conduct surveys for these studies when epidemiological studies that compare the exposed population to a control group have been the standard worldwide to study radiation exposure in communities.[38] Both the independent studies discussed in this essay used epidemiological methodology. They rely on the diseases as observed by the surveying doctors and as reported by the people themselves rather than clinical confirmations.

The first health impact study in the region was done in 1998 after the Environment Committee of the Legislative Council of Bihar (which is now bifurcated into Jharkhand and Bihar states) raised concerns over the health issues being reported in the 1990s. Doctors from the Bihar government and UCIL jointly surveyed residents within 2 km of the mines. About 3,400 people in total and 31 shortlisted cases were clinically examined. Although the entire study is not available, the summary submitted to the court states that the consensus of all the doctors was that "the cases examined had congenital limb anomalies, diseases due to genetic abnormalities like thalassemia major and retinitis pigmentosa, moderate to gross splenomegaly due to chronic malarial infection (as this is hyperendemic area), malnutrition, post-encephalitic and post head injury." The team was convinced and unanimously agreed that "the disease pattern cannot be ascribed to radiation exposure in any of these cases."[39]

The choice of the words "cannot be ascribed" here is particularly noteworthy because even though health studies on uranium miners [40] and residents around uranium mines[41] have found severe health impacts around the world, determining the exact extent of health impact from a source of ionizing radiation is a notoriously difficult task. There is generally no way to distinguish a cancer (or a heart attack) caused by radiation from one caused by smoking or chemical exposures or other factors. Most cancers have several interacting causes. Further, long-term genetic damage and cancer typically manifest years, often decades, later. These factors mean that the effects of radiation are often inadequately recognized, downplayed, or contested. For instance, the estimated death toll from the Chernobyl nuclear disaster has ranged from 4,000 to 985,000.[42]

An independent epidemiological study conducted by the SKV team under Dr. Sanghamitra Gadekar told a different story from UCIL's version. The health status of four villages near Jadugoda was compared to two villages far away from the mines used as a control. The Department of Atomic Energy (DAE) officials had been saying that the deformities in children were because of malnutrition, poverty, and unhygienic practices. So, before they undertook any health study, the SKV team examined the socioeconomic status of the villages around Jadugoda. Their study found that the uranium mine had at least brought relative economic prosperity to the region. Villages around Jadugoda had people working in higher-paying salaried jobs than in the control villages and owned more consumer goods like televisions and refrigerators.

The study found the rate of congenital deformities to be around four times higher in villages near uranium mine: while there were 15 cases among 266 children near Jadugoda, there were only 2 cases among 144 children in the control villages. Similarly, the rate of stillbirths was also found to be higher.[43] The team was also expecting a high incidence of chronic lung diseases in the uranium workers; it has been found in many countries that uranium mine workers are susceptible to diseases like silicosis and lung cancer since they work in an environment of high dust caused by the blasting of rocks.[44] But in Jadugoda, workers did not report any such work-related diseases. However, they did find an anomalously high incidence of tuberculosis (TB). While the incidence of TB was 9 per thousand in control villages, it turned out to be 16 per thousand in Jadugoda villages. This number did not make any statistical sense to the researchers. Further, some of the patients had been on medication for years with no relief for a disease that is easily treatable within six months of medication. Sanghamitra and Surendra suspected what they did not have the resources to prove: the UCIL hospital had been misdiagnosing them because if they were diagnosed to be suffering from an occupational disease, they would have to pay each patient compensation.[45]

Another independent epidemiological study was undertaken by the Indian Doctors for Peace and Democracy (IDPD), an associate of International Physicians for the Prevention of Nuclear War (IPPNW). The study conducted in 2007 found similar patterns as those in the SKV study of higher incidences of deformities, sterility, and cancer in the region as compared to a control population. This study was criticized for reasons such as the lack of clinical confirmations and peer review.

The nuclear establishment has enough resources to conduct detailed clinical confirmations. However, one study funded by the DAE, available publicly, reveals that studies relying on clinical confirmations can also have severe shortcomings. The methodology section shows the scale of the study, which was meant to cover all villages within a 5 km radius of Jadugoda mines and stretched over a period of four years to confirm diseases through a series of clinical tests.

> From 6,900 households covered during survey, 2,693 were called to the health camps for further screening. From 1,523 who attended the health camp, 220 cases were referred for confirmation of diagnosis. Only 91 cases could complete most of the diagnostic tests. Out of these 91 cases, 56 had been called to rule out or confirm cancer and 35 for congenital anomaly. Of the 56 people, only one case of cancer was confirmed. In 38 people, cancer was ruled out and 17 people could not complete the investigation. From 35 people who had been called to rule out or confirm congenital anomaly, six cases of congenital anomaly were confirmed. One of the referred cases died while the investigations were going on. In 28 people, the diagnosis of congenital anomaly was either not sure or the investigations could not be completed.[46]

The investigators note that with no promise of treatment and people asked to come to a hospital 20 kms away multiple times at their own expense and time, most of the people dropped out of the survey at some stage. They also highlight the problem of a small sample due to participants dropping out: "During all the stages patients were missed due to various reasons and so it is possible that there is underestimation of prevalence rates which might lead to presumably wrong conclusion."[47]

1.3 Lessons from Risk Studies

These debates among experts show how measuring risks can be a highly contentious issue. Experts often raise legitimate questions over the limitations of each other's definitions and methodology. The idea of an objective definition of risk then seems rather illusory. Subjectivity permeates risk assessment at every stage of the process, from the initial definition of the risk problem to deciding which results or consequences to include in the analysis, identifying and estimating exposures, choosing appropriate dosage measurements, so on and so forth.

A prominent scholar of risk, Paul Slovic (1999) illustrates this dilemma by presenting nine possible ways to measure something that seems as simple as human fatalities:

- Deaths per million people in the population.
- Deaths per million people within x miles of the source of exposure.
- Deaths per unit of concentration.
- Deaths per facility.
- Deaths per ton of toxic air released.
- Deaths per ton of toxic air absorbed by people.
- Deaths per ton of chemical produced.
- Deaths per millions of dollars of product produced.
- Loss of life expectancy associated with exposure to the hazard.

. . .

> The choice of one measure or another can make a technology look either more or less risky. For example, between 1950 and 1970, coal mines became much less risky in terms of deaths from accidents per ton of coal, but they became marginally riskier in terms of deaths from accidents per employee. Which measure one thinks more appropriate for decision-making depends upon one's point of view. From a national point of view, given that a certain amount of coal has to be obtained to provide fuel, deaths per million tons of coal is the more appropriate measure of risk, whereas from a labour leader's point of view, deaths per thousand persons employed may be more relevant.[48]

Further, institutional structures to which experts belong are involved in creating limits and making definitions of risk. Social contexts of these institutions and competing interests of various actors involved with them inevitably impact these limits and definitions. People consider these factors when making risk decisions. Psychometric studies of risk have consistently found trust (in risk management) as a factor that has a big influence on people's risk assessments.[49] Others argue that trust is a rather reductive measurement for people's past and current social experience of institutions involved in making and regulating risks.

In a study of risk management practices around hazardous wastes in different European countries, Brian Wynne found that "public reactions to 'hazardous waste risks' and associated regulations or proposals are reactions to accumulated experience of 'regulation' as a historical relationship. This includes control of public health and environmental damage, but also intelligibility, competence, trustworthiness, and social identifiability of institutions, which themselves have several public interfaces. These social relationships are the grounding of 'public perceptions of risk.'"[50] Further "the different national institutional settings resolve regulatory definitions of hazard and waste in different ways. And even within a single regulatory framework the precise practical definition of 'fixed' technical criteria is an institutional process. The construction involves actors and groups with divergent interests, perceptions, constraints and rationalities, and thus with competing favoured definitions."[51]

The lack of safety equipment for workers and other negligent practices of UCIL observed in *Buddha Weeps in Jadugoda* make up for part of the social experience that people have of UCIL's management of the mine operations and waste facilities. Further scrutinizing UCIL's operations in other parts of India shows that people's suspicions are not unfounded.

2. A Pattern of Pollution and the Institutional Limitations to the Regulation of UCIL Operations

UCIL's activities have been found to contaminate by three separate government bodies in different parts of India. Here UCIL's pattern of pollution along with a tendency to deny responsibility emerges clearly. Uranium mining can lead to the contamination of water with chemicals, metals, and radionuclides higher than background or preconstruction conditions.[52] The only publicly available study by UCIL on the drinking water surrounding the area of the Jadugoda mines found the uranium levels well within the WHO recommended levels.[53]

In the state of Andhra Pradesh, however, UCIL was questioned for contamination of groundwater around its facility by the Andhra Pradesh Pollution Control Board's Kurnool zonal office (APPCBK). The APPCBK issued a notice to UCIL in March 2018 to explain the increase of uranium in groundwater content in the area. UCIL blamed the natural uranium content of the land for groundwater pollution. AAPCBK replied stating that UCIL's claims of not polluting groundwater were false: uranium values were 1 to 7 ppb in the water samples collected in Mabbuchintalapalle in 2013, whereas uranium values ranged from 690 to 4,000 ppb (against the standard of 60 ppb) when the borewells in the surrounding villages were more recently monitored.[54] The pollution control board also surveyed the residents and confirmed that the residents were indeed victims of contaminated water as they were suffering from skin-related diseases and urged the state government to conduct medical camps at the earliest. AAPCBK has issued directions to UCIL for ensuring safer operation in the area.[55]

In Chhattisgarh, even exploration activities by the Uranium Corporation have been known to cause contamination. People in twenty-two villages in Chhattisgarh have suffered health consequences from arsenic contamination due to the Bodal uranium exploration. The exploration that started in 1976 and ended in May 1989 dug up to 600 meters below the ground level. Kaudikasa, a community situated 3 km away from the mine, has suffered from arsenic contamination for two decades now. The arsenic level in the wells of Kaudikasa was found to be 520 μg per liter, more than 50 times the permissible limit of 10 μg per liter according to the WHO. Whereas in the dug wells (general depth less than 50 m) the arsenic concentration was as high as 880 μg per liter. A medical study done in 2000 in the village indicated that 42 percent of adults and 9 percent of children were suffering from arsenical keratosis, a condition in which changes in skin pigmentation occur from long-term exposure to arsenic. Epidemiological studies indicated high concentrations of arsenic in urine, hair, and nails in 89, 75, and 91 percent of the village population respectively. The government only started supplying uncontaminated water to the village of Kaudikasa eighteen years after this finding.[56]

In Karnataka, the Western Ghats Task Force (a special initiative by the state government to conserve the biodiverse region), also alleged that UCIL has created serious health hazards at Gogi village in Yadgir district, by letting out effluents from its uranium testing site into a lake, which is a source of drinking water to the villagers. Several borewells drilled by the UCIL to ascertain the density of uranium deposits have also, according to the report, caused harmful effects on the ecology.[57]

Further, the institutional structure of the Department of Atomic Energy (DAE) presents serious hurdles to the unbiased safety review of nuclear facilities in India. Until 1972, DAE did not have a separate, identifiable organization or personnel to review the safety of its nuclear installations.[58] There was a suggestion in the 1970s from a senior bureaucrat, Ashok Pastharasathu, to create an independent body located in the department of science and technology, as it had been assigned the national responsibility for ensuring the preservation of the environmental quality.[59] But this was not accepted by the Atomic Energy Commission (AEC).

Currently, the Atomic Energy Regulatory Board (AERB) is responsible for monitoring the safety of all the nuclear facilities operated by the Uranium Corporation of India Limited (UCIL), which fall under the purview of the Department of Atomic Energy (DAE). However, the Board is required to report to the Atomic Energy Commission (AEC), whose chairman is the Secretary of the DAE and which comes under the direct control of the Prime Minister of India. Thus, the regulatory board reports to the very agency it is required to assess and monitor in the interest of public safety.

Moreover, lack of technical expertise outside of DAE puts practical limits to the AERB's autonomy. Dr. A Gopalakrishnan, a former chairman of the AERB observed that "almost 95% of the members in AERB's review and advisory committees are drawn from among retired employees of the DAE, either from one of their research institutes like the Bhabha Atomic Research Center (BARC) or a power generation company like the Nuclear Power Corporation of India Ltd."[60] Other authorities like the Comptroller and Auditor General of India (CAG) (whose function is to enhance the accountability of various public sector organizations and departments to the Indian parliament and state legislatures) and the International Atomic Energy Agency (IAEA) have raised concerns over the subordinate position of AERB to DAE.[61]

The institutional structure reveals that risk regulation of nuclear facilities in India does not occur in an unbiased context. Conflicts of interest and hence politics seep in through organizational structures. Meanwhile, the nuclear establishment also creates narratives about the controversy in Jadugoda that are overtly political. Many claims that are part of this myth narrative do not stand up to rational scrutiny.

3. *Politics of Assigning Blame in Risk Controversies*

Mary Douglas, a cultural anthropologist, is one of the most influential figures in the field of risk. In *Purity and Danger* (1966), Douglas elaborates on her analysis of ideas and rituals concerning pollution and cleanliness in a range of societies. She finds that ideas about order and disorder fundamentally underlie beliefs about purity and pollution. She further sets out to explain how taboos observed in a culture can protect them from behaviors that threaten to destabilize the social order. When taboos are broken, pollution ideas may be used as threats to maintain social order: "At this level the laws of nature are dragged in to sanction the moral code: this kind of disease is caused by adultery, that by incest; this meteorological disaster is the effect of political disloyalty, that the effect of impiety. The whole universe is harnessed to men's [*sic*] attempts to force one another into good citizenship."[62] People who break taboos or pollution rules are seen as wicked both because they have transgressed cultural norms and because they have thus placed others in danger by their actions. Moral lessons are reinforced in the process of blame assignment, and so social order is restored.

Blame assignment is based on the explanation that the community comes up with. This explanation is moralistic—someone died because they broke a taboo, and purification rituals are called for. Another explanation is that which attributes a misfortune to the work of individual adversaries. A third explanation blames the misfortune on an outside enemy, who must be punished. A key insight in Douglas's work is that pollution beliefs, rituals, and logic of danger exist in modern secular societies as well and hold strong resemblances to so-called "primitive" or premodern societies. Only the language has changed to that of risk as it suits the culture of secularized societies that revere science and technology. Risk in Douglas's view is then intimately related to notions of politics, particularly in relation to accountability, responsibility, and blame. It continues to be used as a forensic resource, key to assigning blame.

UCIL time and again invokes science and facts backing their claims in the Jadugoda controversy. It attempts to maintain the gap between scientific reason and premodern logic in the face of risk—a binary that Douglas challenges through her work. In the following sections, UCIL (and the Indian nuclear establishment at large)'s narrative of the risk controversy is probed through Douglas's cultural theory which provides three insights: One, officials and experts belonging to the nuclear establishment also subscribe to myths that support a benign perception of risks from ionizing radiation. Two, even though UCIL (and the Indian nuclear establishment at large) accuse activists of politicizing the risk debate, they are themselves involved in creating overtly political narratives. Three, UCIL's blame assignment for observed health impacts and ostracism of opposition by calling them antinational puts to rest any claim over an unbiased and apolitical position in the controversy.

3.1 Myths about Radiation among Establishment Experts

Although UCIL and the larger Indian nuclear establishment put forth an image of dispelling myths with scientific facts, various important officials have time and again revealed their own myths about radiation. These suggest a benign perception of radiation exposure prevalent among those in the nuclear establishment. In one public meeting, Dr. U. G. Mishra, director of Health, Safety and Environment from Bhabha Atomic Research Center or BARC (the government's premier nuclear research facility) ridicules the standard radiation dose limits for public exposure established by the International Commission on Radiological Protection (ICRP). In a statement recorded in *Buddha Weeps in Jadugoda* he said that "after ICRP 16 came, one scientist from Israel published a very interesting paper. He said that now the limits that are permitted for public exposure are less than the dose you will get by sleeping with your spouse. Because spouse has about 3000 bq of potassium in his/her body. Sleeping close to him/her you get radiation dose and taken on annual basis it works out more than what is permitted under the new regulation."[63]

Independent researchers have been quick to point out the absurdity of such claims. Most people estimate the annual dose from sleeping next to someone to be 1–2 millirem or 0.01–0.02 mSv.[64]

Another article published on the website of the Department of Atomic Energy (DAE) by the Secretary of the Atomic Energy Regulatory Board

(AERB), Dr. K. S. Parthasarathy, stresses the notion that radiation is ubiquitous. He points out various sources of natural radiation and puts out a table of different foods that contain K-40 (without, however, mentioning the unit of measurement). Further in the article titled "Radiation: Perception of People and Reality" Dr. Parthasarathy claims:

> Three exemplary epidemiological studies published in the British Medical Journal involving nearly, 40,000 workers in the UK Atomic Energy WorkForce, 14,000 British Nuclear Fuel Workers and 23,000 Atomic Weapons Research Workers, show the well-known phenomenon called the "healthy workers effect" indicating that these workers are healthier than the national average. But for certain cancers, there is slightly above average incidence rate but none of them is statistically significant. Studies done in USA also indicated similar results.[65]

Others have argued that this is a gross misreading of the "healthy workers effect" which is a type of selection bias, first identified in 1976, common to occupational cohort studies. Workers in professions are healthier and able-bodied. If they fall ill or as they age, they leave the workforce. This becomes a confounder in the results that show workers to be healthier on average than the general population.[66]

The perception of radiation exposure being benign is so prevalent that many in the Department of Atomic Energy have also subscribed to a controversial hypothesis called radiation hormesis, which maintains that chronic low doses of radiation are beneficial because they stimulate repair mechanisms. An early paper on the subject argued that the data available on cancer incidence/mortality rates and environmental radiation levels in various cities and states of India substantiate the hypothesis that low levels of ionizing radiation may be beneficial to humans.[67] It also claimed that where the radiation level is greater, the cancer risk is invariably less. Those who contest such claims have noted that they were based on deficient data and faulty analysis, including the use of only external doses, and ignoring the wide variation in internal doses of radiation, and potentially erroneous estimates of cancer incidence for general populations.[68]

The hormesis hypothesis has been dismissed by most of the scientific bodies that have been constituted to examine the effects of radiation on health. For instance, after examining a large amount of evidence on the hypothesis, the Committee on the Biological Effects of Ionizing Radiation (BEIR) of the US National Research Council concluded that "the assumption that any stimulatory hormetic effects from low doses of ionizing radiation will have a significant health benefit to humans that exceeds potential detrimental effects from radiation exposure is unwarranted at this time."[69]

3.2 Narrative Building 1: The Backward, Unhygienic Adivasi

UCIL does not just critique independent health studies for their methodology, they have also called these the biased work of anti-nuclear activists.[70] Whereas their official health studies have maintained the same conclusion through the years, "villagers suffer from conventional health problems, which could be seen in any village with similar socio-economic condition."[71] Or "No obvious effect of radiation was found in the clinical examination."

The company also plays into the stereotypical backward image of the Adivasis by claiming that their bad habits like smoking, drinking, and bad sanitary conditions are to be blamed for the observed health outcomes. This has helped to shift the blame on to the communities themselves. UCIL's chairman J. L. Bhasin reportedly said in 1999 "that malnutrition is the main cause of the high incidence of tuberculosis and child mortality and the low level of health among the tribal people. The men are generally healthy until the age of 27. He said that after that, they took to drinking. This coupled with their poor economic background resulted in the deterioration of their health."[72] The company has maintained this causal explanation since. In 2014, UCIL spokesperson Pinaki Roy, when questioned about dust from tailing ponds flying over to people's houses, said, "The thing is that in this area their sanitary habits are suspect."[73] In 2018 UCIL's chairman Sundarajan reportedly attributed the diseases to their "economic backwardness, smoking habits and malnutrition," while, in the same breath adding that "the uranium mining in the area brought economic prosperity to the villagers because it offered employment to them and better services."[74]

Ironically, this causal explanation of economic backwardness and malnutrition contradicts the company's claim that it has brought prosperity to the region. Census data from 2001[75] and 2011[76] shows that Jadugoda seems to be more relatively developed. The literacy rate here is 84.06 percent, higher than the state average of 66.41 percent. Independent epidemiological studies as discussed above find that Jadugoda is economically better off than the nearby villages.[77]

Here it is helpful to recall Douglas's insight: notions of order, disorder, and hierarchy in a society underlie ideas about what is clean and dirty. The labeling of Adivasis as unhygienic and backward reveals the hierarchical order of Indian society. Adivasis exist at the lower rungs of this order. Marginalized by a nation-building exercise, Adivasis face social discrimination and violence in their everyday lives. Invoking the image of unhygienic and backward Adivasis then helps in shifting the blame (for the observed health effects) onto them while reinforcing the social order just as Douglas observed in various premodern societies.

3.3 Narrative Building 2: Anti-Nuclear Is Anti-National

Opposition to India's nuclear program has often been labeled as anti-national by the state and the nuclear establishment.[78] UCIL officials have hinted that these cases are brought up as part of a conspiracy against India's nuclear program aimed to thwart the country's progress. P. C. Gupta, who was part of a survey conducted by UCIL in 2017 as a specialist in genetic abnormalities, said that he has come across some cases of congenital deformities in the UCIL hospital in Jadugoda. But he added, "I believe these patients were brought from outside." The chairman of UCIL, Diwakar Acharya, in 2014 when asked about reports about high numbers of physically deformed people reportedly said: "I wouldn't be surprised if a lot of those guys are imported from elsewhere." This is also a standard narrative that UCIL and the nuclear establishment at large have maintained over time. For instance, P. K. Iyengar, former chairman of the Atomic Energy Commission (AEC), when asked about the reports on health effects in Jadugoda said, "These reports are uninformed and in some cases part of a campaign to stop India from pursuing its nuclear research and power generation."[79]

While Adivasis exist at lower rungs of the social order, UCIL and AEC as part of the nuclear establishment enjoy unfettered power at the top. Nuclear technology has been central to the nation-building process in India since the time of Prime Minister Nehru. Adivasis at best remain at the periphery of the nation and at worst are enemies of the state. For instance, the areas of armed struggle by Adivasis for autonomy in the central part of India are often referred to as Pakistan by the police.[80]

The label of anti-national helps in othering them, making them enemies of the Indian society and then easily assigning them blame for the risk controversy. Blaming an outside enemy for a misfortune is an age-old political practice as observed by Douglas in several societies.

Conclusion

In this essay, I have looked at the risk debates on uranium mining in India through the insights from risk studies. It had a twofold aim. The first aim was to challenge the gap between objective assessment of risk by experts and subjective, value or emotion-based perception, of risk by the public. The second was to reveal the politicization of the risk controversy in the uranium mines in India by the Uranium Corporation of India Limited (UCIL) and the nuclear establishment at large, which obscure their politics by using technical language. At first, the essay looked at debates among experts who come to different conclusions regarding the health impacts of the uranium mines by using different methodologies. This showed that objective assessments of risk are often illusory, and disputes about its definition always exist. Then it explored how people's social experience of institutions involved in making and regulating risks affects their risk decisions. It looked at UCIL's pattern of pollution and the lack of independent regulation of uranium mines to argue that these experiences affect people's judgments of risk from the mines. Lastly, it showed, using Douglas's cultural theory of risk, that UCIL and the nuclear establishment are far from unbiased experts merely stating facts and alleviating myths. They politicize the controversy much like any premodern society. A key takeaway from the story of Jadugoda is that the public is not merely reacting to risks based on their emotions and values and experts are not merely imparting unbiased risk assessment. In any risk controversy, politicization is inevitable, the actors claiming to be unbiased perhaps need closer scrutiny.

Notes

1 Robert L. DuPont, *Nuclear Phobia—Phobic Thinking about Nuclear Power: A Discussion with Robert L. DuPont* (Washington, DC: The Institute, 1980); V. T. Covello et al., *The Analysis of Actual versus Perceived Risks* (New York: Plenum Press, 1983); Timur Kuran and Cass R. Sunstein, "Availability Cascades and Risk Regulation," *Stanford Law Review* 51, no. 4 (April 1999): 683, https://doi.org/10.2307/1229439.
2 Itty Abraham, *The Making of the Indian Atomic Bomb: Science, Secrecy and the Postcolonial State* (London: Zed Books, 1998), 34–69.
3 Ibid., 34.
4 "Despite much rhetoric about self-reliance and indigenous development, the AEC sought and received ample help from other countries. Indeed, for the first quarter-century of its existence, until the nuclear weapon test of 1974, the AEC kept on acquiring technologies related to the entire nuclear fuel chain from different countries. Practically all these were then proclaimed as domestically developed by the AEC" (20). For more discussion see M. V. Ramana, *The Power of Promise: Examining Nuclear Energy in India* (New Delhi: Penguin India, 2012), 20–26.
5 Intercultural Resources and George PT, "Return of the Nuclear Shadow: Uranium Mining and the Tribal Community's Struggle for Survival in Jaduguda, India," *ritimo* (blog), August 15, 2011, https://www.ritimo.org/Return-of-the-Nuclear-Shadow-Uranium-Mining-and-the-Tribal-Community-s-Struggle; Xavier Dias, "Uranium Mining—Where the Debate Begins: The Case of Jadugoda," July 6, 2010, https://www.birsa.in/archives/164; Mathew Areeparampil, "Displacement Due to Mining in Jharkhand," *Economic and Political Weekly* 31, no. 24 (1996): 1524–1528.
6 India had 420 MW of nuclear energy capacity by 1974. See NPCIL (Nuclear Power Corporation of India Limited), "Plants under Operation," https://www.npcil.nic.in/content/302_1_AllPlants.aspx (accessed December 6, 2021). Compare this to the figure of 8,000 MW of nuclear capacity that Homi Bhabha, the first chairman of the Atomic Energy Commission, had announced for the year 1980: M. V. Ramana, "Nuclear

Power in India: Failed Past, Dubious Future," Gauging US-Indian Strategic Cooperation (Strategic Studies Institute, US Army War College, 2007), http://www.jstor.org/stable/resrep11995.6.

7 "Nuclear Suppliers Group—About the NSG," https://nuclearsuppliersgroup.org/en/about-nsg (accessed July 23, 2021).

8 A. H. Khan et al., "Assessment of Environmental Impact of Mining and Processing of Uranium Ore at Jaduguda, India," *International Nuclear Information System* 33 (2002): 7.

9 "Uranium Mining: Questionable Decision," *EPW (Economic and Political Weekly)* 35, no. 46 (2000): 3984–3985.

10 Scott Ludlam, "Nuclear India." Maylands, Western Australia: Anti-Nuclear Alliance of Western Australia, 2000, https://slwa.wa.gov.au/pdf/mn/mn2501_3000/mn2867.pdf. Also quoted in Ramana, *The Power or Promise.*

11 Mining Concerns Desk, "Fuel Management in CANDU Reactors," *Annals of Nuclear Energy* 3, no. 7–8 (1996): 359–366.

12 "Uranium Mining: Questionable Decision."

13 "Jharkhandi Organization Against Radiation (JOAR)," http://jadugoda.jharkhand.org.in/2009/10/jharkhandi-organization-against.html (accessed January 22, 2019).

14 C. J. Sonowal and Sunil Kumar Jojo, "Radiation and Tribal Health in Jadugoda: The Contention between Science and Sufferings," *Studies of Tribes and Tribals* 1, no. 2 (December 2003): 111–126, https://doi.org/10.1080/0972639X.2003.11886490.

15 Richard Mahapatra, "Eyewitness: Radioactivity Doesn't Stop at the Mines in Jaduguda," *Down to Earth*, April 30, 2004, https://www.downtoearth.org.in/coverage/eyewitness-radioactivity-doesnt-stop-at-the-mines-in-jaduguda-11118; Adrian Levy, "India's Nuclear Industry Pours Its Wastes into a River of Death and Disease," *Center for Public Integrity*, December 15, 2015, https://publicintegrity.org/national-security/indias-nuclear-industry-pours-its-wastes-into-a-river-of-death-and-disease/.

16 *Frontline* 16, no. 18 (August 28–September 10, 1999).

17 Shri Prakash, *Buddha Weeps in Jadugoda* (Ragi: Kana: Ko Bonga Buru, 1999), https://www.youtube.com/watch?v=FxO_LlHaYvs.

18 Chattopadhayay and Subramanian, "Villages and Woes"; "Jaduguda Operations Safe," *The Hindu*, April 9, 2000; "The Price for Nuclear Capabilities?" *The Hindu*, April 6, 2000; "Living in the Deadly Shadow of Uranium—India's Huge Uranium Mining Complex Threatens Health of Thousands of Villagers," *Toronto Star*, October 10, 1999; "Thousands at Risk of Poisoning from 'India's Chernobyl,'" *Telegraph* (UK), April 25, 1999; "Uranium Mining in Jaduguda, Bihar, Living in Death's Shadow," *SUNDAY Magazine* (Calcutta), April 4–10, 1999; "Radiation from Uranium Mines Hits Bihar Tribals," *Deccan Herald*, January 6, 1999; "Uranium Hits Jadugoda Tribals; UCIL Blamed," *Indian Express* (Bombay), December 28, 1998; "Angry Villagers Take on Uranium Corporation," *Indian Express* (Bombay), September 19, 1998; "Hunt for the Yellow Cake," *Indian Express* (Bombay), June 4, 1998.

19 Xavier Dias, "Testimony: Xavier S. Dias, World Uranium Hearings" (1992), https://ratical.org/radiation/WorldUraniumHearing/XavierDias.html.

20 Prakash, *Buddha Weeps in Jadugoda.*

21 Chattopadhayay and Subramanian, "Villages and Woes."

22 Dias, Testimony: Xavier S. Dias, "World Uranium Hearings."

23 "Myth of UCIL," Uranium Corporation of India Limited, http://www.ucil.gov.in/myth.html (accessed January 16, 2019).

24 Lennart Sjöberg, "Risk Perception: Experts and the Public," *European Psychologist* 3, no. 1 (March 1998): 1–12, https://doi.org/10.1027//1016-9040.3.1.1.

25 "Myth of UCIL," Uranium Corporation of India Limited, http://www.ucil.gov.in/myth.html (accessed January 16, 2019).

26 Allan Mazur, "Disputes between Experts," *Minerva* 11, no. 2 (1973): 243–262.

27 Shyamali Khastgir, *Jadugoda Diary: With the Survey Report on Jadugoda Tragedy from Anumukti* (Monfakira, 2009).
28 BARC is a research facility under the Department of Atomic Energy that oversees UCIL operations as well.
29 Ibid.
30 Chattopadhayay and Subramanian, "Villages and Woes."
31 *Anumukti (Liberation from the Atom)* 13, no. 1 (January 2004).
32 Ibid.
33 Brian Wynne, "Misunderstood Misunderstanding: Social Identities and Public Uptake of Science," *Public Understanding of Science* 1, no. 3 (July 1992): 281–304, https://doi.org/10.1088/0963-6625/1/3/004.
34 Hiroaki Koide, "Radioactive Contamination around Jadugoda Uranium Mine in India," April 27, 2004, Institute for Integrated Radiation and Nuclear Science, Kyoto University, http://www.rri.kyoto-u.ac.jp/NSRG/genpatu/india/JADFINAL.pdf.
35 "Radon and Health," World Health Organisation, February 2, 2021, https://www.who.int/news-room/fact-sheets/detail/radon-and-health.
36 Ibid.
37 R. M. Tripathi et al., "Radiation Dose to Members of Public Residing around Uranium Mining Complex, Jaduguda, Jharkhand, India," *Radiation Protection Dosimetry* 147, no. 4 (November 1, 2011): 565–572, https://doi.org/10.1093/rpd/ncq496.
38 Steven L. Simon and Martha S. Linet, "Radiation-Exposed Populations: Who, Why, and How to Study," *Health Physics* 106, no. 2 (February 2014): 182–195, https://doi.org/10.1097/HP.0000000000000006.
39 Summary from Ali SS (1998) Report of medical survey conducted by a team consisting of Doctors and specialists as submitted to the court: "Court on Its Own Motion v. Union of India and Ors | Jharkhand High Court | Judgment | Law | CaseMine," Jharkhand High Court (2016).
40 Susan E. Dawson and Gary E. Madsen. "Uranium Mine Workers, Atomic Downwinders and the Radiation Exposure Compensation Act (RECA)," in Barbara Rose Johnston and School for Advanced Research, *Half-Lives and Half-Truths: Confronting the Radioactive Legacies of the Cold War*, 1st ed., School for Advanced Research Resident Scholar Series (Santa Fe, NM: School for Advanced Research Press, 2007); Robert J. Roscoe et al., "Lung Cancer Mortality among Nonsmoking Uranium Miners Exposed to Radon Daughters," *JAMA* 262, no. 5 (August 4, 1989): 629–633, https://doi.org/10.1001/jama.1989.03430050045024.
41 G. López-Abente, N. Aragonés, and M. Pollán, "Solid-Tumor Mortality in the Vicinity of Uranium Cycle Facilities and Nuclear Power Plants in Spain," *Environmental Health Perspectives* 109, no. 7 (July 2001): 721–729.
42 Mona Dreicer, "Chernobyl: Consequences of the Catastrophe for People and the Environment," *Environmental Health Perspectives* 118, no. 11 (November 2010): A500.
43 *Anumukti (Liberation from the Atom)* 13, no. 1 (January 2004).
44 National Research Council (US) Committee on the Biological Effects of Ionizing Radiations, *Health Risks of Radon and Other Internally Deposited Alpha-Emitters: Beir IV* (Washington, DC: National Academies Press, 1988), http://www.ncbi.nlm.nih.gov/books/NBK218125/; National Research Council, *Health Effects of Exposure to Radon: BEIR VI* (Washington, DC: National Academies Press, 1999), https://doi.org/10.17226/5499. UNSCEAR, Sources and Effects of Ionizing Radiation: UNSCEAR 2000 Report to the General Assembly, with Scientific Annexes. New York: United Nations Scientific Committee on the Effects of Atomic Radiation, United Nations, 2000.
45 *Anumukti (Liberation from the Atom)* 13, no. 1 (January 2004).
46 Harshad P. Thakur and B. K. Sapra, "Baseline Survey of Health Status of Population in 2006 around a Uranium Mining Site in Jaduguda, India," *Radiation Emergency Medicine* 2, no. 1 (2013): 14–22

47 Ibid.
48 Paul Slovic, "Trust, Emotion, Sex, Politics, and Science: Surveying the Risk-Assessment Battlefield," *Risk Analysis* 19, no. 4 (1999): 689–701, https://doi.org/10.1111/j.1539-6924.1999.tb00439.x.
49 Timothy C. Earle, "Trust in Risk Management: A Model-Based Review of Empirical Research," *Risk Analysis* 30, no. 4 (2010): 541–574, https://doi.org/10.1111/j.1539-6924.2010.01398.x.
50 Brian Wynne, *Risk Management and Hazardous Waste: Implementation and the Dialectics of Credibility* (Springer Berlin, Heidelberg, 1987).
51 Ibid.
52 Committee on Uranium Mining in Virginia, Committee on Earth Resources, and National Research Council, *Potential Environmental Effects of Uranium Mining, Processing, and Reclamation, Uranium Mining in Virginia: Scientific, Technical, Environmental, Human Health and Safety, and Regulatory Aspects of Uranium Mining and Processing in Virginia* (Washington, DC: National Academies Press, 2011), http://www.ncbi.nlm.nih.gov/books/NBK201052/.
53 N. K. Sethy et al., "Assessment of Natural Uranium in the Ground Water around Jaduguda Uranium Mining Complex, India," *Journal of Environmental Protection* 2, no. 7 (2011): 1002–1007, https://doi.org/10.4236/jep.2011.27115.
54 Krishna Shree and Rajesh Serupally, "The Real Cost of Uranium Mining: The Case of Tummalapalle," *Firstpost*, July 16, 2018, https://www.firstpost.com/long-reads/the-real-cost-of-uranium-mining-the-case-of-tummalapalle-4749521.html/amp.
55 U. Sudhakar Reddy, "Uranium Mining: UCIL Ignored Eco Norms," *Times of India*, September 14, 2019, https://timesofindia.indiatimes.com/city/hyderabad/uranium-mining-ucil-ignored-eco-norms/articleshow/71119304.cms.
56 Makarand Purohit, "Arsenic-Affected Village Gets Water after Two Decades," *India Water Portal*, March 12, 2018, https://www.indiawaterportal.org/articles/arsenic-affected-village-gets-water-after-two-decades.
57 "Get Us Out of Here, Say Gogi Residents," *Deccan Herald*, September 10, 2011, https://www.deccanherald.com/content/190077/get-us-here-say-gogi.html.
58 A. Gopalakrishnan, "Evolution of the Indian Nuclear Power Program," *Annual Review of Energy and the Environment* 27 (2002): 384.
59 Ashok Parathasarathi, *Technology at the Core: Science and Technology with Indira Gandhi* (New Delhi: Pearson Longman, 2007), 131–132.
60 A. Gopalakrishnan, "Issues of Nuclear Safety," *Frontline*, March 13–26, 1999.
61 "IAEA Mission Concludes Peer Review of India's Nuclear Regulatory Framework," Text, International Atomic Energy Agency (IAEA) (IAEA, March 27, 2015), https://www.iaea.org/newscenter/pressreleases/iaea-mission-concludes-peer-review-indias-nuclear-regulatory-framework; CAG, "Performance Audit on Activities of Atomic Energy Regulatory Board for the Year Ending March 2012, Union Government, Atomic Energy" (New Delhi: Comptroller and Auditor General of India, August 22, 2012), https://cag.gov.in/cag_old/sites/default/files/audit_report_files/Union_Performance_Atomic_Energy_Regulatory_Board_Union_Government_Atomic_Energy_Department_9_2012.pdf.
62 Mary Douglas, *Purity and Danger: An Analysis of Concepts of Pollution and Taboo* (London: Routledge, 2013).
63 Prakash, *Buddha Weeps in Jadugoda.*
64 Esther Inglis-Arkell, "How Much Radiation Does the Human Body Emit?" *Gizmodo* (July 21, 2015). https://gizmodo.com/how-much-radiation-does-the-human-body-emit-1719085023; Jacklin Kwan, "How Radioactive Is the Human Body?" livescience.com (September 27, 2021). https://www.livescience.com/radiation-human-body.
65 Dr. K. S. Parthasarathy, "Radiation: Perception of People and Reality," *Department of Atomic Energy, Government of India* 35, no. 11–12 (May–June 2002), https://dae.gov.in/node/192.

66 Ritam Chowdhury, Divyang Shah, and Abhishek R. Payal, "Healthy Worker Effect Phenomenon: Revisited with Emphasis on Statistical Methods—A Review," *Indian Journal of Occupational and Environmental Medicine* 21, no. 1 (2017): 2–8, https://doi.org/10.4103/ijoem.IJOEM_53_16.
67 K. S. V. Nambi and S. D. Soman, "Environmental Radiation and Cancer in India," *Health Physics*, 52, no. 5 (1987): 653–657.
68 Arjun Makhijani, "Low Level Radiation and Cancer: Incomplete Data, Faulty Analysis," *Economic and Political Weekly* 21, no. 44 (1987): 1853–1855.
69 National Research Council, *Health Risks from Exposure to Low Levels of Ionizing Radiation: BEIR VII, Phase 2* (Washington, DC: National Academic Press, 2006), 335.
70 Rakteem Katakey and Tom Lasseter, "India's Uranium Boss Says Deformed Children May Be 'Imported,'" *Bloomberg*, July 23, 2014.
71 As quoted in Thakur and Sapra, "Baseline Survey of Health Status of Population in 2006 around a Uranium Mining Site in Jaduguda, India."
72 Suhrid Sankar Chattopadhayay and T. S. Subramanian, "Villages and Woes," *Frontline* 16, no. 18 (August 28–September 10, 1999).
73 Katakey and Lasseter, "India's Uranium Boss Says Deformed Children May Be 'Imported.'"
74 Sagar, "Endorsed by Courts and the Government, Uranium Mining Continues to Create Health Hazards in Jadugoda as the UCIL Expands Its Operations."
75 Literacy rate was about 80 percent, higher than the national average of 64.83 percent (Census 2001), web.archive.org/web/20040616075334/http://www.censusindia.net/results/town.php?stad=A&state5=999
76 Literacy rate of Jadugora city is 84.06 percent, higher than state average of 66.41 percent. In Jadugora, Male literacy is around 92.06 percent while female literacy rate is 75.64 percent. See https://www.census2011.co.in/data/town/363954-jadugora-jharkhand.html
77 Khastgir, *Jadugoda Diary*.
78 For instance see Arun Janardhanan, "8,856 'Enemies of State': An Entire Village in Tamil Nadu Lives under Shadow of Sedition," *Indian Express*, September 12, 2016, https://indianexpress.com/article/india/india-news-india/kudankulam-nuclear-plant-protest-sedition-supreme-court-of-india-section-124a-3024655/.
79 M. V. Ramana, *The Power of Promise: Examining Nuclear Energy in India* (New Delhi: Penguin India, 2012), 236.
80 Arundhati Roy, "Gandhi, but with Guns: Part One," *The Guardian*, March 27, 2010, sec. Books, https://www.theguardian.com/books/2010/mar/27/arundhati-roy-india-tribal-maoists-1.

A Darkened Organ and a Darkened Soul

The Health Effects of Uranium Exposure on a Former Diné (Navajo) Mine Worker

OLIVER GEORGE TAPAHA

On a chilly December day in 1973, *Shimasaní* (my maternal grandmother) introduced my step-grandfather, Joe (he preferred to be called by his birth given name), to *Shimá* (my mother) who had just returned home from Intermountain Indian Boarding School in Brigham City, Utah, for a two-week intercession. *Shimá* described Joe as healthy and quiet. He was forty-three years old at the time and *Shimasaní* was three years younger than he was. They had been secretly courting for months before *Shimasaní* invited Joe to our homestead in Round Rock, Arizona, which is snuggled in the central region of the Diné (Navajo) reservation. Joe's birthplace was in a small community called Rough Rock, Arizona. He had two biological sons when he met *Shimasaní*.

My grandparents' innocent relationship from nearly half a century ago turned into a forty-year marriage. Joe lived a private and simple life. He cared for a large flock of goats and sheep in the daytime, worked on automobiles and makeshift projects in the evenings, and at nighttime, he amused himself with several rounds of solitaire games. He would stay up into the wee hours of the night, sitting near a lantern with dim lighting and shuffling cards after each round of games. Joe and *Shimasaní* were exceptionally close and knew how to pass time. They would bicker, yell, and insult each other one moment, and then they would make eye contact from opposite ends of a room and laugh at their nonsense the next minute. They enjoyed home-cooked meals and never grew tired of hauling water and tending the sheep and goats together. They genuinely loved and took care of each other every single day. But their successful marriage abruptly ended in 2015 when law enforcement uncovered Joe's body at the bottom of a steep rock face in the mouth of Canyon de Chelly, Arizona. He was eighty-five years old.

Joe's death in 2015 was an incomprehensible tragedy that pierced our hearts with great pain. But equally catastrophic was a deadly disease he lived with; a disease he brought back with him from the time he worked in the uranium mines during the early 1960s. That life-threatening illness darkened one of Joe's vital organs, and it also darkened his mind and soul for over half of his life.

According to the stories of *Shimá* and Joe's long-term caretaker (my cousin sister), Joe casually spoke about working in an underground uranium mine near Montezuma Creek, Utah, when he was in his early thirties. He may have worked for the Cottonwood-Montezuma Canyon Mining. He lived with a Diné woman and her three children who resided in that community. Each day he went home to them unaware of the radioactive material on his clothes and tools. For about a year or two, Joe lived with that family and then left and moved north. He secured another mining job with the La Sal Mines near Moab, Utah. He worked there for a number of years. When Joe worked in those mines, he indicated that he did not wear a mask or any protective gear. Like many miners, he also was not educated about the health risks associated with the effects of uranium exposure. He only spoke and understood the Diné language and performed his job duties day by day.

Several years after Joe worked in the mines, he met *Shimasaní,* and they built a life together. Joe was a very active person. He walked long distances herding and tending to a large group of goats and sheep on a high-desert plateau. To seek balance, harmony, and purification, he would conduct a sweat lodge ceremony to relax, pray, and sing healing songs. Around 2003, my family members and I started to notice a change in Joe's health and breathing pattern. After each day's work around the house, he would get tired and display hacking coughs when he rested on his twin-sized bed. And as time progressed, the coughing became more frequent and sporadic throughout the daytime and worsened at nighttime. It concerned us when he began experiencing shortness of breath and coughing up blood and running high fevers. When he visited his primary doctor, Joe was told that he had caught tuberculosis (TB) and had developed pneumoconiosis, commonly referred to as "black lung disease." Black lung disease is a respiratory, job-related disease that is oftentimes connected to the inhalation of ore dusts. When Joe learned of this devastating news, he immediately linked the cause of his black lungs to the radioactive dust particles he was exposed to when mining the *leetso* (yellow dirt). He never smoked cigarettes, cigars, or took any psychoactive drugs.

Joe was outraged, distressed, and had many questions about uranium and radiation, and how such a hazardous chemical and energy could be allowed to cause such serious harm to his body. He also wanted to know why uranium mining companies failed to inform him and other Diné and Indigenous miners about the dangers of uranium mining. Most importantly, he wanted to know how long he was going to live and what treatments, medically and spiritually, he could try to extend his life. He was prescribed only TB medication, which he took daily.

Many Diné uranium mine workers and their families, families of deceased uranium miners, and families who lived close by abandoned uranium mines were also trying to make sense of the same concerns Joe had raised. Many of them also faced severe health problems or had loved ones who returned to Mother Earth after battling various forms of cancer. They demanded answers and justice for the atrocities done to Indigenous miners, their families, Indigenous communities, and sacred lands. In response to this public outcry, many community forums were held in different part of the Diné reservation during the mid and late 2000s to understand and educate others about the negative effects of uranium contamination. Joe and *Shimasaní* attended some of those meetings to get their questions answered. And it was through such spaces that they learned about the process of how to seek radiation exposure compensation.

For Joe to claim reparation for his black lungs, he had to prove that he was employed in the uranium industry and that his illness was a direct cause of high-level radiation exposure. Joe spent nearly two years undergoing numerous medical tests and visiting the Navajo Nation Uranium Workers Office in Shiprock, New Mexico, to meet the requirements of the Radiation Exposure Compensation Act (RECA). In 2010, Joe was finally awarded a one-time lump sum of $100,000. A couple years after receiving his first compensation, he was eligible for another $50,000, and then $10,000 every two years, thereafter. But monetary compensations did not improve his mood and feelings of animosity toward the system that shortened his lifespan.

Joe had an internal war with depression. There were days when he slept for hours with his window curtains closed. Sometimes he would not have an appetite and not eat for days. On a few occasions, he arbitrarily stepped out of the house in the scorching summer heat or in the coldest of days and walked to nowhere. *Shimasaní* was an obsessive worrier and would always ask a family member to drive her around the community, so she could bring her husband home. From time to time, he would go on a drinking spree to suppress his emotions and physical pain. He was carefree, humorous, and kindhearted when he drank. And we know these were the characteristics he desired for himself each day—without medication, without alcohol. Without the black lung disease.

Joe sought mental health counseling for his erratic and ill-tempered behavior, but his mental suffering was too deep and too dark to be healed. To date, we do not know if he took his own life or if someone pushed him off the cliff. And the compensation that was afforded to *Shimasaní* ended the moment Joe was laid to rest. My cousin sister and *Shimá* were informed that Joe's death was not caused by collapsed lungs. But what RECA has failed to recognize is that Joe's mental illness stemmed directly from the emotional turmoil he began to face when he first heard the heart-wrenching news that he had black lung disease. What frustrates us is that we could not prove what really happened to Joe on that spring day when his body parts were found at the base of a deep canyon. As Joe's family, we continue to lose sleep because of this unknown catastrophe. My eighty-nine-year-old *Masaní*, on the other hand, has dementia, and she said she sees him and communicates with him every day. Their married life lives on in her memory and illusions.

I briefly detailed Joe's life story here, on behalf of my family, because it is important to understand that what he went through was real. And our pain was real. By telling this story, *his* story, we begin the healing process. But we also want *his* story to be your teaching, your healing medicine, and your way of seeking justice for those who have been impacted by uranium and radiation exposure throughout the world.

Rest in peace, *Shicheii* (my grandfather), *Nihicheii* (our maternal grandfather), *Nihinálí* (our paternal grandfather), *Nihizhé'é* (our father), Joe Yazzie Tso.

Rocky Flats Health History

Making Risk Visible

SASHA STILES AND EDWARD GRANADOS

On Mother's Day afternoon, May 11, 1969, Ed Martell of the National Center for Atmospheric Research (NCAR), high in the foothills above Boulder, Colorado, saw smoke billowing from a building at Rocky Flats, located sixteen miles northwest of central Denver. He knew that the frequent Chinook winds that roared in excess of 100 mph west to east over the flats into Denver and other communities, sent potentially lethal particles of plutonium toward unsuspecting people in these areas. He asked Rocky Flats officials to sample off-site for plutonium. They refused, so he did his own sampling at various locations east of the plant. He found plutonium deposits in the top centimeter of soil up to 400 times greater than background radiation from global fallout.[1] This now infamous Mother's Day Fire had spontaneously ignited from plutonium in a glove box. Glove boxes are used within the nuclear laboratory to manipulate radioactive, toxic material safely. A glove box is a windowed, sealed container equipped with two flexible gloves that allow the user to manipulate plutonium from the outside in an ostensibly safe environment. The fire burned several hours undetected because of a questionable ventilation system in the glove box. Between 1966 and 1969 the plant's fire department responded to 164 fires, 31 of which were started by plutonium spontaneously igniting.[2]

This essay discusses the health of workers due to the contamination of Rocky Flats, which is just one of many sites in the behemoth nuclear complex. We describe how radiological pollution came to be during the normal functioning of the plant, which produced thousands of plutonium triggers for nuclear warheads in a short time frame, without adequate safety for workers and the surrounding communities. There were several large fires with a high risk of criticality. There were near daily fires and night burning of contaminated debris. Workers were exposed daily to countless toxic chemicals. Cleanup and worker health evaluations were deferred to major production goals. The plant was closed by the FBI due to safety concerns. The cleanup that ensued, as directed by state and federal government agencies, resisted at every turn the concerns and pleadings from local scientists and local citizens. Distrust in government cleanup activities in addition to the veil of secrecy that surrounded the entire production era of the plant continues to breed distrust. The entire issue of past and future harm from radiation is not resolved.

As a physician (Dr. Stiles) and a molecular biologist PhD (Dr. Granados), we see indications in the history of Rocky Flats operations and in our review of past and current epidemiology and scientific papers, the need for a new approach, a paradigm shift to reconsider radiological harm. The major shift requires an understanding of the newer science of epigenetics and genomics. We see a personal responsibility to intervene to create systems of inquiry and protection from radiation harm. We share a responsibility to work to detect radiological bodily damage in order to one day provide better understanding and thus better protection and health care for populations such as the Rocky Flats workers and surrounding communities. Our review underscores a need for unbiased nongovernmental health professionals and affected workers and communities to be included in research design and health protection decisions.

Rocky Flats History

Owned by the US government and operated by contractors, the Rocky Flats Nuclear Weapons Plant produced thousands of plutonium "pits," the fissile plutonium cores for nuclear warheads, from 1952 to 1989. This production process included recycling plutonium metal into plutonium dioxide, conversion of plutonium dioxide to a metal in reduction furnaces, and rolling and machining the metal. Plutonium has toxic and carcinogenic properties. Plutonium metal is pyrophoric, meaning it is liable to ignite spontaneously.[3]

The first explosive fire visible to the surrounding communities occurred in 1957. This fire and explosion destroyed a bank of 620 large filters that existed to protect the public, allowing plutonium particles to escape unimpeded. These filters were caked in plutonium and had not been changed since the plant began operating in 1952. The fire burned unimpeded for thirteen hours, releasing an estimated 500 pounds of plutonium into the air around Denver. The smokestack radiation monitors were not operational from the time the fire began until a week later. When the stack monitors were turned back on the eighth day after the fire, they revealed that stack emissions guidelines were exceeded by 16,000 times greater than accepted background radiation level from global fallout. Soil and property readings adjacent to the plant were read as 225 times normal "background" levels.[4]

There had been no adequate planning for the massive amount of nuclear waste produced at the plant. The square footage of land available on site for waste storage was minimal and far less than usually required for such dangerous nuclear weapons facilities.[5] There were many large thirty-foot-deep trenches where toxic and hazardous waste products were buried in between thin layers of soil. There were many spray fields extending to the perimeters of the plant where all contaminated laundry and other liquid waste were deposited.[6] A waste storage space was compromised when over a thousand barrels of plutonium waste were left exposed to the environment in an outside location called the 903 Pad. It has long been verified that plutonium waste leaked into the water supply and the soil from 1954 to 1968, forcing Broomfield township to switch water supplies. Dow Chemical, the company operating the facility at the end of the plant operation, remediated it by plowing the contaminated soil under, making freed plutonium particles readily available to high winds.[7] The relentless winds uncover the plutonium waste year after year after year. In 2019 Michael Ketterer, PhD, found multiple plutonium dioxide particles along the roadside just east of the 903 Pad right in the path of a proposed highway.[8] A full discussion of soil analyses will follow later in this essay.

Government-funded researchers found no harm to workers and citizens living in the surrounding communities, creating significant distrust in all communities affected by the site. The Colorado Department of Public Health and Environment (CDPHE) conducted studies that many believe are inadequate and methodologically compromised. These studies are often used to counter worrisome findings and discourage further research. For example, when doing epidemiological studies, they included people who did not reside within established wind patterns emanating from Rocky Flats as part of their high-risk cohort, as expressed in a well-known Krey-Hardy Map.[9] This map graphically depicts the prevailing wind flow of airborne contamination from Rocky Flat's Central Operating unit where windblown debris is found to spread out in a wave-like fashion south into Denver, as well as northward spreading in Superior, a township, and Boulder City, both in Southern Boulder County and Broomfield County. The CDPHE used a technique for soil analyses whereby they mixed soil from the most contaminated top centimeter with soil 1–2 feet deeper and averaged the plutonium content.[10]

This method of soil analysis has been questioned by the research community as being unproven and unreproducible. Since then, there has been a concerted effort to utilize new mathematical modeling tools to estimate risk, avoiding the need for massive soil analyses and large epidemiological studies. The Department of Energy (DOE) and CDPHE now base safety on one such risk analysis tool, RESRAD. The RESidual RADioactivity (RESRAD) model and computer code was developed by the DOE's Argonne National Laboratory as a multifunctional tool to assist in developing changing criteria and assess the dose or risk associated with residual radioactive material.[11] One issue addressed in their manual is to assess potential annual or lifetime risks to workers or members of the public resulting from exposure to residual radioactive material in the soil. RESRAD has the mathematical flexibility to manipulate and change dose/exposure data allowing higher levels of plutonium in soil at Rocky Flats or the surrounding communities to be incorporated in their risk calculations.[12] As new soil sampling is done, we have seen RESRAD afford CDPHE the ability to create a new mathematical formula that takes into account a seemingly large particle of plutonium yet concludes that the relative risk of it causing disease is unlikely.

Independent scientists from local and distant universities did their own research, which, like Dr. Ketterer's, disclosed a very different story. In 1975, Harvey Nichols, Emeritus Professor of Biology at the University of Colorado and specialist of airborne transport of pollen, was hired to study airborne particles at Rocky Flats. He did not know at the time that the plant produced fissile plutonium cores. He found that about 14 million radioactive particles per acre were deposited on the site in less than two days of snowfall. He estimated that the snow had scavenged the two days of plutonium particles that were constantly falling over the surrounding area, now ironically designated a wildlife refuge.[13]

Also in 1975, Carl Johnson, director of the Jefferson County Health Department, sampled respirable dust from 25 locations and found that plutonium concentrations on average were 44 times greater than what CDPHE had measured at the same location using their whole-soil samples.[14] Johnson also evaluated leukemia and lung cancer death certificates of census tracts well known to be contaminated and compared this data with the data from census tracts where the soil was clearly known to be near the accepted background radiation. He found a significantly higher (16 percent) incidence of cancer of both types in an area that conformed to the combination of census tracts and wind pattern where past plutonium contamination was documented.[15] Dr. Johnson is well known for his work documenting radiation damage in many sites in the country. His work was contrary to the needs and wants of his superiors at CDPHE. Shortly after the publication of this work he no longer worked at Jefferson County Health Department. His work, cited here, represents one of the only scholarly epidemiological studies of residents' disabilities related to radiation damage from Rocky Flats. Now over fifty years later, the ability to perform classical epidemiological research on the population of the workers and surrounding communities is impossible.

In June 1989, the FBI raided Rocky Flats. Their findings as stated in a Grand Jury trial have been sealed ever since and reportedly now cannot be found, at least by the Court. Jurors were bound to silence as all trial notes were sequestered and unavailable for viewing. The jurors did confirm that they had reviewed very damaging testimony against the DOE and the Rocky Flats governing bodies. However, over time, a significant amount of this testimony has been teased out by resourceful scholars and lawyers.[16] The plant never reopened. The subsequent cleanup was completed in ten years at one-tenth the initially approved cost and time. This expedited cleanup was made possible by taking advantage of new and more flexible risk standards provided to them by RESRAD calculations and a willing set of government paid contractors.[17]

I (Dr. Stiles) have spent years conducting personal interviews with past Rocky Flats workers and reviewing their individual Department of Labor (DOL) client records as a medical consultant for Atomic Workers Advocacy. My work is detailed later in this essay, and my conclusions corroborate onsite reporting such as those by Ted Ziegler, Steelworkers Safety Union Representative at Rocky Flats. Mr. Ziegler states workers labored year after year under substandard safety conditions and massively contaminated air, at the expense of plutonium pit production. Plutonium recovered in the glove boxes was then utilized to create warheads for defense weapons during the cold war. These plutonium-laden warheads were called plutonium pits. Individual workers had significant numbers of major and minor personal accidents exposing them to contaminants. These accidents were underreported and only partially recognized by personnel trained to record such infractions and by their own monitoring instruments. The details concerning each accident, puncture wound, radiation monitor malfunction, lack of appropriate protective garb and respirators are found in each and every record reviewed. Ted Ziegler provided thousands of pages from plant records documenting infractions with no remediation. One such set of records I have reviewed is the signed Safety Team reports with the Rocky Flats plant manager, Dominic Sanchini, during 1988. Some selections from these reports:

> Numerous alpha met probes are found hanging by their wires. . . . Poor glovebox lighting and dirty windows obscure viewing into lower levels of some gloveboxes.

> The inability to control, store and dispose of contaminated waste, obsolete equipment, maintenance supplies and debris efficiently and expeditiously is causing and compounding safety related problems in these buildings.

> The former building manager provided evidence that a large number of safety problems have been identified and documented by area manage-

> ment, remedial action requested formally of Maintenance, and little or no action taken by Maintenance due to workload resource problems.[18]

> Training is inadequate. Lack of written procedures essentially is the basic cause. Asbestos was disposed of into an active landfill - torn bags, covered with dirt and dumped in the wrong place. Improving dosimetry by 10% means nothing. No one insisted on body count on people who might be exposed (states the supervisor) are in extreme danger in 771. Workers have no confidence in dosimetry program.

and finally

> per Mr. Ziegler, the beryllium was sent in large logs to Receiving with toxic notifications removed. They were required to disclose the toxic nature of beryllium. Once in the machine shop without tags, they were sawed up by a heller saw by workers with no respirators.[19]

Many sources led me to realize that you did not want to be a troublemaker. A whistleblower—a glove box operator whose glove was punctured by her coworkers so that when her hand entered the glove box she was immediately massively contaminated—told of regular nightly burnings of nuclear waste that ultimately were uncovered and led, in part, to the closing of Rocky Flats by the Federal Bureau of Investigation (FBI). Kristin Iversen attests to this history in her biography of growing up near Rocky Flats. Similar accounts are documented in *The Ambushed Grand Jury*, a signature account of the legal infractions and government manipulations that speak to the dangers of being a whistleblower at Rocky Flats. Along with the scholarly historical account of legal documents from Rocky Flats by John M. Whiteley, these three books back up Ted Ziegler's deadly picture of disregard for the health of workers and downwind residents.[20]

As a physician, my relationships with such brave workers, especially Ted Ziegler and his massive collection of union safety documents, have afforded me profound insight into what was actually going on at "the Flats." Inspired to learn more, a rigorous review of available documents convinced me to challenge the government that proliferated the view that there is no danger past or present from the Flats. I, Dr. Stiles, know firsthand from my patients and research that there was neglect of safety regulations and neglect of care for injured workers. The numerous accounts of plantwide toxic dust, plantwide asbestos issues, plantwide nonadherence to PPE standards and inaccurate dosimetry counts force me to reexamine any document that is based on plant-derived dosimetry records. I am a physician, and what I learned from the workers I interviewed, one after another, continues to haunt me.

Flaws, Conflicts, and Revisions in Calculating Doses

The dose of acceptable plutonium concentrations at Rocky Flats has been revised repeatedly by the state and federal government over many decades. Whenever a possible contamination event occurs, the government constructs new templates for dealing with problems. The above-mentioned notes and records provided by Ted Ziegler, safety manager at Rocky Flats during the cleanup, describe monitoring practices whereby air monitoring was only conducted after a surface area was cleaned and not during operation when the surfaces were known to be clogged with toxic dusts. Also, with mathematical modeling models as RESRAD, it seems always possible to find a mathematical calculation to show negligible risk. To satisfy a no-risk mandate by mathematical calculations, soil content over multiple depths of soil could be averaged. Exposure is always estimated. In recent reporting of plutonium dioxide content by two reputable soil analysis groups, the CDPHE and Department of Fish and Wildlife were able to develop a new calculation which states that the likelihood of developing disease from such soil analyses is highly unlikely. Of course, if you or your loved one was the one to inhale such a particle known to be on the refuge, you might not appreciate these government dismissals of risk. The language of likelihood is well understood in the field of epidemiology. *Yet in clinical medicine, if your patient is at risk and you do not take reasonable precautions (no matter how unlikely the possibility), well, you just are not a very good doctor.*

Recently, the Fish and Wildlife Service was given the responsibility for caring for and opening the buffer zone around the lethal superfund site as a wildlife refuge. LeRoy Moore, prolific author, former professor at the University of Colorado, Boulder, and head of the Rocky Mountain Peace and Justice Center, confirms a common Department of Energy practice of setting aside contaminated lands around superfund sites as wildlife refuges because the regulations for cleanup are less stringent in such a refuge than for residential use. If people, such as visitors, are estimated to spend only an hour a year, the Colorado Department of Public Health and Environment (CDPHE) estimated the risk for human contamination to be negligible no matter what the exposure dose might have been. LeRoy Moore calls this "reckless and violating the public will."[21]

The standards of cleanup have been a process of negotiation, with state and federal officials resisting advice of citizen advocacy groups. A sanctioned Rocky Flats Citizen Advisory Board and a Rocky Flats Local Impact Initiative, both of which were DOE funded, recommended that the cleanup standard be the average of known background radiation (believed to be around 0.04 picocuries/gram of soil). The DOE and CDPHE countered with a safety number of 651 picocuries per gram to remain in the Rocky Flats soil after cleanup. The 650 pCi/gm was revised by the Rocky Flats Cleanup Agreement (RFCA) to 50 picocuries/gram of soil to remain in the top three feet of soil, and 1000 pCi/gm in the depth of three to six feet of soil, and as much as 6000 pCi/gm left in the area below six feet.[22] Around this time the Institute for Environmental and Energy Research recommended a cleanup level for which a subsistence farmer could occupy the site and eat food grown there—no more than 10 pCi per gram.[23] This concern was rejected by the government. The RFCA numbers were accepted.

Another way that the government obscures risk of contamination is in applying risk calculators such as RESRAD. Simply stated, "risk" is an estimate of the represented duration of time in exposure area times exposure dose. Risk can be calculated without regard for actual epidemiological data of disease incidence. If one spends only a short time in an area, our government can postulate that the level of plutonium in a gram of soil will produce quite a small risk. Based on that, the risk to one Fish and Wildlife Refuge worker (not a resident living on the refuge full time), spending possibly not much more or less time on the refuge than one hour a year is deemed negligible . . . *unless you happen to be at the wrong place and the wrong time.*

I, Dr. Stiles, began this essay considering that the risk to an individual that actually might inhale or ingest plutonium is not trivial and should not be brushed aside in phrases like low risk. If one acquires, or if one's child acquires, a radiation-related disorder, the risk is devastating. Later this essay will bypass the concept of estimating lifetime risk, in the hopes that new epigenetic research can predict what the risk is to the genome, the real dosimeter. Risk measured in generations, is not able to be calculated with any known RESRAD calculation. Individual risk cannot be determined by RESRAD, only probability.

But more on that later. Suffice it to say that when CDPHE/DOE representative Lindsay Masters gave her discussion as if there were no risks at Rocky Flats to the Rocky Flats Stewardship Council in 2019, her titled thesis "Myths and Misunderstandings" was not appreciated.[24] The question of accuracy of dosimeters has long been a sore spot in academic understanding of dosimetry readings. Ted Ziegler, aforementioned Steel Workers Union Safety Representative at Rocky Flats, has documents that question the placement of personal dosimeters, inaccurate recording of personal dosimeters, constant cleaning of building surfaces before any recorded dosimetry is taken, and absence of adequate ventilation of toxic dusts in the plutonium and beryllium buildings.[25] Arjun Makhijani notes that all safe exposure estimates had been based on "reference man" rather than taking into account the very different biological parameters of women, children, and the fetus. New research, such as by the Gender and Radiation Impact Project, calls for more studies on these known disparities.[26]

Local Colorado government officials have continued to resist many new ideas and new soil analyses as late as 2021. In 2019 and 2020 Michael Ketterer, PhD, chemistry and soil scientist, analyzed soil downwind from the 903 Pad along a proposed highway adjacent to the refuge. He found multiple plutonium dioxide particles of 1–3 microns along the roadway destined to be bulldozed and thrown into the ever-growing number of nearby residential developments.[27] The CDPHE and DOE-funded Rocky Flats Stewardship Council and its chairman David Abrams stated (as predicted) that there are already calculations that handle this. He appeared neither surprised nor concerned.[28] A state highway commission contractor, Engineering Analytics, then found an 8.8 micron particle which was eventually dismissed by CDPHE using risk calculations. However, townships along a proposed highway adjacent to Rocky Flats so far have abandoned the project. Science of ingesting or inhaling an 8.8 micron particle is lacking. Much smaller sized particles (2–3 micron) are known to cause severe disability. There is no data on the massive disability suspected from such a large 8.8 micron particle. However, radiation science would suggest that the damage could be catastrophic.

New Worker Study—Ruttenber et al.

It would be incomplete to end our discourse without mentioning one exemplary epidemiological study now long forgotten by A. James Ruttenber and his coauthors. Great epidemiology needs to be applauded! Ruttenber, of the Department of Preventive Medicine and Biometrics, University of Colorado Health Sciences, published an epidemiologic analysis of workers at Rocky Flats employed from 1952 to 1989. The paper identified a total of 23,000 workers who had been hired between 1951 and 1989, including metal workers, chemical process operators, health physicists, chemists, engineers, machinists, radiological protection engineers, guards, and office workers. Over 18,000 workers were monitored at some time for exposure to radiation.[29]

Ruttenber et al.'s analysis leans on SMRs. The Standardized Mortality Ratio (SMR) is defined as the number of observed deaths in the study population divided by the number of expected deaths (calculated from indirect adjustment) and multiplied by 100. "Standardized Mortality Ratios (SMR) for the production era cohort was significantly lower than expected for all causes of death and all deaths with cancer as an underlying cause. The SMR for lung cancer was not elevated. *Only the SMR for unspecified neoplasms of the nervous system was significant ($p<0.05$) when Colorado mortality rates were used to compare the expected number of cases.*"[30]

Concerning brain cancer, Ruttenber et al. reviews other researcher data: "Reyes et al. (1984) first identified elevations in brain neoplasms for plutonium workers. Wilkinson et al. (1987) found a statistically significant elevation in the SMR for unspecified brain neoplasms, based on seven cases. They also found a non-significantly elevated SMR for brain cancer and an elevated rate ratio for unspecified brain tumors in workers with cumulative external penetrating radiation doses greater than 0.01 Sv, as compared with those who had lower doses." Also: "These findings are consistent with a statistically significant dose-response relation between cumulative external radiation dose and cancers of the brain in Los Alamos workers" (Wiggs et al., 1994). Omar et al. (1999) also noted a statistically significant dose-response relation for brain cancer incidence and cumulative external radiation dose for all radiation workers at the British nuclear site at Sellafield, and SMRs for brain cancers have been elevated in other cohorts of nuclear workers (Alexander, 1991).[31] Ruttenber concludes that his findings indicate that the elevated SMR for unspecified neoplasms of the brain has persisted and remained statistically significant.

Ruttenber died shortly after this publication. Had he lived to continue his work, it is probable that his ability to track, using his large database, cancer incidence for cancers that naturally take longer to develop, might have been possibly more reliable than risk analyses.

Going forward, Wilkinson et al. note inconsistencies in the data and made massive efforts to corroborate conclusions using multiple data sources and multiple statistical tools.[32] The analyses of these admirable efforts is beyond the scope of this essay, but they are well appreciated. Acknowledgments of missing data, inaccurate dosimetry data, problems with incomplete cigarette smoking data, and differing data based on date of hire, age of hire, or years of work are appreciated. Radiation technician (RCT) data, as an example given by Ted Ziegler's firsthand reporting, were incomplete at best. Thus the full burden of radiation exposure is not known. *The true importance of data offered by the RCTS at Rocky Flats, in my opinion, is in their ability to consider what is missing.*

ANWAG

Since Ruttenber et al., worker advocacy groups have sprung up supporting past Rocky Flats workers in their quest to gain compensation for illnesses related to their work at the Rocky Flats Nuclear Weapons Plant. One major group, ANWAG (Alliance for Nuclear Workers Advocacy Groups), is quoted below. As recently as 2016, Terrie Barrie, founding member for ANWAG, presented documentation of Safety Violations to the National Institute for Occupational Safety and Health Advisory Board on Worker Safety. These included falsification of plutonium weights on run sheets, falsification of medical records, and falsifications of inspection reports. For example, there was evidence of radioactive materials and contaminated equipment in Building 460 through 1996. Workers in this building wore no protective gear as it was presumed "cold." Later it was clear that it was contaminated. These errors of omission are directly related to NIOSH's questionable ability to reconstruct dose. ANWAG states: *"If the alarm doesn't alarm, then the worker or the Radiation Control Technicians will not know there is a release."*[33]

Dr. Stiles 2021 Clinical Evaluation of Workers applying for Compensation from the DOL

Twenty years after Ruttenber's works, I, Dr. Stiles, reviewed records from a significant number of workers, including those who worked under a health physicist to gather injury-related dosimetry data. I have reviewed their training and their complete work history, building by building. Their ability to collect and process data was severely limited due to the massive needs to produce nuclear weapons and have nothing which might impede the speed of production. I looked at all dosimetry records, which way too often contain glaring missing information. Dose averaging was used to fill in the gaps. Low or high irradiation levels are hidden in the averaging. All workers I spoke with wanted me to know that toxic dust was everywhere, all the time and was never really removed.

As stated previously, Ted Ziegler, safety officer for Rocky Flats, has allowed me to review volumes of Rocky Flats Safety infractions, many of which are related to inadequate and inaccurate dosimetry evaluations. Workers were exposed to plutonium dust daily without adequate protection. Workers were exposed to beryllium and asbestos dust daily with poor ventilation and inadequate cleanup standards. Beryllium and asbestos were documented by workers and Mr. Ziegler as not being limited to some twenty buildings. Due to faulty cleanup and the need for production over safety, toxic dusts were always present all over the plant, from the cafeteria to the areas known to use said toxins. Some buildings required protective gear only later on in plant history. Many other buildings did not require adequate protective gear on a consistent basis. The DOE files later confirmed many buildings to have plutonium, americium, and other deadly toxins that had not been identified initially. Office workers, the great majority of whom were never tested, clearly were at risk of exposure.[34]

I also learned that when workers were involved in an identified incident, assistants to health physicists (the RCTs) would collect badges if worn and distribute urine cups. Often badges were not worn at all. Urine collection was a nuisance and often not fully complied with. There seemed to be, in the records I studied, no rigorous follow up. Often if external dosimetry readings were high, workers were told to go and wash off as many times as needed to bring the surface numbers down. That final number was usually the one recorded to prove that there was no residual contamination. The initial contamination numbers, which would have provided a clearer picture of potential exposures, were not recorded. As there was confusion about the extent of external versus internal damage, such results do not lend clarity. The workers were asked to bring in urine samples on the day of the incident and on subsequent days. Often these were lost or not done more than once and sometimes none at all. If, over time, a urine sampling turned up negative, the workers were told they were clear of any internal radiation. Often incidents (spills and small fires) were not reported due to worker fear of retribution. It is extremely hard to convince myself that the dosimetry records ever give a complete history of plutonium exposure, let alone exposure to the hundreds of other toxic chemicals on site. My patient evaluations, however, would be considered more like an anecdotal study. Although compelling, my information is not considered worthy epidemiologically sound data. My alternative is to explain my clinical observations using new scientific information.

A Paradigm Shift

The linear no-threshold (LNT) model is the most commonly used model to estimate the biological risks from ionizing radiation and to dictate government policy. The National Academies' most recent report on the biological effects of ionizing radiation, known as BEIR VII, states that the linear no-threshold model is based on the biological responses at high radiation doses and dose rates.[35] Since most humans are not willing to expose themselves to harmful levels of radiation, this necessitates that biological harm at lower levels is extrapolated. The initial data came from radiation victims from Hiroshima and Nagasaki, that is to say, very large doses and very obvious immediate health consequences. The effects of radiation at a low dose rate is much harder to monitor initially or over time. For one, low doses of radiation may not be much more than natural background levels. The biological response to radiation comes from a time delay between exposure and developing a cancer or other medical disorder.

Essential to this model is the belief that dosage outweighs biological variables, and dosage is more important than dose rate. The LNT model may claim to understand the effects of high and low dose radiation, i.e., a linear effect per above. However nowhere in the LNT model is there adequate adjustment for a low dose over an extended period of time. The LNT does not account for the difference between initial low dose (which may not be much more than background) from very prolonged effects of such a low dose over time. Clearly, the effects of prolonged exposure to low level doses have been underestimated.

Thus, the LNT government-supported research model did not address the damaging effects suffered by cells (and DNA) that could accumulate over time and that such cumulative damaging effects could occur even after months or years or generations.[36] Now, evidence is clear that genomic instability caused by minimal doses of radiation are often sufficient to produce serious effects. Burgio et al. in a landmark work confirm that the great majority of earlier studies were fraught with assumptions regarding absorbed dose and also clearly noted epidemiological omissions.[37] For example, years later data began emerging regarding thousands of events (abortions and childhood deaths) that occurred in the first weeks/months after Hiroshima and Nagasaki and of congenital malformations that had never been recorded.[38]

H. S. Weinberg et al. identified microsatellite lesions in children exposed to a low dose of radiation, as well as similar lesions in their offspring.[39] These papers and many others marked the beginnings of a far deeper understanding of true genomic instability from a low dose of radiation over an extended period of time. These were consequences unknown during the development of the LNT model, for if they had been known, the model should not have been adopted. These papers are transformative for understanding the importance of supposed unseen damage of ionizing radiation. These papers discussed the clinical conditions we clinicians are now seeing in our patients. They highlight the past disregard for data on women and children. The path forward as physicians and public health advocates is to circumvent the government conclusion that everything is just fine and look to science for the real truth!

To reiterate a central point of this paradigm shift: The mechanism of damage from low dose radiation was stated by the Burgio paper to be fundamentally different from high dose radiation. High doses produced damage consistent with the LNT model and by extension RESRAD mathematical calculations. But effects of exposure to small doses, accumulated over many years, can cause tissue and systemic reactions different in kind from those much more immediate effects from an initial high dose. Exposure to low doses also induce, both in the affected cells and those not directly exposed, a progressive genomic instability which may prove to be even more damaging. I, Dr. Granados, began shifting my paradigm, thinking that the answer was in the perceived difference in low dose emissions.

Fortunately, science is progressing in spite of the DOE faith in RESRAD and adherence to the LNT model and questionable soil testing modalities. Disasters like Chernobyl have proven that exposure at the time of the initial blast may be but a harbinger of worse outcomes to come in the next five to fifty years. It is time for a shift of research: away from faulty dosimetry readings and sloppy accounts of plant malfunctions; away from faulty representations of soil contamination by choosing questionable testing locations and questionable depths; away from choosing cohorts that mix affected with those out of the impact zones; away from being discredited and dismissed by the majority of public health officials in the state. We are now on the verge of following genetic markers from workers and residents in the surrounding communities, through however many generations there may be, to determine the beginnings of genomic vulnerability from ionizing radiation that progresses on to cancer. These techniques, explained in the next section, derived by ever sophisticated biological labs are becoming available now with a power so robust as to place genomic study at our doorstep. It is time for basic science research to defend against further mismanagement of nuclear industries where epidemiology and RESRAD calculations based on the LNT model have, despite good and bad intentions, failed.

Technology Advances for Detecting Genomic Alterations

With the advent of genomic technologies, observing the genetic, epigenetic, and cytogenetic alterations that radiation has caused over generations to the human body can be made visible. By genetic alterations we mean changes to the DNA coding sequences that ultimately result in alterations to the proteins that are encoded in the DNA sequences. In contrast, epigenetic alterations do not involve alterations to DNA coding regions but nonetheless result in alterations to the levels of proteins that are made available for life-supporting functions.

These alterations, that are associated with cancer initiation and progression, can remain unseen for many years prior to the appearance of clinical symptoms of cancer. We will briefly discuss new science techniques that are being used to detect genetic, epigenetic, and cytogenetic alterations that are associated with cancer initiation and progression, prior to the appearance of clinical symptoms.

The Polymerase Chain Reaction (PCR) has been used to detect genetic and epigenetic changes to DNA. Earlier PCR detection involved time-consuming, manual protocols that were often not highly specific for the species being analyzed. Modern Real Time (rt) PCR methods utilize advanced detection reagents and highly automated equipment that results in more accurate detections. Isabelle Miousse et al. have utilized rt PCR to show that an epigenetic modification to DNA, known as methylation, depends on radiation type and dose as well as the type of cell that is irradiated.[40]

Exponential advances have been made in technologies, such as whole Genomic Sequencing of the entire length of human genomes. Recent reports provide examples of studies that describe genetic changes to human and mammal DNA that have been exposed to ionizing radiation. Technologies such as direction Genomic Hybridization (dGH) are advances in the methodologies that can be used to detect chromosomal alterations of the type that are associated with exposure to radiation. These alterations include translocations that result in exchange of genetic material between two different chromosomes and inversions that are rearrangements of genetic sequences that are located on the same chromosome.[41] We hope to introduce these new science techniques to compare the DNA and chromosomal structure of Rocky Flats workers and controls.

Conclusion

Although our local and national government institutions contend that the now closed Rocky Flats Nuclear Weapons Plant has never posed and will never pose any health threats to our environment or our population, the authors have reviewed historical, epidemiological, and emerging scientific data to the contrary. Motivation in the 1950s was to preserve our country in the face of perceived nuclear threats. As the cold war dissipated, we contend that the practices of nuclear proliferation, at the cost of massive personal and environmental health, remained. What once was an act of patriotism at the expense of the individual becomes now a possible cover-up of massive human harm.

Our first objective here is to give the reader a historical understanding of what is known to have happened at Rocky Flats, which is contrary to official government records. The second important objective of our essay is to bring forward the scientific advancements especially of genomics and epigenetics and what these advancements tell us about the health impacts of Rocky Flats. This population has aged, and the effects are often no longer traceable . . . or are they? The genome and epigenome, the finer details of which are currently being defined, offer potential traceable mapping of risk. Our final objective is the hope that this information might sway other governmental and molecular genomics researchers to consider Rocky Flats for future research.

Notes

1 E. A. Martell, *Plutonium Contamination in the Denver Area,* Press Release of the Colorado Committee for Environmental Information, Boulder, Colorado (February 24, 1970). See also LeRoy Moore, "Democracy and Public Health at Rocky Flats," in *Tortured Science: Health Studies, Ethics, and Nuclear Weapons in the United States,* ed. Dianne Quigley, Amy Lowman, and Steve Wing (New York: Routledge, 2012), 69–98.

2 Special Acknowledgment: An overriding historical and scholarly publication of the history of Rocky Flats is *Plutonium and People Don't Mix: Colorado's Defunct Nuclear Bomb Factory,* by LeRoy Moore, PhD (Boulder: Rocky Mountain Peace and Justice Center, 2019). Our thanks to LeRoy Moore and the Peace and Justice Center for their tireless work and their help with this project. Special thanks to Physicians for Social Responsibility Colorado and Rocky Flats Downwinders for their support and guidance. And to the expertise of Cindy Folkers, Beyond Nuclear, who helped me translate the unseen into understandable English. Also see Len Ackland, *Making a Real Killing: Rocky Flats and the Nuclear West* (Albuquerque: University of New Mexico Press, 1999).

3 Arjun Makhijani, Howard Hu, and Katherine Yih, eds., *Nuclear Wastelands: A Global Guide to Nuclear Weapons Production and its Health and Environmental Effects* (Cambridge, MA: MIT Press, 1995).

4 H. Holme, Pre-Trial Statement, Civil Action Nos. 75-M-1111, 75-M-1162, and 75-M-1296 (in the United States District Court for the District of Colorado, Denver Colorado, 1978), 53, 98–101. On these events see also Ackland, *Making a Real Killing,* and Kristen Iversen, *Full Body Burden: Growing Up in the Nuclear Shadow of Rocky Flats* (New York: Crown, 2012).

5 John M. Whiteley, *The Saga of Rocky Flats, Lessons Learned for a Safer Nuclear World in the 21st Century* (Coppell, TX: Gardner Institute for Excellence in Undergraduate Education, 2020), 76.

6 Whiteley, *The Saga of Rocky Flats,* 90.

7 Holme, Pre-Trial Statement, 53, 98–101.

8 M. Ketterer, private communication to the author. See also "Update Regarding Rocky Flats Hot Particle Study," 9 April 2020, https://www.rockyflatsnuclearguardianship.org/single-post/2020/04/09/Update-Regarding-Rocky-Flats-Hot-Particle-Study.

9 P. W. Krey, "Remote Plutonium Contamination and Total Inventories from Rocky Flats," *Health Physics* 30 (1976): 209–214, esp. 210.

10 Colorado Central Cancer Registry, *Ratios of Cancer Incidence Ratios in Ten Areas Around Rocky Flats Compared to the Remainder of the Denver Metro Area 1980–89 with Update for Selected Areas 1990–95: A Report to the Health Advisory Panel on Rocky Flats* (Denver: Colorado Department of Public Health, 1998).

11 C. Yu, A. J. Zielen, J.-J. Cheng, D. J. LePoire, E. Gnanapragasam, S. Kamboj, J. Arnish, A. Wallo III, W. A. Williams, and H. Peterson, *Users' Manual for RESRAD Version 6*: (Argonne, IL: Argonne National Laboratory, 2001).

12 LeRoy Moore, "The Bait and Switch Cleanup," *Bulletin of the Atomic Scientists* (January–February 2005).

13 Harvey Nichols, "Some Aspects of Organic and Inorganic Particulate Transport at Rocky Flats," Final Report on ERDA Contract EY-76-S-02-2736 (US Energy Research and Development Administration, n.d.). See also Moore, *Plutonium and People Don't Mix* and Harvey Nichols, "Rocky Flats: A Detective Story," https://www.rockyflatsnuclearguardianship.org/presentation-by-harvey-nichols.

14 Carl J. Johnson, Ronald R. Tidball, and Ronald C. Severson, "Plutonium Hazard in Respirable Dust on the Surface of Soil," *Science* 193, no. 4252 (August 6, 1976): 488–490.

15 Carl J. Johnson, "Cancer Incidence in an Area Contaminated with Radionuclides Near a Nuclear Installation," *Ambio* 10, no. 4 (1981): 176–182.

16 Wes McKinley and Caron Balkany, *The Ambushed Grand Jury* (New York: Apex Press 2004).
17 Whiteley, *The Saga of Rocky Flats*, chap. 12.
18 Some of these reports are included as documentation in congressional hearings about Rocky Flats. See *Environmental Crimes at the Rocky Flats Nuclear Weapons Facility*, hearings before the Subcommittee on Investigations and Oversight of the Committee on Science, Space, and Technology, US House of Representatives (Washington, DC: Government Printing Office, 1992). Quotations on pp. 708, 711–712.
19 Dominic Sanchini, RF Plant Supervisor, Safety Committee Meetings, as reviewed by the authors.
20 McKinley and Balkany, *The Ambushed Grand Jury*; Iversen, *Full Body Burden*; Whiteley, *The Saga of Rocky Flats*; personal discussion with Ted Ziegler regarding documents in his possession, Book 9 of his personal documents collected during his tenure as head of the Safety Committee of the Steelworkers Union, Rocky Flats. Assessment and Integration of Radioactive Ambient Air Monitoring at Rocky Flats 1993 and Committee documents 1995 with Appendix C Sample Justification.
21 Moore, "The Bait and Switch Cleanup."
22 US Department of Energy, Office of Environmental Management, *Final Rocky Flats Cleanup Agreement, July 19, 1996* (Federal Facility Agreement and Consent Order, State of Colorado Docket # 96-07-19-01).
23 Arjun Makhijani and Sriram Gopal, *Setting Cleanup Standards to Protect Future Generations: The Scientific Basis of the Subsistence Farmer Scenario and Its Application in the Estimation of Radionuclide Soil Action Levels for Rocky Flats* (Boulder: Rocky Mountain Peace and Justice Center, 2001).
24 Lindsay Masters, "Rocky Flats Myths and Misunderstandings," presentation to the Rocky Flats Stewardship Council, February 5, 2019.
25 From documents in Ted Ziegler's possession, Book 2, Joint Company-Union Internal letters on Safety concerns: Discussing lack of Beryllium Surveys. Example: 150 smears in building 144 are the "Tidy Friday level," meaning they are taken only after the area has been cleaned.
26 Arjun Makhijani, Brice Smith, and Michael C. Thorne, *Science for the Vulnerable: Setting Radiation and Multiple Exposure Environmental Health Standards to Protect Those Most at Risk* (Takoma Park, MD: Institute for Energy and Environmental Research, 2006).
27 Michael Ketterer, private communication. See also "Update Regarding Rocky Flats Hot Particle Study," 9 April 2020, https://www.rockyflatsnuclearguardianship.org/single-post/2020/04/09/Update-Regarding-Rocky-Flats-Hot-Particle-Study.
28 David Abelson, Stewardship Council Meeting 2019.
29 A. James Ruttenber, Margaret Schonbeck, Shannon Brown, Timothy Wells, David McClure, Jason McCrea, Douglas Popken, and John Martyny, *Report of Epidemiologic Analyses Performed for Rocky Flats Production Workers Employed between 1952 and 1989* (Boulder: Colorado Department of Public Health and Environment, 2003).
30 Ibid.
31 Ibid.
32 G. S. Wilkinson, G. L. Tietjen, L. D. Wiggs, W. A. Galke, J. F. Acquavella, M. Reyes, G. Voelz, and R. J. Waxweiler, "Mortality among Plutonium and Other Radiation Workers at a Plutonium Weapons Facility," *American Journal of Epidemiology* 125, 2 (1987): 231–250.
33 Terrie Barrie, ANWAG, presentation to NIOSH Advisory Board, November 30, 2016, https://www.rockyflatsnuclearguardianship.org/single-post/2017/01/08/updates-to-the-ambushed-grand-jury-website (accessed November 2, 2022).

34 Records in Ted Ziegler's possession, book 3, Beryllium and Asbestos. Enclosed is a photograph illustrating the common safety infractions: not using correct protective gear. Mr. Ziegler provides documents on inadequate cleanup, inadequate gear, and poor ventilation.

35 National Research Council, *Health Risks from Exposure to Low Levels of Ionizing Radiation, BEIR VII Phase 2* (Washington, DC: National Academies Press, 2006).

36 Ernesto Burgio, Prisco Piscitelli, and Lucia Migliore, "Ionizing Radiation and Human Health: Reviewing Models of Exposure and Mechanisms of Cellular Damage. An Epigenetic Perspective," *International Journal of Environmental Research and Public Health* 15, no. 9 (2018): 1971.

37 Ibid.

38 M. Otake, W. J. Schull, and J. V. Neel, "Congenital Malformations, Stillbirths, and Early Mortality among the Children of Atomic Bomb Survivors: A Reanalysis," *Radiation Research* 122, no. 1 (1990): 1–11.

39 H. S. Weinberg, A. B. Kovol, V. M. Kirzher, A. Avivi, T. Fahima, E. Nevo, S. Shapiro, G. Rennert, O. Piatak, E. I. Stepanova, and E. Skvarskaja, "Very High Mutation Rates in Offspring of Chernobyl Accident Liquidators," *Proceedings of the Royal Society B: Biological Sciences* 268, no. 1471 (2001): 1001–1005.

40 Isabelle R. Miousse, Jianhui Chang, Lijian Shao, Rupak Pathak, Etienne Nzabarushimana, Kristy R. Kutanzi, Reid D. Landes, Alan J. Tackett, Martin Hauer-Jensen, Daohong Zhou, and Igor Koturbash, "Inter-Strain Differences in LINE-1DNA Methylation in the Mouse Hematopoietic System in Response to Exposure to Ionizing Radiation," *International Journal of Molecular Sciences* 18, no. 7 (2017): 1430.

41 On these newer studies, see Manuel Holtgrewe, Alexej Knaus, Gabriele Hildebrand, Jean-Tori Pantel, Miguel Rodriguez de Los Santos, Kornelia Neveling, Jakob Goldmann, Max Schubach, Marten Jager, Maria Coultelier, Stefan Mundlos, Dieter Beule, Karl Sperling, and Peter Michael Krawitz, "Multisite de Novo Mutations in Human Offspring after Exposure to Ionizing Radiation," *Scientific Reports* 8, no. 14611 (2018); Yuri E. Dubrova, "Mutation Induction in Humans and Mice: Where Are We Now?" *Cancers* 11, no. 11 (2019): 1708; and Miles J. McKenna, Erin Robinson, Lynn Taylor, Christopher Tompkins, Michael N. Cornforth, Steven L. Simon, and Susan M. Bailey, "Chromosome Translocations, Inversions and Telomere Length for Retrospective Biodosimetry on Exposed US Atomic Veterans," *Radiation Research* 191, no. 4 (2019): 311–322.

The Town That Fell Asleep

Malignant Infrastructures of Soviet-era Nuclear Ruins in Kazakhstan

MAGDALENA EDYTA STAWKOWSKI

In the summer of 2012, a mysterious sickness struck North Central Kazakhstan. News began circulating that several people in two tiny neighboring villages of Kalachi and Krasnogorskiy, in the country's Akmola region, began to fall asleep, suddenly. They sometimes blacked out while walking and woke up days later with no memory of what had happened. After waking, many experienced debilitating headaches and intense hallucinations that lasted for weeks. Others felt dizzy, slurred their words, and vomited. The mystery sickness would strike some people more than once and affected both children and adults. Local doctors, scientists, and government officials were baffled. At first, they suspected something endemic like alcohol poisoning. When this proved not to be the case, they suggested a seasonal flu as a possible culprit. Once it became obvious that it was not the flu, they hypothesized that the cause was psychological in nature, a mass psychosis of sorts, akin to the "dancing plague" that occurred in communities across Europe in the Middle Ages. As the number of victims grew, "encephalopathy of an unknown origin," a generic term used to describe abnormal brain function of unknown cause, became the default diagnosis. The bouts of sleep lasted for three years, coming in waves during the spring and autumn months. More than 140 people, or about a quarter of the total population, were affected. Residents long blamed the nearby uranium mines, with their network of flooded tunnels and pits, for their ills and for the area's newfound namesake, "sleepy hollow."[1] It turned out the residents were right all along: the mines were the source of their ills. But it took teams of scientists, including virologists, toxicologists, and radioecologists in Kazakhstan three years to come to the same conclusion, even though they still argue whether it was carbon monoxide, radon gas, or some other toxic vapor escaping the mines that was the cause.

The mystery sleeping sickness is more than a biomedical anomaly. Its appearance raises questions about the less visible and often concealed dimensions of environmental histories in Kazakhstan and the controversies that surround them. It exposes the ongoing health effects and ecological damage of "nuclear colonialism," which is broadly understood as a host of disproportionate negative effects on marginalized local populations and lands that once maintained nuclear sites but are no longer considered economically or industrially productive.[2] A close examination of the case reveals Kazakhstan's struggle with "troubled ecologies" that mark a long tale of radioactive contamination from decades of uranium extraction that continues to shape life and environmental politics on its territories since the Second World War.[3] The abandoned mines in Krasnogorskiy were once part of the much greater Soviet nuclear program that was a vast interconnected industry of secret and closed cities housing production facilities, research institutes, extraction sites, nuclear testing areas, and other militarized spaces, that have been abandoned with the breakup of the Soviet Union in 1991. Since the breakup, people have continued to live in far-flung ex-mining communities like Krasnogorskiy and Kalachi, places where others happen to migrate to, in search of affordable places to live. Here, as elsewhere in Kazakhstan, concerns about radioactive pollution and hazardous waste have helped important conversations come to the fore, even as these conversations have been reframed by some scientists and politicians as a reflection of people's cultural "backwardness," leading to a susceptibility to irrational fears of radiation, or "radiophobia."[4]

In this essay I develop the analytic of *malignant infrastructures*. The term describes the ways in which built landscapes that supported things such as uranium mining industries index cases of nuclear violence in the present. Malignant infrastructures, like the abandoned mines of Krasnogorskiy, make people sick. They continue to inflict harm through processes of municipal and industrial ruination which is a direct outcome of key political and economic transformations that occurred in the region since the disintegration of the Soviet Union. I use "malignant" for two reasons. First, toxins from abandoned uranium industries in Kazakhstan are not self-limiting, meaning they will not resolve on their own. Radon gas, heavy metals, and other unsecured "invisible harms" entwined within a complex relationship between the communities, Cold War military investments, and lack of solutions, continue to spread unabated and concentrate in people's bodies.[5] Living atop derelict mines means exposure to radioactive elements and heavy metals, where people's bodies turn into the "living breathing archives of atomic history."[6] In Kazakhstan, the stories of people who live in former uranium mining towns, near the dusty mine tailings or piles of unsecured hazardous waste, are rarely part of the historical narrative about the country's nuclear legacy. What the sleeping sickness shows is how colonial histories and the legacies of the nuclear age are "made visible in negative outcomes."[7]

"Malignant" has a second use here. Like any cancer that is left without adequate treatment, the health effects of malignant infrastructures on the environment and its people spread and get progressively worse. Krasnogorskiy's uranium mines highlight how crumbling and neglected infrastructure shape human experiences of health, through "infrastructural violence," where physical suffering derives from being excluded from social life as citizens.[8] That is, the mysterious sleeping sickness exposes the "routinely violent" former industrial and military geographies that exist as ruins in Kazakhstan. These are places where suffering is experienced in material terms: impaired access to adequate housing, potable water, and health care.[9] These geographies are rooted in large-scale environmental transformations wherein the extraction of natural resources drove the expansion of the Soviet Union's military-civilian nuclear industry, and since the early 1990s, they mark the messy disentanglement from the broader economy and the widespread state divestment in the region and its people.

I first heard about the Kalachi sleeping sickness in summer of 2015. At the time, I was in the eastern part of the country in Semey, talking to physicians, epidemiologists, government officials, and residents about something different. I was there asking questions about the sociobiological effects of forty years of Soviet-era atomic testing that took place on the Semipalatinsk Test Site (or the Polygon, as it is known locally) I've been researching since 2010. I was sitting at a leafy downtown café with Ivan talking about his experiences of atomic testing before our conversation turned to where the uranium for those decades' worth of tests was coming from. A burly man in his early sixties, Ivan once lived and worked in Krasnogorskiy in the late 1970s and 1980s as a foreman and then a supervisor of mine rescue operations for the Tselinny Mining and Chemicals Combine (Tselinny Gorno-Chemicheskiy Kombinat or TsGHK, for short) in charge of uranium production in northern Kazakhstan. "Everyone knows about the Polygon, but no one talks about the uranium towns," he said. Indeed, the history of the Polygon and the human tragedy of forty years of nuclear testing is well known in Kazakhstan. In Semey, near the where we sat, stands a monument, "Stronger than Death." Erected in 2002, the monument and large adjacent park commemorate the many victims of Soviet atomic bombs tested in Kazakhstan. The 100-foot silhouette of a mushroom cloud rises high above a network of walking paths. At the base is a figure of a mother sheltering her infant child. Every year, on August 29, the country celebrates the closing of the Polygon, and "Stronger than Death" is a key focal point. Over the past several decades, scores of newspaper articles appear with stories about people who suffered the devastating effects of radioactive fallout. But there is no such monument to uranium workers and few, if any, stories written about them. So, while nuclear testing is an obvious risk to human safety, heavy metals and radioactivity, especially in the form of invisible radon gas that emanates from the uranium mines, continue to be a serious problem. Though, Ivan said to me, "it's one of a number of secrets among other secrets that no one talks about."

In the pages to come I begin by situating Krasnogorskiy and Kalachi within wider histories of Soviet industrial projects in Kazakhstan. The backstory of the region is crucial to getting a sense of what kinds of processes were under way and exactly how these two towns were brought into existence in the first place. From there, I look at the collapse of the uranium mining industry and the subsequent appearance of the sleeping sickness. Finally, I look at the measures taken by local government officials to study the problem and secure the area within Kazakhstan's post-Soviet nuclear industry.

Secret Towns and Environmental Destruction

Kazakhstan is one of the most heavily polluted countries of the former Soviet Union. Rich in energy resources, like oil, gas, coal, copper, and uranium, among other things, broad swaths of the land and its people were enrolled in the service of centrally orchestrated projects that forced industrial modernity upon them. Like elsewhere in the Soviet Union, economic planners, engineers, and scientists joined forces to transform Kazakhstan's "useless" nature into state commodities, turning the once distant villages and dusty plains into industrial centers where production grew at a breathtaking speed from one month to the next.[10] As in the United States, the Soviet Union's quest for rapid economic growth and military power were affairs that were often cloaked in secrecy. The environmental histories tell of the careless disposal of billions of tons of toxic waste that poisoned rivers, lakes, drinking water, land, air, and people.[11] In these "industrial deserts" of extreme environmental devastation, industry and uncontrolled pollution went hand in hand, while hazardous waste monitoring, environmental standards enforcement, and public health concerns were not a priority.[12] In Kazakhstan, hundreds of industrial towns of various sizes were connected to mining industries that emerged during the early years of the Cold War.

Kazakhstan's key industrial role can be traced to the beginning of the Second World War in the 1940s and the subsequent Cold War that prompted the Soviet Union to evacuate factories and institutes east. As elsewhere, industry and industrial sites were often matters of national security and a growing number of highly specialized and secluded military-industrial-science cities were purposefully funded to provide a technological edge against the United

States.[13] Some were tasked to build and test rockets for the space program and defensive anti-continental ballistic missiles.[14] In others, researchers worked on nuclear and biochemical weapons. These subsidized mono-industry towns were collectively known by the acronym ZATO, or the Closed Administrative Territorial Entities. They were shut off from the world and not featured on publicly available maps. No one could enter without proper documentation. Kazakhstan's nuclear ZATO, developed to serve a specific purpose of Cold War techno-science, and mainly focused on the mining and processing of uranium-bearing ore for nuclear power plants and for the atomic bombs that were tested on the vast open plains of the Polygon.[15] ZATO were paradoxical places. On the one hand, people's health disproportionately suffered from occupational hazards as well as contact with many kinds of toxins, including radioactive elements. On the other hand, those who worked there belonged to privileged communities, where the state paid for everything from housing to schooling to health care and food, incentivizing people to stay despite ecological and health hazards.[16]

Intensive prospecting of uranium deposits in Kazakhstan began in the mid-1940s. The Ministry of Medium Machine Building, in charge of the national nuclear project, directed research and development from Moscow. Northern Kazakhstan was especially important to Soviet economic and military planners because it had substantial uranium deposits. Eight were discovered there in 1956 during a landscape survey ordered by the Soviet Ministry of Geology. This survey's results served as the basis for the creation of TsGHK in the region, headquartered in Stepnogorsk, that ensured the extraction and processing of uranium ores by five regional enterprises it would eventually oversee.[17] In May of 1956, construction of "Ore Management No. 1" (which included the city of Stepnogorsk) began. It was a rapid building project that came together over the course of about four years. According to a personal account of the first director of TsGHK, Sergey Smirnov, in the first year alone, soldiers and prisoners from Gulag labor camps—who made up majority of the workforce—went to work on twelve dwellings (each with four apartments), and built a smith shop for forging metal, and a machine shop for all kinds of repair.[18] Labor conditions were generally abysmal. TsGHK lacked cars, tractors, bulldozers, and other equipment that it was compelled to borrow from nearby collective farms that used them to work on a mammoth agricultural project of the same era—the Virgin Lands.[19] There was no water or electricity, and the work was constantly slowed down by things such as knee-deep mud in spring and violent snowstorms in winter. The challenges of building the site are part of the broader histories of the region as told by managers, directors, and others narrating this chapter in a greater heroic Soviet history.

In the context of the Cold War, landscape and labor were not only key to uranium extraction but there was state-driven competition between sites for producing the most ore regardless of environmental and human cost. Krasnogorskiy and Kalachi were part of TsGHK "Ore Management No. 4" and were located at the far western end of an arc of five regional uranium mining enterprises stretching 500 kilometers (over 300 miles) from Stepnogorsk across north central Kazakhstan.[20] Construction of the mining complex in Krasnogorskiy began in 1964, and the first underground uranium mine went into production in 1969, with a second one soon after. The mines were in the floodplain of the Ishim River, the largest water artery in the region, which flows north into Russia. Eventually, the town population grew to about 7,000 people, and the newly constructed settlement of Kalachi became the local administrative and residential arm of the enterprise. According to Ivan, they were places where "people lived a good life," and in Krasnogorskiy, there was a statue depicting Lenin pointing ahead with his hand showing the way.

Part of ZATO, residents of the two settlements had access to many amenities. Kalachi had several well-stocked grocery stores, a sports complex, and numerous retail shops selling everything from toys, to bicycles, to latest fashions. Workers and their families celebrated birthdays, holidays, promotions, and retirements in a centrally located club built for such things. There was even a hospital, constructed in 1969, with an inpatient facility for 120 beds, and a polyclinic that could accommodate 250 visits per shift.[21] In summer months, people liked to swim in the nearby Ishim River and picnic along its banks. But living the "good life" also meant that residents had to put themselves in harm's way. "There were many accidents in the mines, and people got trapped all the time," Ivan said reminiscing about his rescue work for the TsGHK. Deep underground, he searched for miners in tunnel cave-ins, those injured in falls, or killed in explosions. Although he was periodically tested for radiation exposure, Ivan understood that uranium mining is dirty work, and the priority of the administrators was to keep the industry going. He linked the deaths of his friends and colleagues from cancer to radiation hazards that came with the job, "a price some people paid for having a steady source of income."

As with thousands of previously well-funded and operational monotowns, with the Soviet collapse in the early 1990s, people lost their jobs as plant closures signaled the withdrawal of state resources.[22] Krasnogorskiy's underground mines, once alive with activity, eventually flooded. Many people moved away in search of jobs elsewhere, while others, including pension-aged people, stayed behind with no place to go. Like other highly strategic defense industrial enterprises in Kazakhstan, there was no contingency plan for shutting down operations. The area was not well secured, and more than 1.5 million tons of toxic waste from the towns' mines, laced with radioactive elements and heavy metals, were left in tailings and underground pits.[23]

Economic hardship and the transition to a market economy in the early 1990s relegated cleanup to the peripheries of state action that was tasked with an enormous responsibility for the securitization of 235 million tons of radioactive waste left in Kazakhstan from the Soviet-era uranium extraction alone. Early and ambitious government efforts to rehabilitate and liquidate uranium mining waste in Northern Kazakhstan sought to restore water, soil, and air quality to ensure the maximum protection for public health. This did lead to the closing and sealing of some ventilation shafts, among other things.[24] Government-initiated studies identified Krasnogorskiy as one of thirteen facilities requiring urgent reclamation measures and the elimination of several 15- to 30-meter-wide and 15-meter-deep funnel-shaped sinkholes. Although work was carried out there, the mine reclamation program ended in 2010.[25] State authorities considered areas away from urban centers to pose little danger to human and environmental health.[26] This is even though nearly 600 people lived in the two settlements when the sleeping sickness first appeared.

Kazakhstan's brush with nuclear technologies is complex, ranging from mining and processing to weapons testing, and it is far from over. Malignant infrastructure is somewhat of a banality in a country so littered with hazardous waste and where atomic energy continues to drive the economy. The country is the current leader in global uranium production, accounting for 43 percent of world production in 2019.[27] The Economic Commission for Europe describes widespread radioactive contamination in Kazakhstan from past and current activities as the most "enduring environmental threat for the country."[28] The total dose of artificial and natural radiation per person in the country is one and a half times higher than the global average, and in north Kazakhstan with its significant amount of radioactive waste, it can be three to four times higher than that.[29] Kazakhstan's development of the extractive industry sector, especially its position as the world's number one uranium producer, coupled with serious radioactive waste management issues, means that today Kazakhstan has become the world's largest generator of hazardous waste on a per capita basis.[30]

Post-Soviet political and economic restructuring has had cascading impacts on public health. Working as a researcher and participant observer for nearly a decade, it is clear the state is unable to deal with polluted environments, and the people who live in them, on even a limited scale. Kazakhstan still lacks adequately trained doctors, nurses, or robust chronic disease registers that could help recognize disease clusters.[31] The country has yet to confront the rise of chronic diseases like tuberculosis and HIV/AIDS, and has some of the highest suicide rates in the world.[32] On top of that, scant publicly available records from the Soviet era about the extent of pollution and the lack of government cleanup resources means that enforcing environmental remediation remains nearly impossible. An epidemiologist I spoke with about Kalachi summarized the situation as follows: "we don't do much for people who have found themselves in environmentally difficult conditions." What this means is that the municipal and industrial ruination in places like Krasnogorskiy and Kalachi and their detachment from meaningful economic life (coupled with the lack of state resources) has impaired people's access to adequate housing and health care, not to mention basic amenities. State regulations are well understood by many to be lacking and ineffective.

Sleeping Sickness

Krasnogorskiy and Kalachi are a perfect example of what happens when people end up living among abandoned infrastructures that turn malignant in how they continue to inflict harm through processes of municipal and industrial ruination decades in the making. Many of the boarded-up uranium industries in Kazakhstan lack remediation solutions and adequate funding, while little research exists on the effects of radon or other radioactive substances on people who find themselves living literally on top of toxic dumps.[33] Indeed, residents of the two villages have become the "living breathing archives of atomic history" that no one has ever "opened" until they became visible when a mysterious sickness appeared.[34] What Krasnogorskiy's uranium mines highlight are inhabited geographies of ruination that are experienced in material terms, exposing a highly improvised and unregulated way of life for many who find themselves on the fringes of social, political, and economic orders.

Even before the appearance of the sleeping sickness in 2012, life was unbearable in Krasnogorskiy and Kalachi. According to Ivan, like elsewhere in Kazakhstan, when the local economy collapsed after the mining enterprise shut down, so too did the municipal services. Gas for heating the five-story apartment buildings for hundreds of workers was shut off and so was the electricity. A severe continental climate in northern Kazakhstan can easily see temperatures drop to minus 30 degrees Fahrenheit. Water pipes froze and burst. The subsequent flooding compromised buildings. With no heat and water, some people left. Those who remained, installed makeshift wood-burning stoves in their apartments to survive the winter. Without steady work, they scavenged metal, bricks, or anything useful that could be sold in the city markets hundreds of kilometers away. Gradually, the town fell into ruin this way. Some of the scavenged buildings collapsed as if from an earthquake. Others became too dangerous to occupy, forcing people to move to Kalachi and into single-family homes scattered among the rubble of apartment blocks. I interviewed an epidemiologist who researched the sleeping sickness and likened the area to a "war zone."

Today, Krasnogorskiy and Kalachi have no doctor, no pharmacy, no police station, and no fire department. The school offers only a few primary classes, while the rest of the students are forced to attend boarding schools in faraway towns. There is only one grocery store where residents can buy milk, bread, and fishing tackle. People tend to their small plots of carrots, onions, potatoes, and cucumbers with buckets of water they fill from local wells and cart home to cook and wash with. There is still no central heat source, so locals burn whatever they can, whether it's wood, dried cow manure, or coal. This presence of malignant infrastructure is connected in particular ways with this re-ruralization, and it's nothing unique in either its material or aesthetic characteristic. As elsewhere in Kazakhstan, residents have tried for years to draw government support to Kalachi and with additional emphasis to the abandoned mines. In 2011, a former miner-turned-environmental activist who has lived in Kalachi for decades, organized a hunger strike to highlight the town's plight, but nothing came of the protest.[35] It was not until the mystery sickness appeared that anyone paid attention.

In June 2012, residents began to arrive at a doctor's office in the town of Esil, 40 miles downstream from Kalachi. They complained of drowsiness, memory loss, hallucinations, and the inability to wake up. But doctors couldn't identify the cause of the illness and were summarily sending people home. Things only intensified from there. In 2013, there were more than 30 cases, in 2014 an additional 26, and by January 2015, there were more than 140 people who had been affected.[36] Victims included all strata of the population, including children as young as two years old. As the news spread in the local media, international news outlets picked up on the story, dubbing Kalachi the "village of the damned."[37] It was at the juncture of increasing media attention and concerns of residents that government officials, groups of physicians, and other scientists, including virologists, toxicologists, and radioecologists began to descend on the village. The government authorities went so far as to set up an interdepartmental State Commission and asked international researchers for help. Countless experts ran tests on Kalachi and its residents. They drilled boreholes for soil and water samples; they checked indoor and outdoor air quality, inspected food, collected blood, nails, and hair samples for analysis. Through it all, every possible cause was ruled out, including heavy metal exposure, viruses, bacteria, zoonotic diseases, and alcohol poisoning. People in Kalachi consistently blamed the mines that they were living above and radioactive radon gas making its way into their homes through abandoned infrastructure littering the area.

To link radioactive exposure with a sickness was and remains to be a serious matter, one that unsettles what in many ways has been a political economic structure that has led to a very evident growth in Kazakhstan's national wealth. Malignant infrastructures are part of a continuing cycle rooted in the country's political and environmental history. In a country that produces 43 percent of the world's uranium, government authorities often downplay the harm from present-day exposure to toxic wastes, especially those tied to nuclear industries, past or present. From my own research on the Polygon, I quickly learned that the relationship between ill health and radiation exposure is contested, especially when that exposure is low, slow, and chronic.[38] This is because cancers, cardiovascular diseases, and other illnesses have many causes, from poor diets to unhealthy habits such as smoking or excessive alcohol consumption, which make it difficult for scientists to disentangle cause and effect. It is even harder to show that a new mystery sickness is caused by radon gas. Radon is colorless, odorless, and tasteless. It is difficult to measure it in the air. Although radon has been shown to cause lung cancer, there is no scientific evidence that it puts people to sleep.

To determine whether any linkages could be made between radiation exposure and the sleeping sickness, dozens of scientific teams visited the village. They conducted hundreds of tests, collecting blood, hair, nails, and other data. Scientists from the Institute of Radiation Safety and Ecology (IRSE), part of the National Nuclear Center of Kazakhstan, also got involved. The IRSE is a prominent research body in Kazakhstan that is charged with the enormous task of environmental monitoring of areas of the Polygon, but also those regions throughout the country where radioactive contamination is present. Experts from the institute drilled deeper boreholes than those drilled before and checked air, soil, drinking water, surface water, and food. As for the results from this and other analysis, no radioactive contamination was found at all, and in fact, the scientists concluded that Kalachi is like many other settlements in all of Kazakhstan: generally free from dangerous radiation.[39] Scientists hypothesized instead that carbon monoxide, most likely from wood burning stoves and possibly from mines, was the sleeping agent. One resident I spoke with was adamant that blaming the wood burning stoves was tantamount to blaming the villagers for their problems. But as public concerns were not alleviated and clear causes for sleeping sickness were yet to be found, another high-profile research group in Kazakhstan got involved.

The Research Institute of Radiation Medicine and Ecology in Semey has been an expert on the biological effects of radiation for decades. The institute was once a secret medical laboratory that studied the effects of nuclear testing on people who thought they were getting treated for zoonotic diseases.

In Kalachi, the teams of scientists investigated fifty-nine buildings (residential and non), among other things, and found that 20 percent had average values of radon gas that were well above existing standards.[40] Their conclusion was that a possible salvo of radon emissions (a belching of gases), mixed with carbon monoxide and other gases, released the sleeping agent throughout the community. This view was supported by a Russian team of scientists and physicians affiliated with the Tomsk Polytechnic University.[41] The data they obtained from Kalachi suggest that radon gases escape the mines and concentrate at different levels, depending on the weather, in residential homes for a narcotic effect.

Although scientists continue to fiercely debate the source of the sleeping sickness in Kazakhstan—some suggesting most recently that the cause is underground water poisoned by chemicals—official government opinion has settled on one cause: heightened levels of carbon monoxide and hydrocarbons in the air coming from the mines.[42] The authorities declared the village useless and announced a voluntary resettlement program.[43] They offered compensation and rental apartments in the regional center in the town of Esil. Many people left Kalachi, exchanging their single-family homes and garden patches for apartment blocks. Others found the compensation unsuitable and chose instead to remain. Those who did are ignored once again, living as they were atop the mines, although apparently no one has fallen asleep since 2015 when holes in the ground thought to vent poisonous gases were plugged.

Conclusion

Perhaps because Kalachi is a rural outpost, it was only after almost four years of intense media coverage and pressing demands of residents for state action, that groups of physicians and other scientists, including virologists, toxicologists, and radioecologists finally turned up. What the communities highlight particularly well are two key things. First, there is a spectacular lack of effective policy with which to cope with adverse consequences of environmental and human harm that further obscures the damage done. Today, studies on health risks of residual radioactivity and heavy metals among communities living in the region are either limited or nonexistent despite potentially deadly consequences. Without scientific research to address issues of hazardous waste, radon emissions, and other invisible harms, preemptive securitization of polluted environments becomes impossible. The only solution is to move people out.

Second, these are just two tiny towns out of countless many that have been left with hundreds of millions of tons of toxic waste from the factories, mining operations, or military proving grounds—containing heavy metals or radioactive elements that may still be buried in unmarked pits throughout the country—to deal with the problem on their own. What also becomes clear is that what constitutes toxic harm is not the product of remains and residues but is built into to the entire production cycle of something such as uranium. Only in moments of crisis, as with the case of a mysterious sleeping sickness and the subsequent attention by a variety of media across the world, do we begin to get a sense of the scale of the problem but also how forgettable it can be.

Whether for nuclear weapons or energy, or both, drawing things out of the earth, such as uranium, has been part and parcel of humans changing the planet in irreversible ways. Krasnogorskiy and Kalachi are two bellwether examples of the ways in which municipal and industrial landscapes become malignant infrastructures. Irrespective of scale, these are relics of an intense period of extraction that have produced distinctive environmental histories and are marked by their abandonment from state oversight, their redundancy within new political and economic systems, as well as by their ongoing inhabitation. Malignant infrastructures are those where people live, many if not all of whom lack other options, and where toxic harm does not abate but changes form as it comes through the ground and into peoples' lives.

Notes

1 Rosie McCall, "The Science Behind the Mysterious Town Where Residents Fell Asleep for Days at a Time," *IFLScience*, May 4, 2018, https://www.iflscience.com/the-science-behind-the-mysterious-town-where-residents-fell-asleep-for-days-at-a-time-47455.

2 On uranium mining and human and environmental harm in the US Southwest, see Valerie L. Kuletz, *The Tainted Desert: Environmental Ruin in the American West* (New York: Routledge, 1998); Traci B. Voyles, *Wastelanding: Legacies of Uranium Mining in Navajo Country* (Minneapolis: University of Minnesota Press, 2015); on militarized nuclear landscapes, the consequences of the expansion of peaceful nuclear technology, and the intersections between nuclearism, race, and colonialism, see Jacob D. Hamblin, *The Wretched Atom: America's Global Gamble with Peaceful Nuclear Technology* (Oxford: Oxford University Press, 2021); Laura Pitkanen and Matthew Farish, "Nuclear Landscapes," *Progress in Human Geography* 42, no. 6 (2018): 862–880; Rens van Munster, "On Whiteness in Critical Security Studies: The Case of Nuclear Weapons," *Security Dialogue* 52:S (2021): 88–97; Livia Monnet, ed., *Toxic Immanence: Decolonizing Nuclear Legacies and Futures* (Montreal & Kingston: McGill-Queen's University Press, 2022).

3 Sarah Besky and Alex Blanchette, "Introduction: The Fragility of Work," in *How Nature Works: Rethinking Labor on a Troubled Planet*, ed. Sarah Besky and Alex Blanchette (Santa Fe: School for Advanced Research Press, 2019), 1–22.

4 Magdalena E. Stawkowski, "Radiophobia Had to Be Reinvented," *Culture, Theory and Critique* 58, no. 4 (2017): 357–374.

5 On theorizing invisible environmental toxicity in late capitalist contexts see Donna M. Goldstein, "Invisible Harm: Science, Subjectivity and the Things We Cannot See," *Culture, Theory and Critique* 58 no. 4 (2017): 321–329.

6 Shannon Cram, "Wild and Scenic Wasteland: Conservation Politics in the Nuclear Wilderness," *Environmental Humanities* 7 (2015): 89–105. On human bodies as atomic archives, see p. 90

7 Joseph Masco, "The Age of Fallout," *History of the Present* 5 (2015): 137–168. On radioactive fallout and long-term damage, see p. 138.

8 Dennis Rogers and Bruce O'Neill, "Infrastructural Violence: Introduction to the Special Issue," *Ethnography* 13, no. 4 (2012): 401–412

9 Tania Murray Li, "After the Land Grab: Infrastructural Violence and the 'Mafia System' in Indonesia's Oil Palm Plantation Zones," *Geoforum* 96 (2018): 328–337.

10 On the Soviet Union's approach to the environment and its people, see Paul R. Josephson, *Industrialized Nature: Brute Force Technology and the Transformation of the Natural World* (Washington, D.C.: Island Press, 2002); Paul R. Josephson, Nicolai Dronin, Ruben Mnatsakanian, Aleh Cherp, Dmitry Efremenko, and Vladislav Larin, eds., *An Environmental History of Russia* (Cambridge: Cambridge University Press, 2013).

11 On environmental and human costs of Soviet Union's nuclear weapons development and modernization projects, see Kate Brown, *Manual for Survival: A Chernobyl Guide to the Future* (New York: W. W. Norton, 2019); D. J. Peterson, *Troubled Lands: The Legacy of Soviet Environmental Destruction* (New York: Routledge, 1993); Douglas Weiner, *A Little Corner of Freedom: Russian Nature Protection from Stalin to Gorbachëv* (Berkeley: University of California Press, 1999).

12 Paul R. Josephson, "Industrial Deserts: Industry, Science and the Destruction of Nature in the Soviet Union," *Slavonic and East European Review* 85, no. 2 (2007): 294–321.

13 On a discussion of Soviet closed and secret cities see Robert A. Kopack, *The Afterlives of Soviet Secret Cities: Environment and Political Economy in Kazakhstan's Defense Industry Sites after 1991* (PhD diss., University of Toronto, 2020); Asif Siddiqi, "Atomized Urbanism: Secrecy and Security from the Gulag to the Soviet Closed Cities," *Urban History* 49, no. 1 (2022): 190–210.

14 Robert A. Kopack, "Rocket Wastelands in Kazakhstan: Scientific Authoritarianism and the Baikonur Cosmodrome," *Annals of the American Association of Geographers* 109, no. 2 (2019): 556–567.

15 The Soviet government leaders imagined what is today Kazakhstan, and especially the Polygon, to be a "barren landscape" populated by pockets of "backward" nomadic peoples and peasants; see Francine Hirsch, *Empire of Nations: Ethnographic Knowledge and the Making of the Soviet Union* (Ithaca, NY: Cornell University Press, 2005); David Holloway, *Stalin and the Bomb: The Soviet Union and Atomic Energy, 1939–1956* (New Haven, CT: Yale University Press, 1994); see also Magdalena E. Stawkowski, "I Am a Radioactive Mutant: Emergent Biological Subjectivities at Kazakhstan's Semipalatinsk Nuclear Test Site," *American Ethnologist* 43, no. 1 (2016): 144–157.

16 Kate Brown, *Plutopia: Nuclear Families, Atomic Cities, and the Great Soviet and American Plutonium Disasters* (New York: Oxford University Press, 2013); Stefan Guth, "Oasis of the Future: The Nuclear City of Shevchenko/Aqtau, 1959–2019," *Jahrbücher für Geschichte Osteuropas* 66, no. 1 (2018): 93–123.

17 The eight deposits included six uranium and molybdenum-uranium and two phosphor-uranium bearing ore. Kazakhstan had three industrial plants that extracted and processed uranium: the Kyrgyz Mining Combine (1953) in southern Kazakhstan, the Tselinny Mining and Chemical Combine (1957) in northern Kazakhstan, and the Caspian Mining and Metallurgical Combine (1959) in western Kazakhstan. In total, over twenty uranium deposits were developed in Kazakhstan producing about 40 percent of uranium for the Soviet Union; see Resolution of the Government of the Republic of Kazakhstan, "On Approval of the Program for the Mothballing of Uranium Mining Enterprises and Elimination of the Consequences of the Development of Uranium Deposits for 2001–2010" [In Russian: Ob Utverzhdenii Programmy Konservatsii Uranodobyvayushchikh Predpriyatiy i Likvidatsii

Posledstviy Razrabotki Uranovykh Mestorozhdeniy na 2001–2010 gg.], *Ministry of Justice of the Republic of Kazakhstan*, July 25, 2001 https://zakon.uchet.kz/rus/docs/P010001006_ (accessed July 5, 2021).

18 Sergey Artemovich Smirnov, "It Started Like This . . ." [In Russian: Eto Nachalos' Tak . . .]. In *Our Hearts Are Registered Here: Pages of the History of the City of Stepnogorsk, Akmola Region* [In Russian: *Zdes' Propisany Nashi Serdtsa: Stranitsy Istorii Goroda Stepnogorska Admolinkoy Oblasti*] (Almaty: Atamura, 1994), 25–37.

19 The Virgin Land program in northern Kazakhstan was launched in 1954 to boost agricultural production in the Soviet Union. Within five years of the campaign, the semi-arid plains provided 11.5 percent of the total Soviet grain harvest; see Marc Elie, "The Soviet Dust Bowl and the Canadian Erosion Experience in the New Lands of Kazakhstan, 1950s–1960s," *Global Environment* 8, no. 2 (2015): 259–292.

20 Nikolay Petrovich Petruchin, ed., *Creation and Development of the Mineral Resource Base of the Domestic Nuclear Industry* [In Russian: Sozdaniye I Razvitiye Mineral'no Syr'yevoy Bazy Otechestvennoy Atomnoy Otrasli] (Moscow: Atomredmetzoloto, 2018).

21 Liliya Mihaylova Zamula, "Man and His Health" [In Russian: Chelovek i Yego Zdorov'ye], in *Our Hearts are Registered Here: Pages of the History of the City of Stepnogorsk, Akmola Region.* [In Russian: *Zdes' Propisany Nashi Serdtsa: Stranitsy Istorii Goroda Stepnogorska Admolinkoy Oblasti*] (Almaty: Atamura, 1994), 158–165.

22 On the fate of Soviet-era single-industry towns in the Russian Federation, see Stephen Crowley, "Monotowns and the Political Economy of Industrial Restructuring in Russia," *Post-Soviet Affairs* 32, no. 5 (2015): 397–422.

23 G. V. Fyodorov, "Uranium Production and the Environment in Kazakhstan," *IAEA/OECD NEA International Symposium*, October 2–6, 2000 (International Atomic Energy Agency (IAEA): Vienna, 2002).

24 Resolution, "On Approval of the Program."

25 Resolution, "On Approval of the Program."

26 Fyodorov, "Uranium Production"; UNECE (United Nations Economic Commission for Europe), *Kazakhstan Environmental Performance Reviews* (Geneva: United Nations, 2019).

27 Today, uranium is extracted in Kazakhstan by in-situ leaching (ISL) process only. The ISL method involves recovering uranium ore by first dissolving it underground and then pumping it up to the surface where the ore is recovered. Large amounts of radioactive waste are still generated and must be stored safely. Kazatomprom, Kazakhstan's government nuclear agency, expresses no concern about the legacy of ISL mining, arguing that unique natural environmental processes in the country will clean the mine site; see "WISE Uranium Project," *World Information Service on Energy* (October 20, 2021), http://www.wise-uranium.org; "Decommissioning Projects-Kazakhstan," *World Information Service on Energy* (October 20, 2021), http://www.wise-uranium.org/udkz.html.

28 On an extended discussion of mismanagement of hazardous wastes in Kazakhstan, see UNHRC (United Nations Human Rights Council), *Report of the Special Rapporteur on the Implications for Human Rights of the Environmentally Sound Management and Disposal of Hazardous Substances and Wastes: Mission to Kazakhstan* (Geneva: United Nations, 2015).

29 Rakmetkazhy Bersimbayev and Olga Bulgakova, "The Health Effects of Radon and Uranium on the Population of Kazakhstan," *Genes and Environment.* 37, no. 18 (2015): 1–18.

30 UNHRC, "Report of the Special Rapporteur."

31 S. Adambekov, A. Kaiyrlykyzy, N. Igginov, and F. Linkov, "Health Challenges in Kazakhstan and Central Asia." *Journal of Epidemiology and Community Health* 70 (2016): 104–108.

32 Adambekov et al., "Health Challenges."

33 Bersimbaev and Bulgakova, "Health Effects of Radon."

34 Cram, "Wild and Scenic Wasteland," 90.

35 "Residents of the Sleepy' Village Are Looking for a Meeting with Tokayev" [In Russian: Zhiteli "Sonnoy" Derevni Ishchut Vstrechi s Tokayevym], *K-News,* November 29, 2019, https://knews.kg/2019/11/29/zhiteli-sonnoj-derevni-ishhut-vstrechi-s-tokaevym/; Renat Tashkinbayev, "A Sad Story About How a Paradise Town Was First Built and Then Destroyed in Kazakhstan" [In Russian: Pechal'naya Istoriya o Tom, Kak v Kazakhstane Snachala Postroili, a Potom Razrushili Rayski Gorodok]" *TengriNews,* last modified June 14, 2019, https://tengrinews.kz/article/pechalnaya-istoriya-tom-v-kazahstane-snachala-postroili-1202/.

36 S. N. Lukashenko, V. V. Romanenko., V. I. Surunov, R. A. Sadykov, and S. B. Subbotin, "Study of the Atmospheric Air Contamination in Kalachi Settlement" [In Russian: Issledovaniye Zagryazneniya Atmosfernogo Vozdukha s. Kalachi]," *Bulletin of NNC RK* 4 (2017): 14–22.

37 Will Stewart, "Sex Cravings, Visions of Monsters and Foul-Mouthed Rants: What Force Is Torturing Residents of Kazakh 'Village of the Damned' Hit by Mysterious 'Sleeping Sickness'?," *Daily Mail,* April 30, 2015, https://www.dailymail.co.uk/news/article–3059042/EXCLUSIVE-Sex-cravings-children-seeing-monsters-foul-mouthed-rants-Residents-Kazakh-Village-Damned-hit-mysterious-sleeping-sickness-reveal-sufferers-disturbing-new-symptoms.html.

38 Donna M. Goldstein and Magdalena E. Stawkowski, "James V. Neel and Yuri E. Dubrova: Cold War Debates and the Genetic Effects of Low-Dose Radiation," *Journal of the History of Biology* 48, no. 1 (2015): 67–98; Maxime Polleri, "Post-Political Uncertainties: Governing Nuclear Controversies in Post-Fukushima Japan," *Social Studies of Science* 50, no. 4 (2020): 567–588; see also Jacob D. Hamblin and Linda M. Richards, Introduction to this volume.

39 Lukashenko et al., "Study of the Atmospheric"; V. V. Romanenko, S. N. Lukashenko, M. A. Umarov, Yu. V. Garbuz, A. Yu. Osintsev, and A. N. Shatrov, "Radiation Survey Findings of Kalachi Village in Admolinsk Region" [In Russian: Rezul'taty Radiatsionnogo Issledovaniya Sela Kalachi Admolinskoy Oblasti], *Bulletin of NNC RK* 4 (2017): 23–29.

40 Kazbek Apsalikov, Alexandra Lipikhina, Vladimir Kolbin, Almagul Monsarina, Lyudmila Aleksandrova, Yulia Brait, and Sholpan Zhakupova. 2015. "Radon on the Territory of Kalachi Village in Akmola Region" [In Russian: Radon na Territorii Sela Kalachi Admolinskoy Oblasti], *Universum: Chemistry and Biology* 12, no. 19 (2017), http://7universum.com/ru/nature/archive/item/2811; A. V. Liphikhina, T. I. Belikhina, Sh. B. Zhakupova, V. V. Kolubin, and D. A. Gerasimov, "Consequences of Uranium Technogenesis in Kazakhstan" [In Russian: Posledstviya Uranovogo Tekhnogeneza v Kazakhstane], *Conference Proceedings: Radioactivity and Radioactive Elements in the Human Environment, Tomsk University* (September 13–16, 2016): 394–398, https://elibrary.ru/item.asp?id=28135672&pff=1.

41 Yu. O. Klyuchnikova, N. V. Baranovskaya, L. P. Rikhvanov, "Analysis of Symptoms of 'Sleeping Sickness' and Fatigue Syndrome as the Long-Term Consequences of Gas-Radiation Factor on the Example of the Natural and Technogenic Systems of Kalachi Village (Kazakhstan)" [In Russian: Analiz Simptomatiki Proyavlenia "Sonnoy Bolezni" i Syndroma Povyshennoy Utomlyayemosti Kak Otdalennoye Posledstviye Gazo Radiatsionnogo Faktora na Primere Prirodno Tekhnogennoy Sistemy S. Kalachi (Kazakhstan)], *Materials from International Conference, Tomsk Polytechnic University* (September 13–16, 2016): 303–307, https://elibrary.ru/item.asp?id=28135650&pff=1.

42 Nurlan Ismailov, "NU Scientists Come Closer to Solving the Mystery of the 'Sleepy' Village of Kalachi," *Nazarbayev University,* October 30, 2020, https://nu.edu.kz/news/nu-scientists-come-closer-to-solving-the-mystery-of-the-sleepy-village-of-kalachi.

43 Peter Trotsenko, "Good Morning, Kalachi: How the Village Lives after 'Sleeping Sickness'" [In Russian: Dobroye Utro, Kalachi: Kak Zhivet Selo Posle "Sonnoy Bolezni"], *Radio Azattyk,* July 29, 2020, https://rus.azattyq.org/a/kazakhstan-akmola-regoin-kalachi-reportage/30845602.html.

PART II

International Discourse on Harm

How to Hide a Nuclear Explosion

French Secrets about Saharan Fallout across Decolonizing Africa

AUSTIN R. COOPER

"A complete technical success," one French general described his country's first nuclear explosion, "as much in terms of the construction and detonation of the device as the measurements taken of all sorts."[1] On February 13, 1960, French nuclear planners had conducted this blast in the atmosphere above the Algerian Sahara near the oasis town of Reggane. French forces would complete a total of four atmospheric explosions at Reggane during the Algerian War for Independence (1954–62).[2] The ceasefire terms permitted French use of this site, and another nuclear site built beneath Saharan mountains in Algeria, until 1967. The French general did not want to share all the measurements that French authorities collected in the Algerian desert and in neighboring African territories under French control. In a secret telegram to the French Foreign Ministry, he declined to tell French diplomats the first blast's yield, preferring "to keep that secret for now," meaning more secret than the contents of the secret telegram. He did not want the details that he was willing to disclose to circulate beyond the French Foreign Ministry, either. A handwritten message reminded French diplomats that "this is for your information, and to allow you to respond to questions that could be addressed to you, but it is not intended to be made public."[3] This annotation illustrates how secrecy has shaped and obscured French nuclear history, including technical knowledge about the radioactive debris—known as fallout—produced by nuclear explosions in the Algerian Sahara. It also suggests how this secrecy helped create the social and political orders that continue to perform nuclear weapons governance in France and elsewhere.[4]

Historians of science and technology have shown how nuclear weapons inaugurated a new and paradigmatic regime of national security secrets.[5] This literature has probed questions of democracy, law, and expertise, especially in the United States, the first country to develop the bomb and the only one to use it as an act of war. In his work on the first US satellite reconnaissance system known as CORONA, historian John Cloud has proposed a concept he calls the "shuttered box" to describe the way that Cold War science brought some aspects of classified programs into the declassified realm, often disguised or otherwise transformed.[6] The Cold War arms race and the place that nuclear technologies assumed in decolonization politics meant that French observers were not the only ones paying attention to nuclear explosions in the Algerian Sahara.[7] Radiation exposure from atmospheric detonations had become, since the US bombings of Japan and a botched US nuclear test in the Marshall Islands, an important question for biological sciences and international politics.[8] Declassified documents from many archives on several continents have revealed international institutions, partnerships, and networks for monitoring Saharan fallout. Participants included French allies in the United States, United Kingdom, Canada, and Europe; intergovernmental organizations such as the International Atomic Energy Agency and European Atomic Energy Community; and independent African states like Ghana and Tunisia.[9] Even though fallout levels in some places became public knowledge, French observers worked hard to keep their data to themselves.

Already intrinsic to nuclear weapons development, secrecy continues to mark the French case in profound ways, especially compared to other nuclear-armed North Atlantic democracies in the United States and United Kingdom. French law, for example, does not have the same Freedom of Information principles as its US and UK counterparts.[10] Since 2008, French law withholds "information allowing for design, fabrication, usage, or locating of nuclear weapons," a restriction that tends to receive broad interpretation and prevent access to records that may not threaten nuclear proliferation.[11] In October 2021, French President Emmanuel Macron ordered an unprecedented review of French nuclear archives, but this process focused on Polynesia, still part of France.[12] During the transfer of the two Saharan test sites to the newly independent Algerian government in 1966–67, French forces moved their nuclear weapons development to Polynesia, where nearly 200 atmospheric and underground explosions took place from 1966 to 1996. The recent study *Toxique*, published in March 2021, had garnered media attention and forced Macron's hand by revealing French officials' concealment and underestimation of the radioactive contamination in Polynesia.[13] The declassification review launched by Macron made available some French documents about the Algerian nuclear sites, but access to these records remains partial, uneven, and unpredictable.

This essay uses a broad assortment of French government sources to assess French attention to Saharan fallout.[14] French generals, diplomats, other officials, and technical experts installed radiation detection systems across African territories undergoing complex and diverse processes of decolonization at the turn of the 1960s. At the same time, French officials developed careful procedures for secrecy and compartmentalization to keep this knowledge to themselves, procedures that have persisted in various guises to this day. One key source includes French military reports on the nuclear explosions in the Algerian Sahara that the French Ministry of Defense declassified in 2012–13. This release, which also included the documents about Polynesia that made *Toxique* possible, followed a decade of court battles fought by associations representing victims and survivors of radiation exposure from French nuclear explosions, including French veterans and Polynesian civilians. Patrice Bouveret, founding member of the French anti-nuclear organization Observatoire des armements, which holds these declassified documents in its archives, explains that French military officials refused to release some of the requested documents and redacted others.[15] This essay also relies on documents from the French National Archives made available for research by Macron's recent declassification review, even though it did not focus on Algeria, and documents from the French diplomatic archives and from the personal collections of French officials.

French nuclear planners carefully tracked their own detonations in the Algerian Sahara but shared the results with almost no one. They had confidence in their ability to control these explosions, the fallout they dispersed, and the data produced. Much of this confidence proved to be misplaced, and French nuclear secrecy required negotiation within and among France's national, imperial, and diplomatic agencies. This essay explores three sources of disagreement about French monitoring of the Saharan explosions. First, French military leaders monopolized the fallout measurements in Algeria and other African territories under various forms of French rule. Other French officials tried, with little success, to challenge that monopoly. Declassified reports reveal awareness of the limits of French nuclear expertise, but military analysts dismissed radiation risks in Africa by drawing spatial boundaries, proposing tolerable doses, and ultimately invoking state secrecy. Second, French nuclear planners treated African member-states of the French Community—an inequitable and short-lived federation that characterized the last years of France's sub-Saharan empire—as both political and scientific resources for French nuclear ambitions. African leaders of these member-states requested information from Paris about possible radiation effects from the French explosions in Algeria, but top French officials kept this information to themselves and based this policy decision in part on colonial racism. Finally, French nuclear planners met criticism and commendation in different sectors of the French state. Health officials indicated that they would have taken a more cautious approach to radiation safety, and that they wanted the power to do so, but

French diplomats encouraged and cooperated with their military counterparts to maintain tight secrecy. Measurement and concealment of Saharan fallout became central to French efforts to become a nuclear weapon state.

A Radioactive "Cigar" in the Algerian Sahara

In September 1961, following the fourth and final atmospheric explosion at Reggane, the Atomic Group within the French Army's Technical Section completed a summary report on Saharan fallout.[16] The French Foreign Ministry received a revised version in 1962.[17] These reports revealed that the French military and the French Atomic Energy Commission (CEA) monopolized French fallout monitoring in Algeria and neighboring African territories. High-level responsibility for coordinating and supervising fallout monitoring fell to the Joint-Army Command of Special Weapons and the Technical Bureau of the General Staff of the Armed Forces. Within the Army's Technical Section, the Atomic Group was responsible for fallout forecasts, controlling access to the test site, land and air reconnaissance of potential contamination, and decontamination. The Atomic Group was also responsible for taking radioactivity readings from air, water, soil, and plants. More detailed fallout analysis fell to the CEA's Radioactivity Measuring Services, including studying aerosolized particles and identifying specific radioisotopes. The CEA team conducted a study of radioactivity in foodstuffs, notably milk, based on measurements taken in metropolitan France and overseas territories, including African colonies. The Military Health Service was responsible for testing radioactivity levels of food and water at the test site and in its vicinity. The Military Health Service was supposed to lead collaboration with other French government agencies. This cooperation went well with the National Meteorology service, but poorly with the Central Service for Protection against Ionizing Radiation (Service central de protection contre les rayonnements ionisants, SCPRI), part of the Ministry of Public Health and Population.[18]

The French reports distinguished between Saharan fallout in the "distant zone" (*zone éloignée*) and the "near zone" (*zone proche*).[19] In the distant zone, air samples were taken daily using vacuum pumps, filter paper, and a volumetric counter; water, soil, and plant samples (mainly date palms) were sent "periodically" to Aubervilliers outside Paris, where the Army Technical Section's Atomic Group conducted laboratory analysis.[20] The next section of this essay discusses in more depth the distant zone, which included nearly half of the African continent.

The military reports defined the near zone in terms of "isometric curves" based on radiation measurements taken at ground-level by French Special

Forces. The boundaries of this zone changed for each detonation depending on the expected yield of the nuclear device and the meteorological conditions.[21] Declassified documents do not contain isometric maps of near-zone fallout or detailed charts of radiological and meteorological measurements. This absence makes the Algerian Sahara difficult to assess in the "archeological" way that *Toxique* approached Polynesia. That study used French archival data to reconstruct the path of the radioactive fallout and then to estimate independently the radiation doses to Polynesian populations. Other French sources offer anecdotal insight into possible radiation exposure in the Algerian desert.

The French general's telegram described radioactive contamination from the first French explosion in February 1960, code-named *Gerboise bleue,* after the jerboa, a kind of desert rodent. He reported that *Gerboise bleue* created a "contaminated zone in the shape of a cigar no longer than 300 kilometers long and 30 kilometers wide at most," stretching from Reggane to the Algerian village of Arak. The explosion vaporized the tower suspending the nuclear device 100 meters above the desert floor, scattering "large metal particles" around ground zero. Otherwise, the French general boasted a "relatively clean explosion" that left no crater behind and that kicked up only a "minimal" amount of sand. Further afield, the French general described a "radioactive cloud [that] drifted East-South-East at high altitude," toward a desert fort near Arak occupied by French soldiers. "Airplanes carrying detection devices" followed the cloud across the Algerian Sahara and registered radiation levels that the French general described as "already very weak after several hours." Beyond this radioactive cigar, he continued, "the fallout is too light for us to be able to talk about contamination." As a precaution, French authorities closed the highways they had carved through the Algerian Sahara, but these roads reopened in the days after *Gerboise bleue,* with "radioactivity checks for vehicles crossing the contaminated zone" between Reggane and Arak. The French general estimated that a person who stayed in the cigar-shaped area for a year would receive a radiation dose roughly equivalent to 1 roentgen, about 10 times higher than many current annual exposure recommendations, but more than 30 times lower than the acute threshold for mild radiation poisoning. "Of course," the French general explained, "this zone extends entirely over desert territory."[22] He understood France's atmospheric test site as "empty, valueless, owned by no one, occupied by no one, remote and expendable," in the way that historian Susan Lindee has described many nuclear weapon states' depiction of their test sites.[23] The French general suggested that, in the Algerian Sahara, radiation dosage amounted only to a hypothetical question, presenting no need to worry about human exposure, health effects, or environmental pollution.

Other participants in French nuclear weapons development did worry about radiation exposure in the Algerian Sahara. French mobile spectrometry equipment checked desert residents for "internal contamination" following *Gerboise bleue* and a later atmospheric blast. French military reports attributed this responsibility solely to the CEA Radiation Measuring Services, but other documents suggest that additional French agencies participated, and that these "internal" measurements became a source of disagreement within the French state. The measuring apparatus comprised a lead enclosure with a large sodium iodide crystal inside, sensitive to gamma rays. After *Gerboise bleue* in February 1960, French personnel examined "roughly 125 civilians, Europeans and natives [*indigènes*] living in Reggane and the Touat Valley," and reported normal radiation levels. French teams examined "roughly 70 nomads . . . in the vicinity of Tamanrasset," following the third French explosion in December 1960 code-named *Gerboise rouge*, and reported, "There again no artificial radioactivity could be detected among these people."[24] The CEA Radiation Measuring Services conducted near-zone studies of strontium-90 and cesium-137, radioactive isotopes generated by nuclear fission that scientists knew at the time to cause cancer by accumulating in bone and muscle, respectively.[25] The human and radioisotope studies demonstrated French concern for health risks in the Algerian desert that the French general downplayed, even denied.

When French military leaders received requests in January 1962 from French colonial authorities to loosen movement restrictions in the Algerian Sahara, radiation exposure shaped the decision process. These requests involved the French Minister of the Sahara Louis Jacquinot and his deputies in the desert, whose correspondence suggested that Algerian populations were growing frustrated with these movement restrictions.[26] By early 1962, French forces had suspended atmospheric explosions at Reggane, and moved to the underground site beneath the Saharan mountains, but they did not necessarily see this switch as permanent.[27] Still, French authorities determined that they could allow only a small increase in traffic around Reggane. The chief of staff for the Atomic Department in the Weapons Delegation to the French prime minister advised that French authorities should not eliminate or even shrink the tight military perimeter surrounding the ground-zero points near Reggane, given "the contamination that remains from the previous tests in certain zones of the test site."[28] He implied that residual radiation from the atmospheric explosions could pose health risks to Algerian populations who might travel through this area if not prevented from doing so.

Like their counterparts in other nuclear weapon states, French nuclear planners evaluated these health risks against a "permissible dose." This convention—widely deployed to defend atmospheric explosions against health and safety challenges—did not indicate scientific consensus or even a global standard. These standards varied according to the country developing its nuclear arsenal. Risk tolerance depended not only on the biological effects of radiation exposure, an especially important topic for scientific fieldwork and laboratory studies during the first decades of the Cold War arms race, but also political and economic goals for nuclear technologies.[29] An expert French body, called either the Consultative Security Commission or the Special Commission for Safety, set the French limit at half a rem per year.[30] This commission, and other French nuclear planners, used "permissible concentrations" measured in fractions of Curies per square meter to assess the level of plutonium and various fission products in air, water, soil, plants, and foodstuffs.[31]

The French reports stressed that even the highest radiation doses measured in the Algerian Sahara and neighboring parts of the African desert fell an order of magnitude—measured in dozens of millirems—below the French permissible dose of half a rem. On the African continent, the highest French readings came from the posts at Arak, Amguid, and Ouallen in Algeria; Fort-Lamy (now N'Djamena) in Chad; Ouagadougou in Haute Volta (now Burkina Faso); and Zinder in Niger. The French reports indicated that Arak saw radiation levels in water "three times higher" after *Gerboise bleue* than French standards allowed, but explained that these readings "fell very quickly" to normal levels. "Additionally," the French reports asserted, "the posts at Arak and Ouallen, which were the most 'affected,' do not have a sedentary population, beyond a small military garrison, for whom all protection measures had been taken."[32] Recent testimony by former French soldiers claiming radiation exposure, health effects, and monetary compensation suggests at least some of their protection was inadequate.[33] The small-sample spectroscopy studies near Reggane and Tamanrasset had sufficed to deem nomadic peoples safe.

One Algerian physician has recently called for epidemiological studies, which during the 1960s would have required substantial logistical and diplomatic coordination among the French colonial government and then the independent Algerian state.[34] Scholarship on the Atomic Bomb Casualty Commission, established in Hiroshima after the US atomic bombing, and its successor organization, the Radiation Effects Research Foundation, illustrates opportunities and challenges in conducting large-scale, longitudinal studies of radiation exposure. The effects of low-dose exposure—precisely the scenario created in the Algerian Sahara—remain difficult to assess, in part because the "gold standard" Japanese data excludes many of these radiation pathways.[35] When I asked a US radiation expert who had participated in an international scientific mission to the Algerian nuclear sites in the mid-1990s, he said that epidemiological studies sixty years later would have to make large assumptions, and that he doubted that such a project could deliver much insight.[36] Science became an important tool for making sense of the French explosions in the Algerian Sahara during the 1960s, but science cannot—as it could not back then—resolve all uncertainty.

Monitoring Fallout in the French Community

French officials used the last vestiges of their colonial empire in Africa to develop an elaborate fallout monitoring network, with more than twenty stations in southern Algeria, and nearly a dozen more in African member-states of the French Community. Formed in 1958, the French Community emerged from a decade of discussions about federalizing France's colonial empire in Africa, an idea that many sub-Saharan leaders across the political spectrum supported, even if they disagreed about what shape it should take, and how power should be allocated among Paris and the African capitals. Through its de facto collapse in 1960, the French Community granted equal citizenship and freedom of movement to African citizens of France, and allowed multiple nationalities, but limited the sovereignty of African leaders.[37] The French Community also created an important framework for producing nuclear knowledge during the first French explosions in the Algerian Sahara. African member-states where French nuclear planners installed fallout monitoring equipment included Congo-Brazzaville, Chad, Niger, Haute Volta, Côte d'Ivoire, Mali, Senegal, and Mauritania.

African leaders of French Community member-states, including the president of the Mali Federation Modibo Keïta, asked Paris for reassurance about adverse effects from Saharan fallout. It is not clear how much French officials told Keïta about their monitoring program, which installed equipment in his country.[38] David Dacko, president of the Central African Republic, also asked French officials for clarification. According to one French document, Dacko in August 1959 "confessed still having a certain fear about the consequences of the radioactivity and wondered if the climate and rain patterns would not be modified by these explosions." French officials told Dacko that "there was nothing to fear," citing the two hundred nuclear blasts already conducted in the United States, the Soviet Union, and Australia. These explosions had sparked local resistance near test sites and generated global efforts to ban the practice of atmospheric nuclear weapons testing, but French officials tried to reassure Dacko by telling him that they had taken greater precautions than their predecessors. It appeared to work. Dacko told French officials that they had addressed his concerns and that "there was no need to return to the matter."[39] Like representatives of other nuclear weapon states, French officials spoke as though their decisions about radiation safety rested squarely on scientific grounds, but many factors shaped these decisions.

How did French officials make policy for discussing radiation safety with African member-states of the French Community? Pierre Messmer—writing in August 1959 from Dakar as the French Community's high commissioner general for Senegal, a post he held before joining French President Charles de Gaulle's government as minister of defense—cautioned Paris to limit discussion of this topic in Senegal and neighboring countries. Messmer underscored "the necessity to remain discreet," and outlined his racist rationale. "The reactions of Africans, even *évolués*," Messmer explained, using a colonial term for educated individuals, who often held positions in the colonial state or colonial society, "are in fact fundamentally different from European reactions. For a European, the explanations given about the precautions that will surround an atomic explosion can be reassuring; for an African, talking about the explosion is as impressive as the explosion itself." Messmer's policy proposal rested on nothing more than colonial racism, which portrayed Africans as ignorant, irrational, and childlike. He contended that French officials should, in effect, ignore information requests from African leaders like Keïta and Dacko. His racist logic led to disturbing conclusions: "We would cause less emotion by exploding ten bombs that we talked about only once than by exploding one bomb that we talked about ten times," Messmer advised.[40] He overlooked the corresponding increase in radiation exposure, which carried health risks that were not just talk. Messmer also underestimated African officials' capacity to develop technical systems for radiation detection in their own countries, which recent scholarship on Tunisia, Ghana, and Nigeria has described.

The French general's secret telegram to the French Foreign Ministry reached conclusions similar to Messmer's. Beyond Algeria, the French general had doubted that observers in other African territories would be able to track the radioactive cloud, much less identify it as the result of French bomb development:

> It becomes more and more diffuse, and its radioactivity diminishes as time goes on. It gets mixed up more and more with radioactive residues produced by older explosions that circulate in the same way and fall slowly around the globe.

The era's science demonstrated the sophistication required to match fission products to the explosions that had created them. Either the French general did not understand this science or he assumed that African observers lacked the necessary expertise. Even if a radioactive cloud were to traverse African territories beyond French control, he argued that "measuring this fallout at great distances requires technical equipment and specialists, otherwise crude errors can be committed." Referring to two independent states in West Africa whose Left-leaning, anti-colonial leaders challenged the safety and legitimacy of French nuclear explosions in the Algerian Sahara, the French general declared it "absolutely impossible that any radioactivity [might be detected] days after the explosion in Ghana or Guinea." He preemptively discredited such claims as "fantasy or poor interpretation."[41] He counted on lack of technological access to keep Saharan fallout invisible, but his assumption would prove faulty. Contradicting the general's description of the first French explosion as a "complete technical success," Canadian-Ghanaian and UK-Nigerian monitoring programs detected Saharan fallout that French officials had publicly declared could not travel so far from the Algerian desert as West Africa. These measurements gave the Ghanaian and Nigerian governments a new resource to contest French nuclear ambitions.[42]

The French general did not worry, either, that French atmospheric explosions in the Algerian Sahara might pose health risks to the populations of the African member-states of the French Community. "We can conclude," the French general insisted, "[that there is] no danger, even a minimal one, for Saharan populations even closer in Niger, Chad, Sudan, and other countries to the East and South-East."[43] The declassified French military reports identified radiation levels in Niger and Chad as among the highest readings taken after *Gerboise bleue*. On the one hand, the French general could have been right. Detecting radiation is not the same as demonstrating health risks, and it is likely impossible to prove that *Gerboise bleue* caused adverse health effects in Niger or Chad. On the other hand, he almost certainly spoke too confidently. French officials had not, at the time, announced how many atmospheric explosions they planned to conduct in the Algerian Sahara, or with what yield. They would go on to admit privately that they were not able to predict where the fallout would travel or the amount of radiation it would bring. The French general's confidence rested on little more than guesswork.

French nuclear planners took from the Saharan blasts several "lessons," as they put it, including some about the limits of their own knowledge. One of the declassified reports on Saharan fallout includes a map showing how monitoring stations in French Community territories helped track "long-range fallout" from *Gerboise bleue*. The report warned that this map could give only an "approximate idea of the cloud's progression, [since] the measuring points are far too few in number to allow [us] to trace a precise contour."[44]

Given the rate and range of fallout diffusion in the atmosphere, French nuclear planners concluded that "the term 'cloud' stops making sense very quickly." A map of the long-range fallout trajectories created by *Gerboise bleue* illustrated, on the one hand, that "the configuration of atmospheric currents can explain the presence of radioactivity on a given day in a given place." But French military planners cautioned, on the other hand, that this process "seems very complex," depending on interactions with various atmospheric layers and step-wise descent patterns.[45] They demonstrated real interest in, but limited understanding of, atmospheric science.[46]

French documents reveal stunning confidence despite facing such uncertainty about fallout behavior. Most damningly, French nuclear planners admitted that they were in no position to "go into detail, or especially to make a forecast in advance," even if they could rely on "meteorological circumstances" to explain fallout trajectories after the fact. This forecasting discussion does not appear in the second version of the military report filed with the French Foreign Ministry in 1962, suggesting that it was removed. But declassified documents make clear that French nuclear planners had not known where Saharan fallout would go. They just decided this question did not matter so much after all. On a reassuring note, they concluded that the "considerable diffusion of radioactive residues coincides, in turn, with an even more significant dilution of these residues in the atmosphere and a correlated reduction in the risks that they could pose."[47] This logic assumed that French forces would eventually stop their nuclear explosions in the atmosphere, which did not occur until 1974, more than a decade after the entry into force of the Limited Test Ban Treaty (1963), the agreement that halted the superpowers' atmospheric explosions but that France did not sign.[48]

Public Health Objections and Diplomatic Support for French Nuclear Secrecy

Saharan fallout prompted contests over nuclear secrecy within the French national government, notably among military, health, and diplomatic leaders. A hand-written memo began to make a case in late December 1960 for the SCPRI, then part of the French Ministry of Public Health, to take a permanent, on-site role in radiation detection in the Algerian Sahara. Though unsigned, the memo was probably written by Pierre Pellerin, the French physician and nuclear specialist who played a major role in founding the SCPRI in 1956 and who directed this organization until 1993.[49] Pellerin remains best known in nuclear history for statements that he made—and others that were erroneously attributed to him—downplaying the health risks of the Chernobyl disaster in 1986. France's highest civil law court, the Cour de cassation, dismissed the fraud and homicide charges brought against him after several years of legal battles in 2012.[50] The SCPRI memo from December 1960 was addressed to Jean Charbonnel, the French civil servant and Gaullist politician working at the time as a technical advisor to Bernard Chenot, then French minister of Public Health.[51] In the SCPRI memo, and related documents, Pellerin, Charbonnel, and Chenot challenged French military control of Saharan explosions.

Written days after *Gerboise rouge* in December 1960, the SCPRI memo observed that Saharan explosions were becoming "routine." Regular blasts could pose health risks to Algerian populations, the memo argued, "especially in the region from Adrar to Reggane," referring to another Algerian town roughly 150 kilometers north of the test site. This assertion threatened French nuclear weapons policy because French officials had not announced plans to stop atmospheric explosions at Reggane and because it remained unclear how many of these blasts they could conduct without approaching the limits they had set for radiation exposure. The memo proposed that the SCPRI should intervene by tracing radiation pathways in the desert, given the track record of other nuclear weapon states:

> It is appropriate, in fact, to ensure that no contamination of the water and the food chain occurs in this zone with a relatively dense and poorly resourced population. Now, the American and Russian experiments have shown that accidental contamination was always possible, whatever precautions might have been taken, namely in terms of meteorological predictions.

SCPRI used the earlier US and Soviet explosions, which other French officials had cited to reassure the Central African Republic's President Dacko, to highlight the radiation risks created by French nuclear weapons development. SCPRI stressed the likelihood of nuclear accidents, including ones caused by poor weather forecasting, a kind of expertise that French nuclear planners would admit they had not yet mastered. On the US front, the SCPRI memo alluded to what could have been several instances of fallout contamination known to the public. In 1954, the massive thermonuclear blast in the Marshall Islands that US brass dubbed *Bravo* exceeded pre-shot predictions, and the ashy debris caused acute radiation sickness to Indigenous people, Japanese fishermen trawling nearby for tuna, and US troops. During the second half of the 1950s, US farmers, mothers, and journalists criticized the way that atmospheric blasts at the Nevada Test Site contaminated pastures and milk across the US West.[52] In the Soviet Union, the tight nuclear secrecy maintained through Chairman Mikhail Gorbachev's push for transparency in the late 1980s makes it difficult to say what exactly French observers could have known in 1960 about the Semipalatinsk test site in Kazakhstan. Anthropologist Magdalena Stawkowski has shown that Soviet nuclear planners secretly monitored health effects of radiation exposure from the mid-1950s but lied to Kazakh populations that the pseudonymous clinic in Semipalatinsk was treating "zoonotic diseases."[53] As historian Kate Brown's work on plutonium production suggests, differences in nuclear secrecy between the Cold War blocs came in degree, not kind; US and French nuclear planners withheld and spun information about radioactive contamination like their Soviet rivals.[54]

The SCPRI memo revealed disagreements and tensions within the French nuclear weapons complex. The memo called it "inconceivable" that public health personnel were not alerted well in advance of each Saharan explosion and not authorized to attend. SCPRI participation in fallout monitoring would offer two advantages, the memo argued. One was that SCPRI could take its own readings in the desert and, as the memo put it, "benefit from direct, real, on-site technical information, regarding the conditions of the explosion (yield, nature, geographical and meteorological conditions)." French public health officials did not appear to trust the nuclear planners to deliver accurate and reliable data. To this point, the French general had withheld the yield of *Gerboise bleue* from the French Foreign Ministry. The second advantage would stem from SCPRI's unique expertise, the memo proposed, in "[verifying] if neighboring populations are effectively protected . . . during each test." SCPRI participation could lend an air of objectivity and a united front to French fallout measurements, the memo implied:

> This verification by a Ministry not directly interested in the military or industrial aspects of the tests would have great value in the national and international arenas, politically and diplomatically. One must not forget that CEA authorities assert the Ministry of Public Health's consent each time that a delicate operation is envisaged.[55]

The memo suggested that that French nuclear planners were trying to take advantage of the Ministry of Public Health's trustworthiness without, in fact, earning it. The memo also implied—with good reason—that French officials needed this trustworthiness boost because they were struggling to convince the world that their explosions in Algeria carried no health risks, and because invoking these health risks gave powerful ammunition to France's critics.

Concerns about expertise and authority traveled up the French chain of command. In January 1961, French Minister of Public Health Chenot wrote to French Minister of Defense Pierre Messmer. Chenot complained that Messmer's staff was excluding public health personnel from fallout monitoring.

Chenot indicated that, with Messmer's permission, the SCPRI chief Pellerin had traveled to Reggane for *Gerboise bleue* in February 1960, and that Pellerin had used his mobile lab (*camion-laboratoire*) to measure internal radiation among Saharan populations. (French military reports attributed the responsibility for checking human contamination exclusively to the CEA.) But the situation had changed, as Chenot explained:

> Since then, the nuclear experiments have continued by provoking, whatever precautions might have been taken, an indisputable pollution of the atmosphere. Moreover, in each case, the problem arises of the contamination of water and of foodstuffs in the neighboring oases. Now, my department has no longer been included in the protective measures that you must have had to take to mitigate [*pallier*] the effects of these explosions on the health of these populations.[56]

The Ministry of Defense and the CEA disputed Chenot's assessment that French nuclear explosions were polluting the atmosphere, water, and foodstuffs in the Algerian Sahara. This disagreement helps explain why French nuclear planners did not want public health personnel visiting the desert.

In a memo prepared for Chenot in May 1960, Charbonnel indicated that Pellerin's trip to Reggane with his mobile lab had exacerbated tensions between the SCPRI and the CEA. Charbonnel pinned the blame squarely on the eminent physician Louis Bugnard, one of the creators of the SCPRI, the director of the French National Institute of Hygiene, and France's representative to the International Commission on Radiological Protection.[57] Charbonnel charged that Bugnard's loyalties really lay with the CEA, which Bugnard was advising on radiation safety for the Sahara blasts:

> Very eager not to do anything that might annoy the CEA, where he maintains many friendships, M. Bugnard in short wants the SCPRI, rather than actually monitoring the CEA's activities, to serve only as its outward-facing support.

Charbonnel insisted that Bugnard had a conflict of interest. "While the military had for a long time appeared favorable to Public Health participation [in fallout monitoring]," Charbonnel explained, "the CEA has not hidden its hostility to this arrangement, which would rob it of the monopoly on oversight that it has enjoyed to this point." Charbonnel alleged this conflict was why Bugnard, "after agreeing in principle to [Pellerin's expedition], proceeded to do everything he could to stop it from happening." Bugnard had gone to budget leadership, Charbonnel reported, "to criticize in vehement terms sending the mobile lab to the Sahara, which [Bugnard] presented as a waste of public monies." Saharan fallout played only one part in a larger power grab by Bugnard, who had "been trying as hard as he could to retake control of the SCPRI," according to Charbonnel.[58] By May 1964, Bugnard had quit his public health post and started working officially for the CEA.[59]

French leaders continued to jockey over fallout monitoring. In his letter to Messmer, Chenot admitted that "the responsibility for public health in the Saharan departments" technically fell outside his purview.[60] This responsibility likely belonged to the Ministry of the Sahara, an agency created in 1957, when Reggane and neighboring desert communities became an integral part of France, after several decades under a different territorial status than the rest of Algeria.[61] Nonetheless, Chenot insisted that the Ministry of Public Health should take an on-site role in measuring fallout near Reggane, and he borrowed heavily from the earlier SCPRI memo to make his case. Chenot reiterated how "useful" it would be for Public Health personnel to know the schedule for future blasts, confirming that his ministry had been kept in the dark. Chenot contended it was "also necessary" that his staff attend "each explosion" and participate on a local basis in radiation protection. He assured Messmer that public health personnel would maintain the "absolute secrecy necessary for this kind of operation," suggesting that Chenot understood his government's priorities.[62] Like the SCPRI memo, Chenot insisted that the entire French government, including military leaders, would benefit from public health involvement, given the domestic and international perception that Chenot claimed his staff enjoyed as objective experts, divorced from nuclear-weapons and nuclear-power interests.

Chenot listed other ways that the Ministry of Public Health felt left out. Implying that he did not know that the French military and CEA had been taking fallout measurements in the Algerian Sahara and in African member-states of the French Community, Chenot proposed creating a "sampling network" around the blast zone and under his control. Chenot also noted that public health personnel had not been included in "biological experiments always rich in lessons for public health and radiation protection," which Chenot knew the Military Health Services had carried out "on a large scale" during *Gerboise rouge* in December 1960. These experiments involved rats and goats that French troops placed in cages, exposed to the blast radiation, and shipped back to Paris for examination.[63] Chenot complained to Messmer that the SCPRI "had not been invited to participate in these tests," and that he "could not help but regret this lack of coordination."[64] Chenot worried that factional squabbles over fallout monitoring might cannibalize not only his turf but also valuable scientific knowledge about the biological effects of radiation exposure.

Unlike Ministry of Public Health officials, French diplomats regularly participated in monitoring Saharan fallout across the African continent, but even they did not have full knowledge of this operation. Following a request from French military officials, the French Foreign Ministry asked in December 1959 that its embassies and consulates in Nigeria, Ghana, Guinea, Kenya, and Sudan begin taking radiation measurements using air pumps and paper filters. The Foreign Ministry also hoped that its diplomatic posts could take samples of "staple foods and milk" and return these for analysis.[65] In March 1960, following *Gerboise bleue*, the French Embassy in Sudan indicated that its staff had been sending fallout samples "regularly" back to Paris by diplomatic bag.[66] A message from the military attaché in Khartoum confirmed that his embassy had been taking "daily" samples using air filters as well as "periodic" samples of soil, plants, and milk. The results, however, were kept secret, even from French diplomats in Sudan.[67] Declassified military reports confirm that these diplomatic accounts from Sudan described routine measurements. Cooperation with the French Foreign Ministry extended the reach of French fallout monitoring beyond the French Community and into African territories whose leaders opposed French use of the Algerian Sahara for bomb development.

French diplomatic leadership ultimately supported French nuclear secrecy, even though the military reports drew confident and reassuring conclusions about Saharan fallout:

> Analysis of the results discussed above shows that, in all the cases studied, *the levels of radioactivity released were notably lower than the standards established* for population safety, whether it was a matter of the air, drinking water, or foodstuffs. . . . The large number of measuring posts and their distribution across a vast geographic area, for one thing, [and] the continuity over time of this monitoring, for another, guarantee that *no harmful effect could have resulted,* in terms of global health, from the French nuclear experiments of 1960–61.[68]

French officials considered publishing the fallout analysis, and some French diplomats preferred that option, at least at first. In September 1961, the Ministry of Defense asked the Foreign Ministry what its diplomats thought of the French military report on Saharan fallout and how widely France should share it. The Service of Atomic Affairs, the Foreign Ministry's nuclear experts, initially recommended publication: "[The document] clearly establishes the innocuousness of our experiments, and despite its technical nature, it constitutes a good response to the accusations that have been brought against us."[69] But Atomic Affairs changed its mind.

Just weeks later in September 1961, on behalf of Foreign Ministry leadership, Atomic Affairs recommended that Defense keep the fallout results secret:

> However, the publication [of this report] appears to us, in the current circumstances, hardly appropriate. *Despite its objectivity, our fear is, in fact, that the report might have an effect contrary to the desired one,* by attracting attention once again to our experimental program, and thus giving certain parties the occasion to express, at our expense, the frustration that they have felt since the resumption of Soviet testing. Hence we figure that it would be preferable to defer publication of this report, its distribution being limited, for the moment, to the French organizations involved in its production, and our principal Embassies, for the use of their leadership.[70]

French officials considered if the reassurance and rigor of French fallout measurements—should these become public—could deflect criticism of French nuclear ambitions. But they chose not to distribute the report and blamed the Cold War superpowers for forcing a more secretive posture on France's part. Soviet authorities had just ended a three-year test moratorium, in place since an agreement in 1958 with the United States and United Kingdom that also paused their testing, by resuming atmospheric blasts in September 1961; US forces followed suit, though at first underground.[71] Worried that France might serve as a scapegoat for Soviet policy, French diplomats encouraged what appears to have become an indefinite deferral in releasing this information, at least until activists and researchers began requesting documents about Saharan fallout in the first decades of the twenty-first century. In 1961, the Service of Atomic Affairs recommended that only ten copies of the French fallout report be made and sent to the embassies in London and in Washington, and to New York for use by France's United Nations delegation.[72]

Conclusions

Saharan fallout mapped institutional, epistemological, and personal disputes about French nuclear secrecy. But compartmentalization and classification of nuclear knowledge have obscured these debates and the extent to which they roiled factions within the French government. French nuclear secrecy did not only face outward; it constituted and complicated the internal order of the French nuclear weapons complex. Declassified military reports show

how French nuclear planners maintained control over Saharan blasts despite fallout measurements that contradicted their forecasts. African member-states belonging to the French Community mattered crucially for French efforts to monitor and trace Saharan fallout; it remains unclear how much the African leaders learned about their countries' role in the French bomb project and the extent of the radioactive debris that reached their territories. Medical and environmental concerns about Algerian populations spurred demands by French health officials to attend the Saharan explosions and participate in radiation monitoring, demands that intensified a turf war brewing with French military leaders. The health officials did not always agree with the military's claims that atmospheric explosions would not risk radiation exposure for desert residents, but the health officials did not manage to make French nuclear testing policy more cautious or even to serve as a more independent check on the military fallout monitoring program. French diplomats had, for their part, always had some role in the fallout measurements, and they drew reassuring conclusions from the results much like their military counterparts. When asked for advice on releasing fallout reports, the French Foreign Ministry advocated maintaining nuclear secrecy rather than using French data to back the French argument that the French explosions had done no harm. Presented in 1961 as a deferral—ostensibly temporary—that decision to withhold the fallout report proved quite durable, cementing French policy on the matter for more than forty years. This insight into French nuclear weapons policy-making, and which state agencies have had the power to institutionalize their priorities, represents one of the main payoffs of the newly accessible records. That power determined who knew what about Saharan fallout and what about it French authorities made knowable in the first place.

Notes

1 Military telegram, 1809INVA/306, Fonds Questions atomiques et spatiales (QAS), Archives diplomatiques, La Courneuve, France (ADLC).

2 On the connection to the Algerian War, see Roxanne Panchasi, "'No Hiroshima in Africa': The Algerian War and the Question of French Nuclear Tests in the Sahara," *History of the Present* 9, no. 1 (2019): 84–112.

3 Military telegram, 1809INVA/306, Fonds QAS, ADLC.

4 This argument about the "co-production" of nuclear knowledge and nuclear order, domestic and international, draws on Science and Technology Studies; see Sheila Jasanoff, ed., *States of Knowledge: The Co-production of Science and the Social Order* (London: Routledge, 2004).

5 Alex Wellerstein, *Restricted Data: The History of Nuclear Secrecy in the United States* (Chicago: University of Chicago Press, 2021); M. Susan Lindee, *Rational Fog: Science and Technology in Modern War* (Cambridge, MA: Harvard University Press, 2020), esp. chaps. 8–9; Eric

Schlosser, *Command and Control: Nuclear Weapons, the Damascus Accident, and the Illusion of Safety* (New York: Penguin Books, 2013); Peter Galison, "Removing Knowledge," *Critical Inquiry* 31, no. 1 (September 2004): 229–243.

6 John Cloud, "Imaging the World in a Barrel: CORONA and the Clandestine Convergence of the Earth Sciences," *Social Studies of Science* 31, no. 2 (April 2001): 231–251.

7 Jeffery T. Richelson, *Spying on the Bomb: American Nuclear Intelligence from Nazi Germany to Iran and North Korea* (New York: Norton, 2007), 195–218; Abena Dove Osseo-Asare, *Atomic Junction: Nuclear Power in Africa after Independence* (Cambridge, UK: Cambridge University Press, 2019), 19–48; Christopher Robert Hill, "Britain, West Africa and 'The New Nuclear Imperialism': Decolonisation and Development during French Tests," *Contemporary British History* 33, no. 2 (April 2019): 274–289.

8 Toshihiro Higuchi, *Political Fallout: Nuclear Weapons Testing and the Making of a Global Environmental Crisis* (Stanford, CA: Stanford University Press, 2020); Alison Kraft, "Dissenting Scientists in Early Cold War Britain: The 'Fallout' Controversy and the Origins of Pugwash, 1954–1957," *Journal of Cold War Studies* 20, no. 1 (April 2018): 58–100; Laura J. Martin, "Proving Grounds: Ecological Fieldwork in the Pacific and the Materialization of Ecosystems," *Environmental History* 23, no. 3 (July 2018): 567–592; Jacob D. Hamblin and Linda M. Richards, "Beyond the *Lucky Dragon*: Japanese Scientists and Fallout Discourse in the 1950s," *Historia Scientarium* 25, no. 1 (August 2015): 35–56; Sean L. Malloy, "'A Very Pleasant Way to Die': Radiation Effects and the Decision to Use the Atomic Bomb against Japan," *Diplomatic History* 36, no. 3 (June 2012): 515–545; M. Susan Lindee, "The Repatriation of Atomic Bomb Victim Body Parts to Japan: Natural Objects and Diplomacy," *Osiris* 13 (January 1998): 376–409.

9 See, e.g,. Austin R. Cooper, "The Tunisian Request: Saharan Fallout, US Assistance, and the Making of the International Atomic Energy Agency," *Cold War History* 22, no. 4: 407–436.

10 Benoît Pelopidas, "France: Nuclear Command, Control, and Communications," NAPSNet Special Reports, June 10, 2019, https://nautilus.org/napsnet/napsnet-special-reports/france-nuclear-command-control-and-communications/.

11 "Loi no. 2008-696 du 15 juillet 2008 relative aux archives," Légifrance, https://www.legifrance.gouv.fr/jorf/id/JORFTEXT000019198529.

12 Austin R. Cooper, "A New Window into France's Nuclear History," *Bulletin of the Atomic Scientists*, September 16, 2022, https://thebulletin.org/2022/09/a-new-window-into-frances-nuclear-history.

13 Sébastien Philippe and Tomas Statius, *Toxique: Enquête sur les essais nucléaires français en Polynésie* (Paris: Presses universitaires de France, 2021).

14 The classic study on "radiation safety"—monitoring fallout and managing the health risks from nuclear testing—focuses on the US case; see Barton Hacker, *Elements of Controversy: The Atomic Energy Commission and Radiation Safety in Nuclear Weapons Testing, 1947–1974* (Berkeley: University of California Press, 1994).

15 Patrice Bouveret, "Petit historique sur l'obtention des documents déclassifiés," communication to author, April 27, 2021.

16 Groupement Atomique, Section Technique de l'Armée, "Synthèse sur les Enseignements tirés des quatre premières expérimentations nucléaires, 1. Retombées radioactives provoquées par les premières explosions nucléaires françaises, 2e partie: Retombées lointaines," Rapport technique no. 108/B, September 1961, document déclassifié no. 27/154 (April 2013), Observatoire des armements (OBSARM) (hereafter referred to as "1961 fallout report").

17 Note sur les retombées radioactives dûes aux premières explosions atomiques françaises, 1962, Calcul de la radioactivité à la suite de la campagne (February 15, 1960–September 22, 1961), 1809INVA/307, Fonds QAS, ADLC (hereafter referred to as 1962 fallout report).

18 Longer description in 1962 fallout report, 1–4.

19 1961 fallout report, 4; 1962 fallout report, 4–5.
20 1961 fallout report, 5-6; 1962 fallout report, 6–7.
21 1961 fallout report, 4; 1962 fallout report, 4–5.
22 Military telegram, 1809INVA/306, Fonds QAS, ADLC.
23 Lindee, *Rational Fog*, 18.
24 1961 fallout report, 13; 1962 fallout report, 13.
25 Angela N. H. Creager, "Radiation, Cancer, and Mutation in the Atomic Age," *Historical Studies in the Natural Sciences* 45, no. 1 (February 2015): 14–48; on strontium-90, see Jeffrey C. Sanders, "History Uncontained at the B Reactor," this volume.
26 Jacquinot to Ouargla and Colomb-Béchar, January 23, 1962; Préfet des Oasis to Jacquinot, January 30, 1962; both in Levée des restrictions apportées à la liberté de la circulation au Sahara, Sécurité du CEMO, 19940390/60, Fonds du Cabinet militaire du ministre ou du secrétaire d'État chargé de l'Outre-mer, Archives nationales, Pierrefitte-sur-Seine, France (AN).
27 Thomas Fraise and Austin R. Cooper, "France Struggled to Relinquish Algeria as a Nuclear Test Site, Archives Reveal," *The Conversation*, August 3, 2022, https://theconversation.com/france-struggled-to-relinquish-algeria-as-a-nuclear-test-site-archives-reveal-187712.
28 R. Lévêque, Directeur de Cabinet, Département "Atome," Délégation Ministériel pour l'Armement, to le Vice-Amiral, Chef du Cabinet Militaire du Premier Ministre, "Restriction apportée à la circulation autour de Reggane, Règlementation de l'entrée des Français et Étrangers dans les Départements Oasis et Saoura," February 6, 1962, no. 002279/DMA/DAT, 19940390/60, Fonds du Cabinet militaire de l'Outre-mer, AN.
29 See Higuchi, *Political Fallout*; Creager, "Radiation, Cancer, and Mutation in the Atomic Age"; Soraya Boudia, "From Threshold to Risk: Exposure to Low Doses of Radiation and Its Effects on Toxicants Regulation," in *Toxicants, Health and Regulation since 1945*, ed. Soraya Boudia and Nathalie Jas (London: Routledge, 2013), 71–87; Soraya Boudia, "Global Regulation: Controlling and Accepting Radioactivity Risks," *History and Technology* 23, no. 4 (December 2007): 389–406; Jacob Darwin Hamblin, "'A Dispassionate and Objective Effort:' Negotiating the First Study on the Biological Effects of Atomic Radiation," *Journal of the History of Biology* 40, no. 1 (February 2007): 147–177.
30 "Communication au sujet des conditions techniques des expériences atomiques au Sahara," September 10, 1959, enclosed with Defense Ministry to Foreign Ministry, no. 8054, September 21, 1959, Préparation de la campagne et réactions internationales à une éventuelle expérience nucléaire française (September 1–30, 1959), 1809INVA/304, Fonds QAS, ADLC.
31 1961 fallout report, 8–13; 1962 fallout report, 7–12.
32 1961 fallout report, 10–11; 1962 fallout report, 11.
33 Yannick Barthe, *Les retombées du passé. Le paradoxe de la victime.* (Paris: Le Seuil, 2017); Louis Buildon, *Les irradiés de Béryl* (Paris: Thaddée, 2011).
34 Mostéfa Khiati, *Les irradiés algériens. Un crime d'Etat.* (Algiers: ANEP, 2018), 287–289.
35 Susan Lindee, "Survivors and Scientists: Hiroshima, Fukushima, and the Radiation Effects Research Foundation, 1975–2014," *Social Studies of Science* 46, no. 2 (April 2016): 184–209; Sumiko Hatakeyama, "Let Chromosomes Speak: The Cytogenetics Project at the Atomic Bomb Casualty Commission (ABCC)," *Journal of the History of Biology* 54, no. 1 (April 2021): 107–126.
36 Steven L. Simon (staff scientist, Trans-Divisional Research Program, US National Cancer Institute), in discussion with the author, October 16, 2019, recording on file with the author.
37 Frederick Cooper, *Citizenship between Empire and Nation: Remaking France and French Africa, 1945–1960* (Princeton, NJ: Princeton University Press, 2014).
38 Draft letters to Modibo Keïta, Notes, AG/5(F)/2640, Fonds Foccart, AN.

39 Memorandum of Conversation with President Dacko (Centrafrique), "Explosion de la bombe atomique française," August 5, 1959, AG/5(F)/2640, Fonds Foccart, AN.

40 Pierre Messmer, Haut-Commissaire Général (Dakar) to Paris, "Explosion atomique au Sahara," August 6, 1959, no. 1897 DIR/CAB, Notes, AG/5(F)/2640, Fonds Foccart, AN.

41 Military telegram, 1809INVA/306, Fonds QAS, ADLC.

42 Osseo-Asare, *Atomic Junction*; Hill, "Britain, West Africa and 'The New Nuclear Imperialism.'"

43 Military telegram, 1809INVA/306, Fonds QAS, ADLC.

44 1961 fallout report, 14–16; absent from 1962 fallout report.

45 Ibid.

46 On nuclear weapons testing and the development of atmospheric science, see Joshua Howe, *Behind the Curve: Science and the Politics of Global Warming* (Seattle: University of Washington Press, 2014), esp. chap. 1; and Paul Edwards, *A Vast Machine: Computer Models, Climate Data, and the Politics of Global Warming* (Cambridge, MA: MIT Press, 2010), esp. chap. 8.

47 1961 fallout report, 14–16; absent from 1962 fallout report.

48 William Burr and Hector L. Montford, "The Making of the Limited Test Ban Treaty, 1958–1963," National Security Archive, August 8, 2003, nsarchive2.gwu.edu/NSAEBB/NSAEBB94.

49 "Nucléaire: Le professeur Pierre Pellerin est décédé à l'âge de 89 ans," *Le Monde*, March 3, 2013.

50 Cyrille Vanlerberghe, "Pr Pellerin : 'L'injustice de Tchernobyl est réparée,'" *Le Figaro*, November 21, 2012.

51 "Charbonnel, Jean," Archives d'histoire contemporaine du Centre d'histoire de Sciences Po, https://www.sciencespo.fr/histoire/fr/fonds-archive/charbonnel-jean.html.

52 See, e.g., Hacker, *Elements of Controversy*, 131–210.

53 Magdalena E. Stawkowski, "'I Am a Radioactive Mutant': Emergent Biological Subjectivities at Kazakhstan's Semipalatinsk Nuclear Test Site," *American Ethnologist* 43, no. 1 (2016): 148–149.

54 Kate Brown, *Plutopia: Nuclear Families, Atomic Cities, and the Great Soviet and American Plutonium Disasters* (Oxford: Oxford University Press, 2013).

55 Note pour M. Charbonnel, "Participation du SCPRI aux Tests Nucléaires du Sahara," December 31, 1960, Box 3 ("Débuts en politique 2"), Fonds Jean Charbonnel, Archives d'histoire contemporaine, SciencesPo centre d'histoire, Paris, France. Kind thanks to Thomas Fraise for sharing this document during a global pandemic.

56 Ministre de la Santé Publique et de la Population (Chenot) to Ministre des Armées (Messmer), January 30, 1961, Box 3, Fonds Charbonnel, SciencesPo. Document courtesy of Thomas Fraise.

57 "Louis Bugnard," INSERM, October 16, 2013, https://www.inserm.fr/portrait/histoire/louis-bugnard/.

58 Jean Charbonnel, Cabinet du Ministre, Ministère de la Santé publique et de la population, "Note à l'attention de Monsieur le Ministre sur les problèmes actuels du S.C.P.R.I.," May 9, 1960, Fonds Charbonnel. Document courtesy of Thomas Fraise.

59 Raymond Marcellin, Ministre de la Santé publique et de la population, au Ministre d'État chargé de la recherche scientifique et des Questions atomiques et spatiales (Gaston Palewski), May 5, 1964, 19760161/19, Fonds de la Direction générale de la santé, AN. Document courtesy of Thomas Fraise.

60 Chenot to Messmer, January 30, 1961, Fonds Charbonnel, SciencesPo.

61 Sarah Abrevaya Stein, *Saharan Jews and the Fate of French Algeria* (Chicago: University of Chicago Press, 2014).

62 Chenot to Messmer, January 30, 1961.

63 Jean-Marie Colin and Patrice Bouveret, *Radioactivity Under the Sand: The Waste from French Nuclear Tests in Algeria; Analysis with Regard to the Treaty on the Prohibition of Nuclear Weapons* (Heinrich Böll Foundation, July 2020), 17–19. https://www.boell.de/sites/default/files/2020-08/Under%20the%20Sand_english.pdf.

64 Chenot to Messmer, January 30, 1961.

65 Direction Politique, Services des Affaires Atomiques, Note, très secret, December 30, 1959, Préparation de la campagne et réactions internationales à une éventuelle expérience nucléaire française (November 21–December 31, 1959), 1809INVA/305, Fonds QAS, ADLC.

66 Jacques Dumarcay (Khartoum) to Direction Politique, Service des Affaires Atomiques, no. 166/QA, re : Dépêche de l'Attaché Militaire concernant l'explosion de Reggane, March 7, 1960, Soudan, Suite, 49QONT-71, Fonds Afrique-Levant, ADLC.

67 Le Lieutenant-Colonel François, Attaché Militaire et de l'Air près de l'Ambassade de France au Soudan (Khartoum), to Division du Renseignement (Paris), Etat-Major Général de la Défense Nationale, re: Réactions soudanaises à l'explosion de la bombe atomique française, March 7, 1960, Soudan, Suite, 49QONT-71, Fonds Afrique-Levant, ADLC.

68 1961 fallout report, 14; 1962 fallout report, 13; emphasis added.

69 Direction Politique, Service des Affaires Atomiques, Note, September 9, 1961, Calcul de la radioactivité à la suite de la campagne (February 15, 1960–September 22, 1961), 1809INVA/307, Fonds QAS, ADLC.

70 Atomic Affairs, for Foreign Ministry, to Defense Ministry, September 22, 1961, no. 60/QA, Calcul de la radioactivité à la suite de la campagne (February 1960–September 1961), 1809INVA/307, Fonds QAS, ADLC, emphasis added.

71 Robert A. Divine, *Blowing on the Wind: The Nuclear Test Ban Debate 1954–1960* (New York: Oxford University Press, 1978), 310–317.

72 Atomic Affairs to Defense Ministry, September 22, 1961.

Exposing Contested Sovereignties

Morocco and French Atomic Testing in the Sahara

MATTHEW ADAMSON

France tested its first atomic weapon on February 13, 1960, in southern Algeria, in what was commonly considered at the time French territory. Two maps serve to illustrate what was at stake in this territory as testing began. The first map, from late 1959, comes from the French Foreign Ministry. Composed of three sections, it depicts three atomic weapons testing ranges: the US Nevada range, the Soviet Union's Semipalatinsk range, and the first French Sahara testing range. Rings denote distance from ground zero, and solid circles represent population size; the larger the circle, the more inhabitants. On the map labeled "USA," there is a small circle for Las Vegas (according to the map, 45,000 inhabitants) less than 100 kilometers from the testing site. According to the same map, the four million inhabitants of the Los Angeles area are just under 500 kilometers away. In sum, by drawing a contrast to US and Soviet testing ranges and population centers, this first map was meant to suggest the remoteness of the French test site at Reggane in southwestern Algeria and imply a territory devoid of inhabitants, contents, or political markers.

A second map, from 1963, offers another perspective. From the front page of the newspaper of the Moroccan Istiqlal political party, the map shows the location of Reganne as well as the other French atomic bomb testing site in Algeria. Rather than an empty field overlaid with distance markers and circles representing population centers, this map provides political information. Two mushroom clouds denote the two French Sahara weapons test locations. Both fall within a shaded territory that today comprises much of southern Algeria, Mali, and Mauritania. This is the area Istiqlal claimed as Moroccan, including the desert area the map labels "*Sahara marocain,*" where the French Sahara tests unfolded.

These maps point to something crucial concerning the historical circumstances of French atomic testing in the Sahara at the beginning of the 1960s. French atomic tests were one element in the extended, complex process of decolonization of North Africa and occurred amid the claims of sovereignty of newly independent countries. Exposure to the fallout of French testing in the Sahara between 1960 and 1963 thus illuminated the status of Moroccan claims concerning the Sahara in the wake of Morocco's independence gained in 1956. Moreover, fallout tested what the very notion of sovereignty really meant and what purposes claims of sovereignty might serve. This was true not only for the Kingdom of Morocco but also, as of March 1962, newly independent Algeria.

Throughout the period of atomic tests in the Sahara, French authorities claimed that the testing grounds and the larger region around them constituted a *terra nullius*. Scholars have problematized these claims—made not only for France's Sahara test site but by nuclear powers for many other test sites[1]—and have shown how depiction of "putative 'empty' spaces" and the "memorialized absence of humans" has importantly influenced narratives of the nuclear age.[2] In the case of the Sahara, the irradiated spaces were devoid neither of life nor of politics, but in fact were full of objects and meaning that made them crucial sites in the process of decolonization, just as the French atomic tests themselves were important events in that contingent process.[3] In fact, in their exposure of "vulnerable national sovereignties,"[4] the atomic bomb tests can be seen specifically to be "radiopolitical events," events that left a radiological marker with both political intent and effect.

Historians have by now detected many examples of the link between exposure to weapons testing fallout and the process of decolonization. Opposition to nuclear weapons and weapons testing was one of the pillars of postcolonialism.[5] This opposition was especially significant in postcolonial Africa.[6] What requires further attention are the many specifics of this opposition and the reaction to it. Opposition came differently in different countries, and French authorities responded differently to the anger and tension triggered by atomic testing according to precisely how they were attempting to steer their postcolonial interests in a variety of places and contexts. The specifics of the radiation exposure of the Sahara Desert demonstrate how fallout's political geography could be one of postcolonial linkages, connecting not only Paris and Rabat to the Sahara but both to the United Nations in New York and ultimately to Algiers as well.

This essay therefore views radiation exposure as a source for illuminating a novel perspective of decolonization, the first years of Moroccan independence, and the geographies and sovereignties clarified by that exposure. After a brief reflection on the nature of the historical sovereignties overlapping in the French Saharan testing grounds, it considers the specific reactions of the Moroccan government and major political parties in 1958 and 1959 to the prospect of nuclear testing in the Sahara. The essay then turns to the French effort to persuade the Moroccan people and government that the tests posed no danger. This campaign was bound to fail. Because of the imperial position the French could not help but assume and because of the historical and symbolic significance ascribed to the Sahara by Moroccans, the bomb tests were irretrievably understood as political acts regardless of the minimal radioactive fallout claimed by the French.

The essay then turns to examination of the reaction of the Moroccan authorities and political parties to the first French tests. While the French, Moroccans, and others made efforts to detect and measure radioactive fallout, in Rabat and elsewhere in Morocco attention centered not on these measurements but on the reaction of Moroccan political factions to the atomic tests. Both the French embassy and the Moroccan authorities worried about how acutely the tests upset populace and political parties. French officials and the Moroccan king came for different reasons to try to minimize the stresses these radiopolitical events caused. The last section examines the situation after Algeria's independence and notes that Moroccan King Hassan II and Algerian head of government Ahmed Ben Bella faced similar pressures from impending French tests—pressure that, on at least one occasion, Hassan turned to his advantage.

Decolonization, Sovereignty, and the Geopolitical Nature of Nuclear Tests in Newly Independent North Africa

Belying the impression given by the French foreign ministry map, the Sahara atomic testing grounds were not empty. They were full—of human communities, of resources, and of historical import for different countries and factions in Algeria, Morocco, France, and elsewhere—something brought to light by French atomic testing. For the Kingdom of Morocco, the Sahara was a place of importance if also ambivalence. On the one hand, Moroccan historiography has emphasized the contrast between the *Bilâd al-Makhzen,* "the realm of governance," and the *Bilâd as-Sibâ,* "the anarchic mountain and desert edges of the realm."[7] On the other hand, the most nationalist elements of the Moroccan polity placed great weight on the historical Moroccan claim to sovereignty over that part of the desert in which France's atomic testing sites were found. This was of acute importance to the Moroccan Palace. As Tony Chafer has noted, leaders of newly independent African countries—the Moroccan monarchy included—were anxious not only about communist infiltration but also "the spread of radical nationalist ideas" that might threaten their own claims to national leadership.[8] For the Palace, French atomic testing illuminated not only sovereignty disputes but also the threat of the domestic opposition.

Contemporary accusations in Francophone Africa of France's "nuclear imperialism" underline the significance of various sovereignty claims, not only from various forces in Morocco but from other countries as well. After all, at the end of the 1950s, many borders were still disputed and the war in Algeria ongoing.[9] Moroccan sovereignty claims on the Sahara shed new light in particular on the nature and study of Moroccan nationalism. Much of the study of the latter has focused on cultural and linguistic expression and assertion,[10] but it is important to bear in mind that, as Montserrat Guibernau has argued, national identity can also be strongly linked to "imagined" territories unknown or at least unvisited by most in the nation.[11] For this reason, more attention could go to nuclear technology and objects, including weapons tests and their fallout,[12] as radiopolitical events around which competing claims about the nation can be framed. This attention becomes all the more important given Larmer and Lecocq's important recommendation to make the *construction* of nationalism the object of study. Here, bomb testing and fallout on a disputed territory become resources for competing claims about the Moroccan nation.[13]

Finally, it is crucial to understand that the fury in Morocco and other newly independent African countries against France for its plan to test weapons in the Sahara was not, as Colette Barbier contended in 1998, irrational.[14] Not only did Moroccans and others have reason to doubt French assurances and view with apprehension the uncertainty of the health effects of atomic testing. Concern with the sovereignty of the Sahara Desert was not just symbolic. France's Algerian war effort involved the creation of a physical barrier, the Morice Line, that extended to Beni Ounif, partway to the ostensible border. South of this, French military forces waged a campaign of interdiction to stop supplies and men from adding to the Algerian independence forces. Furthermore, France maintained its claims to the vast area of the test site knowing full well that its geologists had discovered not only petroleum but many strategic mineral resources under the surface.[15] That is why this history is ultimately a diplomatic one, making use of previously unexamined sources from the archives of the French foreign ministry: because the stakes were ultimately geopolitical, involving sovereignty as well as the resources that undergird geopolitical power and sovereignty.

The Narrative Gets Away (July 1958–January 1960)

When Charles de Gaulle announced on July 22, 1958, that he would press ahead with plans to test a nuclear weapon in the Sahara in the first half of 1960, his decision might have seemed simple enough. The previous year, the French political and nuclear leadership had already indicated this to be their aim. De Gaulle, faced with the collapse of the Fourth Republic, was simply confirming that as policy. In point of fact, his announcement came at a moment of great historical complexity in and outside of France.

One dimension of that complexity stemmed from the most recent actions of the nuclear powers. During the summer of 1958, the US, the USSR, and the UK—the three nuclear weapons states whose ranks France would soon join—had agreed to a voluntary moratorium on nuclear weapons testing while entering negotiations for a test ban treaty.[16] France's announcement disappointed all who argued for an end to the testing that had lifted radioactive fallout into the upper atmosphere around the globe. It caused particular discomfort to the US and UK, since the USSR could now use France's planned actions in propaganda against the Western Alliance.[17]

There were other reasons why the matter of nuclear weapons testing in the Sahara became internationalized. Morocco, Tunisia, Sudan, and Ghana were all recently independent, soon to be joined by several other countries, all of which were former (mostly French) colonies. All of these countries had reason to hold their former colonizer in suspicion when France, fighting a brutal colonial war in Algeria, declared Algeria's southern reaches a nuclear weapons testing zone. That zone, the southwestern Sahara desert, was hardly empty. Native populations as well as migratory groups passed through regularly, and in fact testing brought more people to the area as Nigerian Tuareg and others gravitated to the good wages their labor brought them in test site construction.[18] Meanwhile, underneath the rocks and sands, French prospectors had struck rich petroleum reservoirs as well as discovered several valuable mineral deposits.[19] Those resources underscored the significance of the stakes involved in the contested sovereignty of the Sahara. The Alaouite dynasty ruling Morocco, for one, claimed that much of what would eventually become Algeria, Mali, Mauritania, and the Western Sahara had in the past been ruled by, or been loyal to, the Alaouite family and therefore should be part of independent Morocco. Fighting for these regions had entered into Morocco's national story—occasional skirmishes between Moroccan forces and various tribal groups still erupted outside Morocco's borders—and for the older, most powerful elements of Morocco's leading political party, Istiqlal, the notion of "le Grand Maroc," "al-Maghreb al-Aqsâ" (Greater Morocco) were central to their ideology.[20]

This mattered for the Moroccan domestic political scene. At the end of the 1950s, King Mohammed V was in a running battle with Istiqlal for political supremacy in Morocco's constitutional monarchy. Istiqlal's insistence that the country do more to assert its right to Greater Morocco constituted one front in this fight, and by 1958 Mohammed V had begun in response to assume a more irredentist posture in his public statements. His concerns were wider, as he and Morocco's leaders were aiming to lead Africa's newly independent countries in an alliance that could assert their interests against the privileges of the former colonial powers. This new alliance-building did not lack for rivalry, witness the French diplomatic corps' interest in and hopes to exploit the competition between Mohammed V and Gamal Abel Nasser to claim priority in leading North Africa. All the while, the US and USSR regarded North Africa as a front in their own global conflict, the US administration being particularly sensitive to Morocco's posture because of US air force and naval bases there. Nouasseur Air Force base was used at the time for staging Strategic Air Command bombers, while naval installations included the Port Lyautey Naval Air Station and a nearby communications center.[21]

Protests against the planned nuclear tests arose first in pan-African forums, echoing the anti-nuclear sentiments expressed in previous meetings like the 1955 Bandung Conference.[22] The Afro-Asian Solidarity Conference that concluded at the beginning of January 1958 in Cairo passed a motion condemning the tests. So, too, did the Conference of Independent States in Accra in April, and the Panafrican Student Conference in July.[23] These sorts of international displays of protest continued into 1959, when the French tests appeared on the agenda of the Conference of Independent States in Monrovia in August, again resulting in a consensus motion against testing. The press in these countries recorded these protests; in Morocco, where Istiqlal's papers were highly influential, the Istiqlal organ *Al-Alam* spoke of the coming tests as "a hostile act . . . in an independent country" and alluded to the need to "recover the country's natural borders," thus establishing a link between borders and atomic tests in its earliest communications on the topic.[24]

Twice in 1958 in Morocco, governing coalitions broke down, and at the beginning of 1959, Istiqlal split, a younger progressive faction leaving it to form the Union Nationale des Forces Populaires (UNFP). Nevertheless, all sides, including the Palace, agreed on the need to protest the planned French tests. During the first half of 1959, the Moroccan foreign minister delivered several notes of protest to the French embassy in Rabat. The ambassador refused them because the sovereignty of the Sahara was evoked, declaring "what happens in the French Sahara does not concern the Moroccan government."[25] It clearly concerned the French government. The cover of the August 17, 1959, issue of *Time* magazine illustrated it: below a banner reading "France Finds Wealth in the Desert" are oil derricks and pipelines, visually dominating a traditional caravan. Before the scene hovers the visage of Jacques Soustelle, who, significantly, served at the time as both the state minister for the Sahara and overseas territories, and as the state minister for atomic affairs.[26]

French assertions notwithstanding, at least one petition against the planned French test was already circulating in Morocco, and all of Morocco's dailies, regardless of party affiliation, objected to the planned French tests.[27] While the UNFP-leaning *Al-Tahrir* cautioned that "Africa's atmosphere will be contaminated and the African peoples, our children, our vegetation, our civilization . . . are threatened with annihilation." *Al-Alam*, in which the words of Istiqlal's ideological leader Allal al-Fassi often appeared, warned that "Your life is in danger. . . . [T]he risks of contamination are not small, the water and plants will be polluted in ten years, if similar tests are carried out, the atomic clouds will invade the atmosphere of the African continent to such a degree that its inhabitants will be badly harmed."[28] All the while, the Istiqlal-leaning newspapers continued to remind their readers of Morocco's claim to the area in which the tests were to be conducted.

As the critical tone in the press rose in intensity, France's ambassador to Morocco, Alexandre Parodi, suggested the Ministry of Foreign Affairs "publish an official communiqué meant to ease the worries that this [press] campaign stirs in the countries bordering the Sahara, and in particular to show the senselessness of the allegations concerning the risks . . . to the populations of North Africa."[29] We should see that the concerns of Parodi, shared by the Quai d'Orsay (French Foreign Ministry headquarters), reflected the threat of the above tensions to disrupt French postcolonial strategy in Morocco and in other former French colonies. Historians have shown that France hoped in its former colonial possessions and protectorates, now independent African states, a system of "*coopération*" could be established in which France's influence and interests could be maintained.[30] In this way, by threatening this strategy, by illuminating and therefore amplifying the intensity of sovereignty disputes, French atomic testing became an important part of the process of decolonization.

The record shows how the French Foreign Ministry painted a picture of minimal risk. Their principal rhetorical tool was comparison to previous nuclear weapons tests. The French foreign ministry declared that upcoming French tests would add infinitesimally little to the atmospheric fallout already deposited by over 200 US and Soviet tests, "some ten thousandths." (This statement, however true in its own terms, exposed the degree to which tests had already deposited massive doses of radiation into the atmosphere.) As for local fallout, French diplomats emphasized what they considered the relative isolation of the Sahara test site—exactly the purpose of the map described at the opening of this chapter. Las Vegas, Nevada, was several times closer to the US testing center than Marrakech (1,000 kilometers) or Gao (1,200 kilometers) were to the French test site. French diplomats were instructed to report that French meteorologists were constantly monitoring the weather, ready to trigger the first test at the optimum moment when any local fallout would be swept into an unpopulated area. Finally, French officials reminded African countries that overseeing all tests was an ostensibly disinterested

Safety Commission of eminent scientists and physicians who had determined an admissible yearly limit of exposure to radiation twice as strict as their US equivalents (i.e., 1.5 versus 3.9 roentgens/year).[31]

French diplomats described these factors to anyone making inquiries as to the dangers of the testing to come and they implored their US and British colleagues to amplify their assurances of safety.[32] African governments, including Morocco's, were skeptical. At the meeting of the IAEA General Assembly in Vienna, not only did the Moroccan delegate declare in the process of denouncing all nuclear tests that the French test, to take place "on our territory," must too be denounced, but he urged the IAEA to send a technical assistance mission to Morocco to set up systems for detecting and measuring fallout. This put France in a bind. A blanket refusal might suggest that its assurances were not reliable.[33] However, Morocco could not force the issue: responsibility for measurement of fallout fell solely on UNSCEAR, the United Nations Scientific Committee on the Effects of Atomic Radiation, and any appeal to the IAEA was thereby deflected away.[34]

In the end, though, French assurances were fated to fail. Not only did Moroccan references to fallout risks differ from the examples provided by the French—the Moroccan press tended to give particular attention to previous atomic tests, especially the March 1954 Bikini test that irradiated a Japanese fishing vessel, as well as the bombings of Hiroshima and Nagasaki.[35] More to the point, for Moroccans, the assessment of risk was less important than the geopolitical meaning. France continued to insist on conducting nuclear tests on territory claimed as Moroccan. The Quai d'Orsay was prepared to counter these claims directly, mostly in noting that virtually every time French officials invited their Moroccan counterparts to discuss arrangements for common planning of the Sahara region, the Moroccans refused, even as they protested rocket tests, atomic tests, or the granting of research permits to petroleum and mineral prospectors. If the French thought the Moroccans categorically obstinate, Moroccan diplomats understood any acknowledgment of a common effort to manage Saharan affairs might give the French reason to claim Saharan borders settled, at least between Morocco and French Algeria.[36] In the end, the Moroccan foreign minister chose not only to protest nuclear tests "within the confines of Moroccan territory," but more fundamentally "the decision taken to transform the Sahara into a nuclear testing ground."[37] The number of roentgens released mattered, but, for Moroccan officials, the primordial offense was the nuclear despoilment of the contested Sahara.

Politically, Morocco found success not in Vienna at the IAEA, but in New York, at the United Nations, where it succeeded in having the matter of French testing in the Sahara put onto the autumn agenda for the General Assembly.

Starting November 4 to 13, 1959, the United Nations General Assembly took up debate about French testing via a draft resolution to demand that France stop test preparations. Despite a UK attempt to draw attention to an alternative, less categorical resolution, and US support for the UK effort, Morocco's work as the lead sponsor of the draft resolution succeeded, and it was adopted on November 20, 1959.[38] During the debate, the Moroccan representative refused to permit his French counterpart "to depict the Sahara as a desert, as it appeared in past romances or as certain film producers would seek to describe it." Rather, in an area he described as "contested territory," the Moroccan representative noted that "there are hundreds of villages which are in some instances only fifteen to twenty kilometers apart."[39]

This success in the UN became a touchstone for Moroccan officials and the Moroccan press. Especially in the latter's view, a majority of the world's nations had agreed with Morocco that the real meaning of the French tests was not a minimal risk of harm to the peoples of North Africa. It was a signal that, as an announcer stated on Radiodiffusion Nationale du Maroc, "the horrors of the explosions which have taken place in certain countries" were enough to "bring a shudder to anyone." With its "criminal operation," France was simply signaling to the world that it "considers the Sahara as its own property, an integral part of its national territory."[40] When Mohammed V met with Dwight Eisenhower in December, he reminded the US president that France was still exploiting countries that had gained their independence.[41] For the countries of Africa, Morocco included, the upcoming test was a sign, not that France was bettering its security with a new generation of weapons, but that France maintained a violent imperial presence in Africa. Try as they might in the halls of the UN or in the diplomatic corridors of Rabat, French officials could not deter this narrative, at least not in Morocco.

Diplomatic and Political Fallout (February 1960–April 1961)

From Rabat, French ambassador Alexandre Parodi reported to Paris that in the weeks before the nuclear test, the "propaganda campaign" in Morocco intensified, "in newspapers, by radio, and by oral propaganda in the medinas." The French diplomat thought it "essential that secrecy be maintained for as long as possible from the moment of the explosion and even for some time after in order to avoid all danger of protests and to deny anti-French propaganda." (Upon receiving this note, a functionary in the Quai d'Orsay wrote in the margin, "*Mais c'est impossible!*"[42])

French diplomats thought they were observing a "striking contradiction." The Moroccan government appeared to be playing a double game, on the one hand permitting alarming messages on the radio and in the press while on the other hand assuring French officials that all was under control and there was no hostility to France in Morocco. That noted, the French embassy in Rabat witnessed genuine anxiety on the part of Moroccan officials.[43] The declaration of a no-fly zone for hundreds of kilometers around Reggane distressed them.[44] Meanwhile, the Moroccan minister of economy and finance, Abderrahim Bouabid, requested to know whether desert locusts might bring contamination into the country. The Service of Atomic Affairs reported back to Rabat that of the two desert locusts known in the Reggane area, *Dociostaurus maroccanus* was not migratory, and *Schistourca gregaria* only reached Morocco from migratory routes originating in Sudan or Mauritania, not the Algerian desert.[45]

Ambassador Parodi was not exaggerating when he stated that the tone of the Moroccan press had grown harsher. The early evening January 18, 1960, broadcast on Radiodiffusion Nationale du Maroc gives a sense of this. The announcer accused the French of high arrogance and linked the tests to the colonial war in Algeria:

> So much for humanity if the atomic test benefits me. . . . Who cares about the cries of the African and Asian peoples, when it is a question of mounting the summit of glory. This is why [France] insists on turning African territory into a laboratory and its inhabitants into guinea pigs, without worrying about the consequences. . . . France has exterminated more than a million Algerians without worrying about the anger of world public opinion. It massacred these martyrs in an odious manner and wiped out entire tribes. Who would stop it from persisting in this massacre of innocents and in this crime?[46]

French officials interpreted these texts as provocation. However, it appears that there was, on the part of the Moroccan government, real worry. In the last week of January, Bouabid went from embassy to embassy in Rabat, pleading with the US, Portugal, Italy, the UK, the USSR, Spain, and Belgium to find a way to persuade the French to postpone the test.[47] When in Paris the Moroccan ambassador called on Maurice Couve de Murville to protest the coming test, the French foreign minister accused the Moroccan government of knowingly feeding "the agitation now brewing in Morocco" and coldly questioned why Morocco should have an opinion about whether France should develop nuclear weapons.[48]

In the event the test could not be stopped, what the Moroccan government was most anxious to know was the date—*when* would the explosion occur? The Quai d'Orsay reported that it could not reveal this because it did not know: it depended on the weather.[49] The Palace and the Moroccan government worried, test date unknown, that the mounting anger and protests might explode. As rumors of an impending test circulated, the UNFP declared January 31 a day of national protest and general strike. Crossing party lines, Istiqlal announced it would join the protests. The Moroccan authorities added extra protections around the French embassy and consulates. From a confidential source, Parodi learned that the Minister of the Interior had authorized force against any protests that threatened to get out of hand.[50] It appears that the Palace was as worried as the French embassy that protests might physically harm French interests, further suggesting the degree to which French atomic tests and the anxiety they stirred had the potential to interfere with French influence in Morocco.

In the end, there was no such escalation. Istiqlal withdrew and insisted only on national prayer. The UMP, the UNFP's labor arm, suspected that the protests were as much to agitate for the release of a political prisoner as to demand French renunciation of tests and stepped back from the call for a strike. Not that the day lacked drama: in Casablanca, 2,000 protesters took to the streets, while in Rabat, a crowd four times as big marched from the medina toward the French embassy. Police deployed fire hoses and discharged guns into the air to bring them to a halt.[51]

Ultimately, the Moroccan government was powerless to impede the French. Even their supposition as to the day of the test—February 6—was off the mark. A week later, on February 13, 1960, *Gerboise bleue* ("blue jerboa") exploded in the early morning in Reggane, yielding approximately 70 kilotons. Moroccan radio announced the news of the test later that morning. French consuls around the country reported the situation. In Casablanca, there was some unrest in the poorer neighborhoods and a scuffle between students and police. In Oujda "a certain feeling" permeated the streets, and reinforcement of the guard was made around the French consulate and the consul's home. In Souk El Arba, unrest in the street came with rumors that the fallout would cause blindness and sterility—there the consul thought the worst threat might come to French landowners in the countryside. Meanwhile, in Rabat, the situation remained calm despite what the French embassy regarded as provocation on the radio.[52] While some French diplomats were dismissive, even contemptuous, of Moroccan fears—one diplomat wrote from Rabat that he could not count the number of times Moroccan mothers had claimed to give birth to "two-headed infants"—real effects were observable.[53] At Hassi Zerzour, just 175 kilometers from the test, a bright light illuminating the whole region was observed in the early morning, and residents in Taouz and Zagora (375 kilometers) heard a sound like rumbling thunder.[54]

The Moroccan Council of Ministers met. According to French intelligence, debate concerned what to do, and what harm might come if relations with France were disrupted. The most extreme reaction came from the minister of education, a member of Istiqlal, who proposed breaking off all relations, confiscating French goods, obliging French troops to leave Morocco, and ending all technical assistance from France. An unnamed minister thought such a reaction would worsen Morocco's situation; the king hoped the debate would produce less drastic views.[55] While Moroccan representatives in New York tried and failed after the test to produce an intervention on the part of the UN Security Council, the Moroccan ambassador in Paris stepped into the French Foreign Ministry to make a dramatic announcement: Morocco was breaking off the March 28, 1956, accords—the agreement reached by Paris and Rabat at the moment of Moroccan independence—and he, the ambassador, was being recalled.[56] Paris found this only mildly disconcerting, as many of the elements of the accords had lapsed into "lettre-morte," and the Moroccan gesture was a unilateral one.[57]

It might have appeared at first glance that the Palace and Istiqlal were in harmony. In reaction to the French nuclear weapons test, both made reference to the sovereignty of the Sahara. But the French intelligence service thought a large Istiqlal protest organized on February 21 in Casablanca was meant in point of fact "to demonstrate the power of Allal al Fassi's movement," and that, in general, protests against the test provided all opposition forces a pretext "to mobilize their troops."[58] Parodi in Rabat observed arrests of leading UNFP figures—there was pretext for the Palace to counter the mobilization as well.[59] In this way, the *Gerboise bleu* test can be understood as an intensely political event, provoking and casting light on the various Moroccan factions who believed the test took place on contested territory. Under those circumstances, the French tests inevitably alarmed the Moroccan government and exposed fault lines in the Moroccan polity.

The political fragility of Morocco's situation was evidenced in March, when Parodi informed the Crown Prince Moulay Hassan that another test would soon take place. Hassan appealed to Parodi to have the test delayed: there were elections coming, and he wondered aloud whether a new Moroccan government could survive this stress. The Moroccan government's lack of agency—its inability to have the slightest influence on the timing or in any way mitigate the dangers of the French tests—was all the more exposed when all sides of the political spectrum were making claims about the sovereignty of the Sahara and putting increasing pressure on the Palace and the government to do something about it.[60] In fact, the "poisonous atmosphere for Franco-Moroccan relations" stirred by the tests threatened French interests and goals as well.[61] Several French citizens working in Morocco addressed a letter directly to de Gaulle warning that another test could have "disastrous effects on the life and property of members of the French colony" in Morocco. Even if the Foreign Ministry minimized this warning by identifying the authors as having "liberal tendencies," their observations could not be ignored, especially with the Moroccan government applying increasing pressure for the withdrawal of all French troops.[62]

Anxiety in the face of French claims of the conditions of safety in which *Gerboise bleue* unfolded only rose after the catastrophic Agadir earthquake of February 29, two weeks after *Gerboise bleue.* A tremor originating in the shallows of the earth's crust killed almost 15,000 people, nearly a third of the entire city. Dailies in the Arab world asked whether the French nuclear test had not somehow set off the earthquake; in Tripoli, *Al Talia* accused France of choosing to test atomic weapons in "a fragile zone of the Earth's crust" and predicted more earthquakes.[63]

Moroccan officials do not appear to have credited these rumors, but after the rejection of their request for technical aid from the IAEA, the record shows that they did seek other means to establish a fallout detection and measurement capability. In February 1960, Moroccan diplomats made an unsuccessful attempt to gain US aid.[64] In March, Italian physicist Giampietro Puppi received an invitation directly from Moroccan Prime Minister Abdallah Ibrahim not only to give a series of lectures in the country but to set up a series of atmospheric fallout detection sites around the country.[65] The latter apparently never came off. French intelligence identified another attempt. Bernhard Witt, a German national living part-time in Switzerland, told French officials he had been under contract during the second half of 1959 with the Moroccan government to establish a detection system. Witt claimed that this system detected no additional, dangerous radiation, but that, nevertheless, he was approached in Germany and Switzerland by Moroccan officials who asked him to return for a second venture, this time to join a motorized convoy in the direction of Timbuktu meant to uncover "depots of atomic bombs and nuclear arms, spreading dangerous radioactivity on Moroccan soil." Witt's impression was that this was subterfuge, meant among other things to enter French territory and plant evidence. He declined.[66]

If the French resented this purported Moroccan plot, they hardly had a leg to stand on. Unbeknown to the Moroccans, the French Foreign Ministry was helping the French Commissariat à l'Energie Atomique gather rain samples for measurement of fallout in Agadir and Rabat, something they continued to do in Agadir until 1962 and Rabat until an unknown later date.[67] In fact, monitoring of this and subsequent French atmospheric tests took place in a number of sites, including US Atomic Energy Commission stations in Libya, Italy, Egypt, Pakistan, Lebanon, Thailand, Taiwan, and Japan, as well as a US naval research laboratory facility in Morocco itself.[68]

The French foreign office began to realize—despite the secret rainwater collection—that the most immediate concern of the Moroccan government was to know *when* tests would take place. On March 29, 1960, de Gaulle opened a new French effort at selective disclosure by informing Mohammed V by personal communication that a second explosion "much less powerful than the previous" would take place sometime after March 31.[69] While *Al-Alam* called for moving "from words to acts" to stop French testing,[70] when the test ("*Gerboise blanche*") came on April 1, 1960, the French embassy in Rabat reported the atmosphere in the streets much more subdued than after the February 13 test and the Moroccan government more cautious about calling for retaliatory measures.[71] Indeed, the Moroccan government had observed just how little changed after its dramatic post–*Gerboise bleue* diplomatic moves. Beyond general condemnation, Morocco's government took no retaliatory steps.[72] Privately, Mohammed V responded to de Gaulle's letter with a reminder not only of the "great danger for all humanity" brought by the nuclear arms race but of Morocco's claim of sovereignty to the Sahara: "We condemn these tests all the more as they take place in inhabited regions and on territory which We consider an integral part of Our Kingdom."[73]

By the time of France's next atomic test in December 1960, King Mohammed V had moved to increase his power. In May 1960, he dismissed the Abdallah Ibrahim government and appointed himself prime minister and his son deputy prime minister. Two months later the king appointed the notorious Mohamed Oufkir director general of security, adding to the autocratic nature of the regime. There was also a notable deterioration in Franco-Moroccan relations. Moroccan willingness to ship arms to the FLN, to welcome Algerian refugees, to oppose the independence of Mauritania (on the grounds that it was Moroccan territory), and to gather the so-called Casablanca Group (Morocco, Egypt, Mali, Guinea, Ghana, and the provisional Algerian government), all infuriated de Gaulle.[74] Nevertheless, after France's December 27, 1960, test ("*Gerboise rouge*"), the Moroccan government refrained from any unilateral measures, despite calls for such from the opposition.

On January 13, 1961, France's new ambassador to Morocco, Roger Seydoux, sat down for lunch with the heir to the throne, Prince Moulay Hassan. Hassan warned Seydoux that if France conducted another nuclear weapons test, "France would have no more diplomatic representation in Africa." Seydoux judged it a bluff and responded that French citizens in Africa could not be deprived of the protection of their embassy. When Seydoux tried to put Hassan's mind at ease by explaining that a future test would be of a very small weapon "without dangerous radioactive effects," Hassan made clear that the problem was political: it was the principle of the test he was questioning. After all, reports had it that the mayors of Corsica had through protest dissuaded France from using the island as a site for underground tests. Moroccans, and other newly independent Africans, could not remotely achieve the same effect.[75]

The point revealed the Palace's geopolitical discomfort. De Gaulle's own declarations suggested that Algerian independence was increasingly likely. Hassan claimed to Seydoux that Morocco's concern was not sovereignty but fair distribution of the Sahara's resources. While the Moroccan government sympathized with the FLN, it recognized any Algerian regime to come would be a republic, ideologically hostile to Morocco's monarchy (not to mention Morocco's territorial claims in Algeria). Tunisia was a republic, and Spain was, according to Hassan, a "feudal regime" with a "crusade spirit"—Morocco was surrounded by potentially hostile powers. In inter-African forums such as the January 1961 Casablanca conference, Morocco continued to make territorial claims (in that case, concerning Mauritania).[76] Bomb tests, Hassan correctly understood, exacerbated *political* tensions. At the core of the Moroccan quandary was lack of agency amid assertions of sovereignty.

Radiopolitical Events in a Changed Landscape (May 1961–March 1963)

In February 1961, Mohammed V died, to be succeeded by his son; meanwhile France announced its next weapons test would be its last atmospheric one. A new phase of Moroccan-French conflict over testing began. However, when France conducted its next test, *Gerboise Verte*, on April 25, 1961, it did not dominate the news. The Generals' *putsch* did, as conspirators in Algiers attempted to stop de Gaulle's government in its steps to cede independence to Algeria. French diplomats in New York, Tripoli, Accra, Beirut, and Rabat reported that there was little to no reaction to the test. The attempted coup had seized all the attention.

The French government again warned the Moroccan head-of-state of the imminent test. The day before the explosion, the French ambassador sought but could not find the new king. He spoke instead with Hassan's closest counselor, Ahmed Guedira. Guedira immediately phoned the king in the countryside, who thanked the French government for the message but who "did not wish to receive the ambassador to avoid the obligation to raise a vehement protest against this new explosion, protest that he did not wish to raise at this moment when General de Gaulle 'needed the support of all of his friends.'"[77] It was a sign of improvement in Franco-Moroccan relations, as Hassan II stepped away from the non-aligned position his father had increasingly entertained in his final months. Moroccan denunciation of testing was this time general rather than particular.[78] In a further turn of events, in early 1962 the French Commissariat à l'Energie Atomique hosted a Moroccan delegation, taking them to several of their nuclear research centers.[79] Nevertheless, the Moroccan government was determined to keep its options open: Guedira angered de Gaulle when he reported that Morocco would be willing to turn to the US for military aid if France didn't provide it.[80]

Meanwhile, the political changes which the rebellious generals feared came to pass in Algeria. The Évian Accords were signed in March 1962. Four months later, Algerian residents overwhelmingly voted for full independence. Algeria's government, soon to be headed by Ahmed Ben Bella, would now experience the same difficulty Morocco's government had faced, only more acutely. This was for two reasons. First, Ben Bella and the Algerian government quickly became a leading voice speaking to issues confronting the Global South, including opposition to nuclear testing. Second, French nuclear testing posed a direct challenge to claims of territorial sovereignty in the Sahara. To be certain, testing would continue: included in the Évian Accords were statutes granting France rights to a testing ground until 1967. Testing moved from Reggane 600 kilometers east to In Ekker, where two tests had already taken place: *Agathe* in November 1961 and *Béryl* in May 1962. The bombs were now exploded underground (which did not prevent a disastrous venting of radiation during the *Béryl* test), but this could not mask atomic testing's capacity to illuminate geopolitical conflict and rivalry.[81]

Relations between Morocco and the new Algerian state were fraternal but tense. With Ben Bella's assumption of power in September 1962, Hassan II had an interlocuter who in many respects he sympathized with—after all, Morocco had supported the FLN during Algeria's long war for independence. However, the borders were a sore spot. In 1962, there was a series of skirmishes between pro-Moroccan and pro-Algerian militias around Tindouf.[82] The Palace continued to feel the heat from Al-Fassi and Istiqlal, and as Hassan had told Roger Seydoux, the Moroccan and Algerian regimes were ideologically divergent.

The Commissariat à l'Energie Atomique and French Army made final preparations for the next French atomic test in March 1963. As it approached, French diplomats in North Africa were on alert to evaluate the potential Algerian reaction. Georges Gorse, ambassador in Algiers, noted that many in the diplomatic corps thought the moment bad for a French nuclear test. Late the previous year Ben Bella had publicly denounced French nuclear testing.[83] Now he had just overcome a political crisis and still faced stiff opposition.

Furthermore, the scheduled test would come at the moment the Algerian nation was celebrating the first anniversary of the signing of the Évian Accords. At the very least, thought Gorse, the French ought to provide Ben Bella with the exact date of the upcoming test as well as dates of all the tests to come.[84] Meanwhile, the mood in Rabat was similarly tense. In the words of a French diplomatic report, the king, recently returned from a trip to Algiers, "was profoundly worried about what his Algerian interlocutors told him about the possible repercussions of this test."[85] Ben Bella appeared agitated, and, out of solidarity, Rabat would be obliged to follow Algiers in strong protest, even if reluctantly.[86] The Moroccan foreign minister, Ahmed Balafrej, wondered whether Hassan's upcoming voyage to meet de Gaulle in Paris might have to be canceled.

In Algiers, on the 16th, Algerian Foreign Minister Mohamed Khemisti was able, somewhat, to calm Ambassador Gorse. Khemisti reported to Gorse that his political bureau had drafted a "very moderate" response to the approaching test. He then cut to the chase. Algerian territory could not serve as "a testing ground" for atomic devices without Algeria losing face internationally. When Gorse protested that, after all, such use was written into the Évian Accords, Khemisti returned to the real problem. It was not one of international law, he said, but one of politics, "the need to reinforce the authority of a newly created state, and not to feed the forces of subversion and division active in Algeria." It was in France's best interest, Khemisti insisted, that it have a strong, unified Algerian partner, not a weak, divided one. The pending test, illustrating Algerian lack of agency, would problematize its sovereignty and harm its government.[87]

The Algerian warning did not deter France from the planned test. The *Émeraude* nuclear device exploded inside the mountain of In Ekker on the morning of March 18, 1963, with a yield of 10 kilotons. French diplomats were now willing to account more carefully for the stress these tests placed on their Moroccan and Algerian interlocutors. If the tests could not be stopped, then the more information about their timing, the better, so as to maximize chances of shaping the flow of information afterward. Late that morning, the Quai d'Orsay instructed its ambassadors in Rabat and Algiers to inform Hassan II and Ben Bella of the test, and to report that for the moment, the test was to remain a tightly held secret.

The ambassadors moved to make the disclosure. In Rabat, French ambassador Pierre de Leusse called on Balafrej after he found the king away and unavailable. In Leusse's presence, Balafrej phoned King Hassan. The king thanked Leusse for informing him of the explosion and agreed to keep the knowledge secret. Balafrej added that, once knowledge of the test became public, the government of Morocco had no choice but to lodge a protest, out of solidarity with Algeria. Balafrej assured Leusse that order would be kept in the streets.[88] Meanwhile, in Algiers, Gorse met directly with Ben Bella and found him unsettled. Gorse judged his reaction as that of someone whose "pride was hurt." Ben Bella wondered why France wanted to weaken his government. He assured Gorse that any protests would be kept in check, but he predicted calls from the opposition to revisit the Évian Accord statutes pertaining to the testing. Khemisti added in a separate meeting with Gorse that, given its non-aligned stance, which included opposition to nuclear testing, Algeria would have no choice but to protest.[89]

As of the evening of March 18, 1963, the morningtime 10-kiloton explosion of *Émeraude* was unknown outside a small circle of French leaders, officers, and diplomats, as well as King Hassan, Ben Bella, and their closest confidants. The silence provided time to the leaders and diplomatic corps of all three countries to hone their strategies for minimizing the political stress caused by the revelation of the test. In fact, Hassan II and Ben Bella had had the opportunity to confer by phone. Their dilemmas were similar, if of different intensities: how to treat a fait accompli outside their control, one that problematized their various claims to the sovereignty of their respective countries over given territories and therefore their respective agency to do something effective to reinforce such claims.[90]

To the surprise of all, that evening Hassan took matters into his own hands. The king released a communiqué announcing the test. It was a protest against the explosion in the "Arab Maghreb," and it expressed Moroccan solidarity with Morocco's "Algerian brothers."[91] It was most of all an act to demonstrate Moroccan agency, especially the Palace's. It came as a complete surprise to the French and, it appears, to the Algerians.

The French wondered why Hassan had done it. When pressed, the king told the French ambassador that it was all he could do to control the situation in Morocco, to keep a step ahead of Al-Fassi and Istiqlal, who, Hassan noted, not only attacked France for once again defiling the Moroccan Sahara but blasted the Palace for its failure to adequately address the matter of the sovereignty of the Sahara in his remarks about the test.[92] As for Ben Bella and his government, they believed the purpose of Hassan's maneuver was to embarrass them.[93] On the 19th, Ben Bella had avoided any reference to the nuclear weapons test the previous day, preferring to celebrate the first anniversary of the Évian Accords. By that time, however, the Algerian press had gotten wind of Hassan's statement and spread the news themselves. Gorse thought Ben Bella "personally wounded" by the whole affair.[94]

At the very least, the atomic test had made the incapacity of the Ben Bella government to influence the test's timing and the dissemination of information about it visible for his opponents to see. The protests in the streets of Algiers that followed, though aimed at the French atomic test, were a signal of Ben Bella's vulnerability in the face of his government's incapacity to stop the testing.[95] This was a painful political experience for Ahmed Ben Bella, as it already had been for Mohammed V and Hassan II. Just as prior atomic tests had exposed the tensions of French-Moroccan relations and the important place of the Sahara in those tensions, *Émeraude* briefly and conspicuously illuminated the triangle of interests and claims on the sovereignty and uses of the Sahara which followed Algerian independence.

Conclusion

As James McDougall has remarked before, colonial violence is marked by its unaccountability, a sort of colonial impunity many Moroccans had experienced as a French Protectorate and many more Algerians had faced firsthand.[96] With this in mind, it is easier to understand why protests in Africa to French nuclear testing were as passionate as they were, why charges of nuclear imperialism reverberated deeply, and why tests affected the politics of newly independent postcolonial states. However, diplomatic and geographic specificities are crucial. In terms of reactions to exposure to French testing fallout, an account of Morocco must refer to the historical and political specifics of the Saharan testing grounds and Morocco's claims to them if we are to understand atomic tests and the fallout as radiopolitical events, distinctly political in motivation and effect.

The contingency of borders and the contingency of atomic testing and fallout overlap and illuminate one another. Frederick Cooper has observed that, in the 1950s and 1960s, as new states appeared in Africa, borders were highly contested, but have since become "perhaps the most stable element of African history."[97] Indeed, these borders were contested, sometimes violently: witness the so-called "War of the Sands" between Morocco and Algeria in October 1963. For Morocco, the defining of the nation was tied to the Sahara in and outside its official borders. The dynamism and contingency of those borders were illustrated not only by the fallout of French bomb tests but also by other events, such as the Moroccan invasion of the Western Sahara in 1975.

The fallout from atomic tests in the Sahara forced a continued expansion of the conflict over the meaning of those tests, as the struggle over them spread to Rabat, to Algiers, and elsewhere. It is interesting to note that the French strategy to minimize conflict by treating the tests as harmless radiological events without any political import remained for years to come. In a 1973 White Paper on testing in Oceania, French officials again chose to depict a bare, featureless space indicating only distances between the ground center of tests and populated areas—and again attempted to compare this favorably with other testing sites (in this case the Maralinga site in Australia, the US Nevada test site, and the Soviet test site in Kazakhstan).[98]

Peoples and countries in the region, however, could not treat these tests without political reference. For the peoples of Oceania, the testing represented a threat to their homeland. And even for more distant populations, tests in the Pacific counted as radiopolitical events. In the mid-1990s, when France resumed testing in the Pacific, there was uproar in Australia and New Zealand, though the countries were several thousand kilometers away from the testing grounds. Trevor Findlay finds, however, that "in the geography of the imagination," the Pacific island test site was close to both and of great significance.[99] One wonders if the New Zealand prime minister, Jim Bolger, was aware of the degree to which his accusation that France was acting like an anachronistic colonial power matched the criticisms levied against France by Moroccan leaders fifty years before. At the very least, he would have affirmed the intractably political nature of atomic testing and the fallout that results.

Notes

1 M. Susan Lindee, *Rational Fog: Science and Technology in Modern War* (Cambridge, MA: Harvard University Press, 2020).

2 Joe Lockhard, "Desert(ed) Geographies: Cartographies of Nuclear Testing," *Landscape Review* 6, no. 1 (2000): 3–4.

3 Tony Chafer underlines the contingency of the process of decolonization in "Decolonization in French West Africa," in *Oxford Research Encyclopedia of African History*, 2017.

4 Jeffrey James Byrne, *Mecca of Revolution: Algeria, Decolonization, and the Third World Order* (New York: Oxford University Press, 2016), 185.

5 Vincent Intondi, "The Dream of Bandung and the UN Treaty on the Prohibition of Nuclear Weapons," *Critical Studies on Security* 7, no. 1 (2019): 83–86.

6 Jean Allman, "Nuclear Imperialism and the Pan-African Struggle for Peace and Freedom: Ghana, 1959–1962," *Souls* 10, no. 2 (2008): 83–102; Rob Skinner, "Bombs and Border Crossings: Peace Activist Networks and the Post-Colonial State in Africa, 1959–62," *Journal of Contemporary History* 50, no. 3 (2015): 418–438; Christopher Hill, "Britain, West Africa and 'The new nuclear imperialism': Decolonisation and Development during French Tests," *Contemporary British History* 33, no. 2 (2019): 274–289; Roxanne Panchasi, "'No Hiroshima in Africa': The Algerian War and the Question of French Nuclear Tests in the Sahara," *History of the Present* 9, no. 1 (2019): 84–113.

7 Baz Lecoq, "Distant Shores: A Historiographic View on Trans-Saharan Space," *Journal of African History* 56 (2015): 30.

8 Tony Chafer, *The End of Empire in French West Africa: France's Successful Decolonization* (Oxford: Berg, 2002), 4.

9 Camille Evrard, "Mauritanie 1956–1963: Les multiples dimensions d'une indépendance contestée," *L'Année du Maghreb* 18 (2018): 149–167. Douglas Ashford made the interesting observation that, as these borders became increasingly a fait accompli, and actions

that might lead to change became decreasingly possible, irredentism increased, becoming more effective as a domestic political rhetoric. Douglas Ashford, "The Irredentist Appeal in Morocco and Mauritania," *Western Political Quarterly* 15, no. 4 (1962): 641–651.

10 Eva Cantat, "Imaginaire nationale et territoire: la construction nationale marocaine après l'indépendance," *Revue de l'Institut des langues et cultures d'Europe, Amérique, Afrique, Asie et Australie* 30 (2018), http://journals.openedition.org/ilcea/4468 (consulted April 23, 2023); Rocío Velasco de Castro, "La monarquía alauí, símbolo identitario de la nación marroquí: Legitimidad histórica e instrumentalización política," *Diacronie: Studi di Storia Contemporanea* 16, no. 4 (2013): 1–17.

11 Montserrat Guibernau, *The Identity of Nations* (Cambridge: Polity Press, 2007). The affinity with Winichakul's notion of a "geo-body" is unmistakable and significant. Thongchai Winichakul, *Siam Mapped: A History of the Geo-body of a Nation* (Honolulu: University of Hawai'i Press, 1997).

12 For local effects of the first French weapons tests, see IAEA, *Radiological Conditions at the Former French Nuclear Test Sites in Algeria: Preliminary Assessment and Recommendations* (Vienna: International Atomic Energy Agency, 2005); and A. Chebli et al., "First Approach to Studying the Impacts of the Nuclear Tests on Insects in Reggane, Algeria," *Ciência e Técnica* 31, no. 3 (2016): 119–132.

13 Miles Larmer and Baz Lecocq, "Historicizing Nationalism in Africa," *Nations and Nationalism* 24, no. 4 (2018): 893–917.

14 Colette Barbier, "L'Afrique face aux premières expérimentations nucléaires françaises," *Centre d'études d'histoire de la Défense* 8 (1998): 113.

15 Baz Lecocq, "L'histoire d'un coq qui grattait le sable: Conflits frontaliers et nationalisme au Sahara à la fin de l'époque coloniale," presented at CEMAF, Séminaire Frontières et IndépendancesenAfrique,May21–222010,Paris,France.Anais...abr.(2010),https://www.academia. edu/925275/Lhistoire_dun_coq_qui_grattait_le_sable_Conflits _frontaliers _et_nationalisme_au_Sahara_a_la_fin_de_lepoque_coloniale (accessed April 23, 2023). See also Roberto Cantoni, *Oil Exploration, Diplomacy, and Security in the Early Cold War: The Enemy Underground* (New York: Routledge, 2017).

16 Lawrence S. Wittner, *Resisting the Bomb: A History of the World Nuclear Disarmament Movement* (Stanford, CA: Stanford University Press, 1997); Robert A. Divine, *Blowing on the Wind: The Nuclear Test Ban Debate, 1954–1960* (New York: Oxford University Press, 1978).

17 Mervyn O'Driscoll, "Explosive Challenge: Diplomatic Triangles, the United Nations, and the Problem of French Nuclear Testing, 1959–1960," *Journal of Cold War Studies* 11, no. 1 (2009): 28–56.

18 Julien Brachet, "Movement of People and Goods: Local Impacts and Dynamics of Migration to and through the Central Sahara," in *Saharan Frontiers: Space and Mobility in Northwest Africa*, ed. James McDougall and Judith Scheele (Bloomington: Indiana University Press, 2012), 240. Lecocq adds that, since the French did not trust Algerians to carry out the construction work necessary around France's two testing sites at Reggane and In Ekker, they had to bring in labor from elsewhere. Baz Lecocq, *Disputed Desert: Decolonization, Competing Nationalisms and Tuareg Rebellions in Mali* (Leiden, The Netherlands: Brill, 2010), 172.

19 Lecocq, "L'histoire d'un coq qui grattait le sable."

20 For more on the main elements of identity formation in Morocco and the Maghreb, see Mimoun Hillali, "Identité ou identités au Maghreb: du poids de la culture, de l'idéologie et de la religion (éléments de problématique géopolitique)," in Olivier Lazzarotti and Pierre-Jacques Olagnier, *L'Identité, entre ineffable et effroyable* (Paris: Armand Colin, 2011). Lecocq notes that as early as 1956, Allal al-Fassi, one of the founders of Istiqlal, proposed a map suggesting that the Algerian testing range, the Spanish Sahara, and Mauritania were all "occupied areas" requiring further liberation. Lecocq, *Disputed Desert*, 62.

21 El-Mostafa Azzou, “Les relations entre le Maroc et les États-Unis: regards sur la période 1943–1970,” *Guerres mondiales et conflits contemporains* 1 (2006): 105–116. El-Mostafa Azzou, “La présence militaire américaine au Maroc, 1945–1963,” *Guerres mondiales et conflits contemporains* 2 (2003): 125–132.

22 Vincent Intondi, “The Dream of Bandung and the UN Treaty on the Prohibition of Nuclear Weapons,” *Critical Studies on Security* 7, no. 1 (2019): 83–86.

23 SDECE (Service de Documentation Extérieure et Contre-Espionage) Sommaire pour le Primier Ministre, July 11, 1958. Maroc 1956–1958, 160, Folder 1959. Archives Diplomatiques, La Courneuve, France. (Hereafter: AD.)

24 SDECE, Expériences à Colomb Bechar, July 11, 1958. Maroc 1956–1958, 160, Folder 1959. AN. Colomb Bechar was the site of French rocket tests. French diplomats thought it had been confused with the coming atomic tests, to be staged at Reggane. Perhaps this was the case, but the authors of these articles understood they were commenting on the upcoming atomic tests.

25 Protestations Marocaines au Sujet des Essais Balistiques et des Expériences Nucléaires Françaises au Sahara. Undated summary of all protests delivered to the French embassy in Rabat, probably from mid-1960. Maroc 1956–1958, 160, Folder 1959. AD. See also Panchasi, “‘No Hiroshima in Africa,’” 93.

26 See Kelsey Suggitt, *Impossible Endings? Reimagining the End of the French Empire in the Sahara, 1951–1962*. Diss., School of Languages and Area Studies, University of Portsmouth, 2018.

27 Reported in *L'Humanité*, July 28, 1959.

28 Telegram, French embassy in Rabat to Paris, July 24, 1959. Maroc 1956–1958, 160, Folder 1959. AD.

29 Parodi, Rabat to Paris, July 23, 1959. Maroc 1956–1958, 160, Folder 1959. AD.

30 See especially Guy Martin, “Continuity and Change in Franco-African Relations,” *Journal of Modern African Studies* 33, no. 1 (1995): 1–20. Martin suggests that one reason France encouraged the creation of very small, non-federated postcolonial states was to maximize the maintenance of its influence. See also Ryo Ikeda, *The Imperialism of French Decolonisation: French Policy and Anglo-American Response in Tunisia and Morocco* (Basingstoke: Palgrave Macmillan, 2015).

31 Circular from Paris to all French embassies, composed August 5, 1959, dispatched August 6, 1959; for meteorologists, see Circular from Paris to all embassies in Africa, composed August 5, 1959, dispatched August 6, 1959; for details concerning the Safety Commission and exposure limits, see Conseil Executif de la Communauté [i.e., Communauté Français]. Communication au sujet des conditions techniques des expériences atomiques au Sahara, presentée par le Ministre des Armées chargé pour la Communauté des Forces Armées.
Maroc 1956–1958, 160, Folder 1959. AD.

32 Memorandum of Conversation, August 24, 1959. Herve Alphand, ambassador of France; Claude Lebel, minister counselor; Livingston T. Merchant; Robert H. McBride, *Foreign Relations of the United States, 1958–1960, Western Europe,* vol. 8 (Washington, DC: US Government Printing Office, 1993), 242.

33 See Bertrand Goldschmidt, Vienna to Paris, September 25, 1959, and Crouy, Vienna to Paris, September 25, 1959. Maroc 1956–1958, 160, Folder 1959. AD.

34 Néstor Herran, “‘Unscare’ and Conceal: The United Nations Scientific Committee on the Effects of Atomic Radiation and the Origin of International Radiation Monitoring,” in *The Surveillance Imperative* (New York: Palgrave Macmillan, 2014), 69–84. The most the IAEA did in this regard was to create a committee dedicated to examining typical radiation levels in all aspects of the biosphere, including descriptions for reliable techniques for such detection and measurement. See IAEA, *Radioactive Substances in the Biosphere* (Vienna: IAEA, 1961).

35 These reports and articles illustrate Roxanne Panchasi's argument about the elements informing African reaction and protest to the French tests. Panchasi, "'No Hiroshima in Africa.'"

36 See Note pour la Direction des Nations–Unies et des Organisations Internationales, October 20, 1959. Maroc 1956–1958, 160, Folder 1959. AD.

37 Rabat to Paris, November 4, 1959. Report of visit from Moroccan foreign minister. Maroc 1956–1958, 160, Folder 1959. AD.

38 O'Driscoll, "Explosive Challenge."

39 Panchasi, "'No Hiroshima in Africa,'" 95.

40 French translation of Arabic text broadcast on November 25, 1959, 19h45, Radiodiffusion Nationale du Maroc, "La Voix du Sahara du Maroc." Rabat to Paris, November 30, 1959. Maroc 1956–1958, 160, Folder 1959. AD.

41 *Foreign Relations of the United States, 1958–1960*, vol. 8.

42 Parodi, Rabat to Paris, January 19, 1960. For reference to images of Hiroshima, see Parodi, Rabat to Paris, January 16, 1960. Maroc 1956–1958, 160, Folder 1960. AD.

43 Rabat to Paris, January 20, 1960. Maroc 1956–1958, 160, Folder 1960. AD.

44 Parodi to Paris, January 21, 1960. Maroc 1956–1958, 160, Folder 1960. AD.

45 François de Rose, Paris, to Alexandre Parodi, Rabat, January 26, 1960. Maroc 1956–1958, 160, Folder 1960. AD.

46 Text of Radiodiffusion Nationale du Maroc, from January 18, 1960. Rabat to Paris, January 21, 1960. Maroc 1956–1958, 160, Folder 1960. AD.

47 Parodi, Rabat to Paris, January 21, 1960; Parodi, Rabat to Paris, January 22, 1960 (received at 11:20 pm). Maroc 1956–1958, 160, Folder 1959. AD.

48 Couve de Murville to Parodi, January 23, 1960. Maroc 1956–1958, 160, Folder 1960. AD.

49 François de Rose to Rabat, January 22, 1960. Maroc 1956–1958, 160, Folder 1960. AD.

50 Parodi, Rabat to Paris, January 29, 1960. Maroc 1956–1958, 160, Folder 1960. AD.

51 Jean Le Roy, Rabat to Paris, February 1, 1960. Folder 1960. AD.

52 Rabat to Paris, retransmission from Oujda, Souke El Arba consulates, February 13, 1960; retransmission from Casablanca consulate, as well as report from Rabat, February 14, 1960. Maroc 1956–1958, 160, Folder 1960. AD.

53 Jean Le Roy to Jean Basdevant, February 26, 1960. Maroc 1956–1958, 160, Folder 1960. AD.

54 SDECE report, February 15, 1960. Maroc 1956–1958, 160, Folder 1960. AD.

55 Rabat to Paris, February 15, 1960. Maroc 1956–1958, 160, Folder 1960. AD.

56 Couve de Murville, Paris to Rabat, February 15, 1960. Maroc 1956–1958, 160, Folder 1960. AD.

57 Note au sujet des reactions marocaines à la suite de l'explosion de la bombe atomique française. February 17, 1960. Maroc 1956–1958, 160, Folder 1960. AD.

58 SDECE reports, February 15 & February 21, 1960. Maroc 1956–1958, 160, Folder 1960. AD.

59 Parodi, Rabat to Paris, February 16, 1960. Maroc 1956–1958, 160, Folder 1960. AD.

60 Parodi, Rabat to Paris, March 16, 1960. Maroc 1956–1958, 160, Folder 1960. AD.

61 Parodi to Couve de Murville, March 1, 1960. Maroc 1956–1958, 160, Folder 1960. AD.

62 Letter of Lefebvre, Jouannet, Faure (Pasteur), Moreau, Denis, Nataf, to the President of the Republic. March 18, 1960. Maroc 1956–1958, 160, Folder 1960. AD.

63 Tripoli to Paris, March 8, 1960. The rumor also appeared in the French-language press, apparently started by a mentally unstable naval officer stationed in Morocco. See LeRoy, Rabat to Paris, March 28, 1960. Finally, two months later, the notion also appeared in a German technical journal. See Bonn to Paris, April 23, 1960. Maroc 1956–1958, 160, Folder 1960. AD.

64 See Austin Cooper, "The Tunisian Request: Sahara Fallout, US Assistance and the Making of the International Atomic Energy Agency," *Cold War History* 22, no. 4: 1–30.
65 Le Roy, Rabat to Paris, March 28, 1960. Maroc 1956–1958, 160, Folder 1960. AD.
66 Alexis Hartel, French Consul in Mainz, to French Ambassador in Bonn, May 20, 1960. Maroc 1956–1958, 160, Folder 1960. AD.
67 J. Labeyrie to Randet, September 15, 1961; Roger Seydoux (ambassador) to Couve de Murville, February 24, 1962; Jean Renou to Jacques Martin, April 12, 1962. Affaires atomiques-380-1955–1972, Folder: Relations avec la France, coopération scientifique, 1959–1971. AD.
68 Edward Gamarekian, "French Atom Cloud Floats Over Mideast," *Washington Post*, February 16, 1960. A6. Notably, other monitoring stations, including ones stationed in Italy, detected fallout from the test. For what the archives have recently divulged on the Commissariat à l'Energie Atomique, atomic testing, and fallout, see Austin Cooper's chapter in this volume.
69 De Gaulle to King Mohammed V (transmitted through Moulay Hassan), March 29, 1960. Maroc 1956–1958, 160, Folder 1960. AD.
70 Rabat to Paris, March 18, 1960. Maroc 1956–1958, 160, Folder 1960. AD.
71 Parodi, Rabat to Paris, April 4, 1960. Maroc 1956–1958, 160, Folder 1960. AD.
72 Rabat to Paris, report of 1:00 pm broadcast of Radiodiffusion Nationale du Maroc, April 1, 1960. Maroc 1956–1958, 160, Folder 1960. AD.
73 King Mohammed V to Charles de Gaulle, April 19, 1960. Maroc 1956–1958, 160, Folder 1960. AD.
74 Maurice Vaïsse, *La grandeur: Politique étrangère du général de Gaulle (1958–1969)* (Paris : Fayard, 2014), 474.
75 Roger Seydoux, Report on Lunch with Prince Moulay Hassan, Ahmed Guedira, and General de la Chenlière. January 13, 1961. Maroc 1956–1968, 156. Situation Politique Expériences Nucléaires au Sahara. AD.
76 Margaret Roberts, "Summitry at Casablanca," *Africa South* 5, no. 3 (April 6, 1961): 68–74.
77 Seydoux, Rabat to Paris, April 24, 1961. Maroc 1956–1968, 156. Situation Politique Expériences Nucléaires au Sahara. AD.
78 See the Moroccan Minister of Information's statement reported in Seydoux's note, Rabat to Paris, October 21, 1961. Maroc 1956–1968, 156. Situation Politique Expériences Nucléaires au Sahara. AD.
79 See P. Fouchet, Direction Générale des Affaires Culturelles et Techniques, January 16, 1962. Maroc 1956–1968, 156. Situation Politique Expériences Nucléaires au Sahara. AD.
80 Vaïsse, *Le grandeur*, 475. See also Edouard Moha, *Histoire des relations franco-marocaines, ou, Les aléas d'une amitié* (Paris: J. Picollec, 1995).
81 Bruno Barrillot, *Les Irradiés de la République: Les victimes des essais nucléaires français prennent la parole*, vol. 269 (Brussels: Editions Complexe, 2003).
82 Byrne, *Mecca of Revolution*, 185.
83 Alain Raymond, "No Atom Tests, Ben Bella Says," *Washington Post*, November 4, 1962, K7.
84 Gorse, Algiers to Paris, March 14, 1963; Gorse, Algiers to Paris, March 16, 1963. Maroc 1956–1968, 156. Situation Politique Expériences Nucléaires au Sahara. AD.
85 Note pour le ministre, March 16, 1963. Maroc 1956–1968, 156. Situation Politique Expériences Nucléaires au Sahara. AD.
86 Leusse, Rabat to Paris, March 16, 1963. Maroc 1956–1968, 156. Situation Politique Expériences Nucléaires au Sahara. AD.
87 Gorse, Algiers to Paris, March 16, 1963, received at 11:00 pm, three pages. Maroc 1956–1968, 156. Situation Politique Expériences Nucléaires au Sahara. AD.
88 Leusse, Rabat to Paris, telegrams received at 5:30 and 7:00 pm. Maroc 1956–1968, 156. Situation Politique Expériences Nucléaires au Sahara. AD.

89 Gorse, Algiers to Paris, March 18, 1963, received at 8:20 pm. Maroc 1956–1968, 156. Situation Politique Expériences Nucléaires au Sahara. AD.
90 Leusse, Rabat to Paris, March 18, 1963, Telegram no. 1321. Situation Politique Expériences Nucléaires au Sahara. AD.
91 Leusse, Rabat to Paris, March 18, 1963, Telegram no. 1321. Situation Politique Expériences Nucléaires au Sahara. AD.
92 Leusse, Rabat to Paris, March 20, 1963. AD.
93 Gorse, Algiers to Paris, March 20, 1963. AD.
94 See Gorse, Algiers to Paris, March 16, 1963. AD.
95 See Benjamin Stora, *La gangrène et l'oubli* (Paris : La Découverte, 2005), 325 as well as Byrne, *Mecca of Revolution*, 154.
96 James McDougall, *A History of Algeria* (Cambridge: Cambridge University Press, 2017); Austin Cooper, this volume.
97 Frederick Cooper, "Possibility and Constraint: African Independence in Historical Perspective," *Journal of African History* 49, no. 2 (2008): 169.
98 Joe Lockhard, "Desert(ed) Geographies: Cartographies of Nuclear Testing." *Landscape Review* 6, no. 1 (2000): 3–20.
99 Trevor Findlay, "Explaining Australasian Angst: Australia, New Zealand and French Nuclear Testing," *Security Dialogue* 26, no. 4 (1995): 374.

"Carrying the Can for Chernobyl"

Visualizing Radioactive Contamination in Sheep in North Wales

JOSHUA McMULLAN

On April 26, 1986, an estimated 14,000 petabequerels (PBq), or roughly 14 billion-billion becquerels (Bq), of radiation was emitted from reactor 4 at the Chernobyl Nuclear Power Plant, the worst nuclear accident to date. In the immediate aftermath of the disaster, radiation was detected over the United Kingdom. A notable amount of this radiation was unevenly deposited across the country, with especially high levels of radioactive contamination in North Wales and Scotland.[1] This proved especially troubling for sheep farming communities in North Wales who faced growing concerns over the safety of foodstuffs produced there. For these communities there was significant anxiety over whether their sheep would become too contaminated for human consumption, with many farmers fearing that they would be unable to sell their sheep at market value, or that the slaughter of their sheep would be prohibited.[2]

Public perceptions of nuclear power in turn became the focal point of the response for Prime Minister Margaret Thatcher and pro-nuclear policy makers such as Lord Walter Marshall, chairman of the nationalized Central Electricity Generating Board (CEGB).[3] To calm public anxiety, the Thatcher government directed the Ministry for Agriculture, Fisheries and Food (MAFF) to initiate a program to mark sheep with paint to indicate whether they were over the limit of radioactive contamination. These animals, grazing in the open air, were then visible to the local public, and to the wider population, through media reporting. Changes in color over time marked the decline in contamination and served as constant reminder of the system in place to prevent contaminated sheep meat entering the human food chain, a system called the Mark and Release scheme. For farmers, the government also initiated a compensation scheme, to recompense them for financial damages arising from the decision to allay public anxiety by taking meat out of the food chain and, in doing so, communicating a commitment to the long-term future of the industry.

Together, these government measures in the wake of the Chernobyl disaster can be seen as different forms of science communication, underpinned by the visual articulation of what was deemed safe. The government sought support for its policies from an anxious public, who were already wary of nuclear weapons and energy and radiation. The Chernobyl incident occurred after a period of intensified Cold War tensions, leading to increased support for nuclear disarmament, with the Labour Party calling for unilateral disarmament in their 1983 manifesto, a revitalized Campaign for Nuclear Disarmament (CND), and the development of grassroots anti-nuclear movements.[4] This included local authorities in cities such as Manchester, Leeds, and Leicester declaring themselves "nuclear [weapon] free cities." Local authorities in Wales similarly declared the country as the first European "nuclear [weapon] free zone" in 1982.[5] However, as this example attests, the common specter in the 1980s British nuclear debate was the cataclysm of nuclear warfare more than the perils of a nuclear accident.[6] The essay shows that the debate was wider ranging than just weapons or even the impact of a nuclear disaster on human health.

Scholars of this era in UK nuclear history have similarly focused on nuclear weapons and war, though debates about nuclear power are gaining greater historiographical attention.[7] In the case of Wales, scholars have linked nuclear power and weapons to welfare discourse and Welsh nationalism.[8] Sean Martin's recent intervention on the politicization of Chernobyl, in a Welsh context, for instance, provides a useful insight into the political impact Chernobyl had on the debates arising from civil nuclear power in Wales after Chernobyl.[9] Other scholars, such as Jonathan Hogg and Christoph Laucht, have also broadened their analysis away from nuclear weapons discourse toward exploring "British nuclear culture."[10] This essay builds on such work to show the role of government bodies in shaping public discourse about nuclear power, telling us more about how the government reacted to technological disasters.

This essay discusses the regional case study of North Wales, showing how the British government tried to make visible the radioactive contamination of sheep flocks in Wales.[11] Brian Wynne, in particular, has drawn attention to the failure of government planners to appreciate the expertise of farmers and to accommodate their local, contextual knowledge into containment plans.[12] Similarly, this essay focuses its attention on the impact of the Chernobyl fallout on sheep farmers, but also places a greater emphasis on understanding the interactions between farmers and the state. In North Wales, officials navigated a complex situation in which they needed to ensure consumer confidence in foodstuffs while also ensuring the continued viability of the North Wales economy, which was heavily dependent on sheep farming. This drove them to try to bolster public confidence in the future of nuclear power as a safe means of producing electricity that would not negatively impact human health, but also crucially, the suitability of the region around the nuclear plants for agriculture. As the fallout from Chernobyl spread far and wide, it reinforced the notion that fallout from nuclear power would not respect or recognize traditional borders of nations.[13] What in other circumstances would have been seen as a national problem, or easily dismissed in the West due to its occurrence behind the "Iron Curtain," was very quickly acknowledged, in the words of Ursula K. Heise, as a European-wide, "transnational risk scenario."[14] This blurring of national borders also posed an additional challenge to the UK government, as it attempted to evaluate the potential health impacts. By making sheep the focal point of their monitoring program, the sheep become the "canary in the mine."[15] Consequently, Welsh sheep flocks became not only a visual indicator of radioactive contamination but also a visual indicator of the risk of nuclear fallout to public health.

I point out here that by focusing attention on contaminated sheep, the government's role served to shift the public narrative about Chernobyl from a debate over the future of nuclear power to a very different set of concerns focused primarily on socioeconomic considerations related to financial losses incurred by sheep farmers. First, I explore the initial response to the crisis to show how the issue was framed by the farmers and by government. Second, I show how the Mark and Release scheme visualized radioactive contamination and later the decontamination of Welsh sheep. And third, I analyze the purpose of the compensation policy, showing how it shifted the narrative away from the health impacts of radiation to socioeconomic concerns. In turn, by focusing so much attention on the impacts of radiation on sheep, the political narrative moved away from the question of nuclear power itself and instead toward the financial difficulties faced by farmers. This is significant as it is a clear attempt by government to alter public discourse and perception of radiation away from narratives that placed it in opposition to government policy. This work is an effort to reveal public discourse and the way it was shaped by government actors such as the Welsh Office and the CEGB.

The Initial Response: Complacency, Paranoia and Restrictions

Although the Chernobyl "cloud" that reached the UK at the beginning of May contaminated many parts of the country, the topography of some regions meant that radioactive fallout was retained in the ground, and thus the food chain, for a longer period. Some of the uplands of northwest Wales, especially in parts of Snowdonia (bordering on Trawsfynydd nuclear power station), the Denbigh Moors, and the isle of Anglesey (which housed the Wylfa nuclear power station), were particularly hospitable to this radioactivity due to the climate, soil type, and drainage factors.[16] This meant that the contamination, and the need for monitoring, remained for decades. The initial reaction from the British government was one of calm, describing the contamination level as "very low" and posing "no health risk to the public."[17] The radiation was reported in North Wales on May 5, 1986, with local newspapers reporting claims that some people felt sick. This included Brian Rearden, a former mayor of Builth Wells and his wife, Anne, who claimed to be sick with headache and nausea caused by the radiation.[18] It is not clear whether these illnesses were caused by radiation, but the very existence of these articles illustrates the growing concerns produced by fears of incidental irradiation.

Monitoring in North Wales began on May 16, 1986, following a Welsh Office announcement. The data was collected by the CEGB, UK Atomic Agency, and the National Radiological Protection Board (NRPB) across multiple sites (including Trawsfynydd and Wylfa), although their preparedness for monitoring radiation had been brought into question. Dr. J. A. V. Pritchard, a scientific adviser at the Welsh Office, wrote to civil servants that pre-Chernobyl he believed that the CEGB had whole-body radiation counters at both power stations. This proved to be inaccurate, and since none were available at Ysbyty Gwynedd, the regional hospital, either, there were clearly insufficient monitoring preparations in place if a comparable incident emerged at either of the nuclear power stations.[19] Friends of the Earth (FOE) also disputed the accuracy of the data presented to the Welsh Office by these groups. They claimed that despite a close agreement between their data and official data there were "in several cases . . . discrepancies which raise questions about the sampling technique employed by the official monitoring organizations."[20] They also claimed to have found high levels of contamination outside of restricted zones. However, the FOE does not support the claim that Trawsfynydd was a major source of contamination, given their claims that all their samples produced a "Chernobyl signature."[21]

The government's reassurances failed to qualm the anxieties of the public, who, when looking at the wider reaction, were frustrated by the indifference of Thatcher's government toward the Chernobyl disaster. This was especially so, given the reactions of other countries to the disaster. Poland, for instance, had given out iodine tablets to children, while countries including Germany, Italy, and the Netherlands had imposed restrictions on the sale of agricultural produce.[22] The actions of other countries were also noticed by print and audio media outlets. However, some representatives within the UK industry questioned the actions of other countries. For example, John Dunster, director of the NRPB, questioned the validity of the images coming from the Soviet Union at the time, showing Geiger counters being used to show vegetables having no radiation on them. He argued that this was a poor way of showing radioactive contamination, for, if a counter could pick up radiation on the surface of a crop, then it "was in serious trouble."[23] Nonetheless, public information telephone lines set up in response to the crisis were "inundated," which "seriously hampered communications."

Meanwhile, internal memos presented to the Prime Minister's Office claimed that the "ill-coordinated nature of the information and advice aroused rather than calmed public anxiety."[24] Government scientists also created further confusion on May 6, 1986 by advising the public not to drink rainwater. Dunster claimed that for those who did drink rainwater, it would be "not bad at all" and it was not "very serious" but that they should still avoid doing so.[25] Nevertheless, some questioned that if rainwater was contaminated, would other water sources not be contaminated as well? As time passed, monitoring of areas and foodstuffs impacted by the fallout continued to reveal high levels of radioactivity, seemingly confirming such anxiety, and it became clear to government ministers that they had underestimated the problem.

Thatcher's government perceived public anxiety as the significant problem arising from the Chernobyl crisis, realizing that action was urgently needed to counter growing concerns around the country. On June 20, 1986, seven weeks after the incident, Michael Jopling, the minster for MAFF, thus announced that livestock with radiation levels over 1,000 Bq/Kg would be prevented from entering the food chain. Restrictions were also placed upon the movement and slaughter of sheep in regions that were registering a cluster of high measurements, such as Cumbria, North Wales, and parts of Scotland. Portrayed as a short-term measure, Jopling initially placed these restrictions for just twenty-one days, which would be "reduced as soon as the monitoring results . . . confirm the expected fall in levels."[26] However, these levels did not fall as quickly as envisioned, and the initial groundswell of concern was in part due to the government's complacency over, first, admitting there was a problem with radiation levels and, second, identifying how to deal with it.[27] When radiation levels did not fall in areas, the need for a longer term policy of restriction, monitoring, and gradual derestriction—including a policy to reimburse farmers for the financial losses incurred—was identified as a necessary economic intervention. Indeed, even Thatcher complained of the government's apparent complacency, stressing that action was required. In response to a briefing on whether to publish contingency plans for potential overseas nuclear accidents in the future, for instance, Thatcher noted that "I do not like this plan—it sounds far too complacent." Thatcher instead wanted to hold off on publishing the plan as it was an election year and there were fears anti-nuclear groups would "have an opportunity to make the government look silly."[28]

As the radiation level did not decrease, and fears over its longevity and impact on public health persisted, sheep in the UK became the embodiment of this fear of radioactivity. They were portrayed as the British victim of Chernobyl, while farmers were the ones who would carry the financial burden. This view was aptly surmised in a press release from the Farmers Union of Wales (FUW), which stressed that Welsh farmers were "carrying the can for Chernobyl."[29] This attitude was also evident in a letter sent to the MAFF. A Mr. M. P. Watkins described requesting New Zealand lamb in the knowledge that "lamb from North Wales was not fit for consumption."[30] However, Mr. Watkins's butcher sought to refute this belief, claiming Welsh lamb was in excellent condition, and increasingly cheap because of discriminatory consumption, handing Mr. Watkins an explanatory pamphlet detailing the quality and cheapness of Welsh lamb.[31] The existence of this pamphlet shows how local businesses thought it necessary to launch their own information campaign to highlight that local sheep were safe for consumption and that the government was the regulatory body that ensured this safety. Keith Best, local Conservative MP for the constituency of Ynys Môn, also called for the rehabilitation of Welsh lamb and argued that it was the responsibility of government to do so.[32]

Sheep took the brunt of nuclear fear and, consequently, the character of Welsh lamb as a high-quality meat suffered. The entire sheep population came to symbolize the impending threat of nuclear power, leading to growing disdain throughout the UK. In turn, it was this disdain that served as the genesis of the Mark and Release scheme. Through this scheme, the government sought to work with farmers to make visible to the public that the sheep were, in fact, safe to consume. The Welsh Office, however, was in a conflicted position. On the one hand, it was responsible for implementing the restrictions on animals under the direction of MAFF out of a public health rationale. On the other, these measures were equally aimed at retaining and restoring market confidence in British agriculture after concerns were raised from producers.[33] Officials therefore needed to minimize fears about the impact of the Chernobyl fallout on food supply.

In balancing these conflicting demands, officials in Wales at times found themselves in contradictory positions. For example, in September 1986, a Welsh Office press statement announced that there was "no hazard to the health of sheep and cattle grazing the most heavily contaminated pastures in the UK," while simultaneously affirming the continuation of restrictions on the sales of sheep that had grazed over swaths of North Wales.[34] Although farmers understood the necessity for the restrictions due to health concerns, the contradictory language at times caused some confusion. If the sheep themselves were safe, why was there a ban in the first place?[35] Welsh Office officials responded by saying that the limit of 1,000 Bq/Kg was a "conservative figure" that "incorporates a substantial safety margin."[36]

The Mark and Release Scheme: Visualizing Radioactive Contamination

From June 20, 1986, all movement of sheep beyond their home pasture was forbidden on irradiated farms in North Wales, Cumbria, and the Scottish Highlands. However, on discovering that irradiated sheep quickly became decontaminated after grazing on clean pastures, moving sheep to clean pastures became routine policy. Within this monitored movement to unpolluted pastures, sheep exceeding the government's prescribed irradiation limit were distinguished from those under the limit by green paint. These green-marked sheep were identified as irradiated and unsuitable for slaughter, whereas unmarked sheep were deemed suitable for consumption. The government continued to monitor the sheep every three months. If a sheep exceeded the official irradiation limit a second time, it was marked with apricot paint, and thus the sheep marked green were no longer the contaminated bodies and were released from restrictions. This system of re-monitoring continued seasonally, working from green to apricot, apricot to blue, and blue back to green, to establish visually which sheep were supposedly unsuitable for human consumption.[37] Marked sheep could be sold but only for further feeding and not for slaughter outside of restricted areas, an action that some farmers hoped would be profitable. For example, David Ford, a former civil servant, mentions how sheep from the Lake District were sold to farmers at a "knockdown price" in Devon in the hope that the sheep would become decontaminated to then be sold.[38]

The restricted areas were also divided into low (LDA) and high (HDA) deposit areas. These were initially identified by green and blue paint respectively. As sheep were not individually tagged for identification, so that their exact place of origin could not be determined, no sheep from any LDA or HDA could be slaughtered until the entire category was derestricted.[39] This decision not to identify the exact place of origin also had political ramifications. Dafydd Wigley, an MP for Plaid Cymru, questioned a press release from the Welsh Office over the location of flocks with the highest radiocesium levels, as well as over an "alleged delay" in obtaining samples of sheep meat in Wales. He also sought advice about the action level for contamination over 1,000 Bg/KG. The first part of Wigley's question was problematic for Welsh Office officials, who did not want to identify individual farms, and so only published results at a county level. They did this because they wanted to "protect individual farmers and to ensure their continued cooperation in the monitoring programme."[40]

The implication here is that if people were able to pinpoint exactly which farms were the most contaminated, then those farms would be discriminated against when they took their produce to market. The government wanted to protect the livelihoods of farmers in contaminated areas, possibly because farmers in the region were a key Conservative voting demographic.[41] Yet by doing this, the radiation within contaminated areas would be obscured and therefore, invisible at a detailed level. The Welsh Office believed that they could make the data "more meaningful" by publishing it on a district level as well, so that it was more precise. However, this data was published in its raw state, with little to no scientific analysis attached. In turn, the Welsh Office seemed to assume that people would be able to read and understand what these large datasets meant, although anti-nuclear groups such as the Welsh Green Party believed that "most of what had been collected was being withheld from the public."[42] This is why the color coding of the sheep scheme is particularly significant because it created for farmers, and for the public, a visual semi-catalogue of where the radiation was and partially what it was impacting. They would not be able to recall which sheep were contaminated, when, by how much exactly, but what they could see is that the numbers of contaminated sheep were dropping as time went on.

The Mark and Release system was introduced due to consumer concern, and there was certainly precedent for such concerns. Although the memory of the Windscale fire in 1957 was distant for many, there were those who would remember the images of milk being poured away in the aftermath of the fire as a health precaution. The foot-and-mouth outbreak of 1967 also provided some precedent for the restrictions that were put in place. During this outbreak, the movement of sheep and cattle were restricted if a farm was found to be infected with the disease, with regular monitoring from veterinary experts.[43] Both episodes gave British officials precedent for putting in place restrictions to maintain public confidence in the food supply in times of crisis. In doing so, however, the Mark and Release scheme placed a great emphasis on ensuring the confidence in the food supply, rather than addressing the broader issue of the safety of nuclear power.

The government and businesses were aware of the concern in the British public over the safety of eating foodstuffs coming from areas where radiation was deposited. For example, South Caernarfon Creamery expressed concern about consumer reaction to milk and milk products. This concern was backed up by drops in sales when nurseries stopped purchasing milk, while other consumers switched from fresh bottled milk to alternative milk products, although in this case, this did not cause a drop in the overall sale of milk products.[44] Despite this stability in sales, anxiety over agricultural produce remained. As more information became available the concern shifted to focus solely on sheep as they grazed on high pasture grounds where most of the radiation was deposited.

One useful impact for the UK nuclear industry and government of the Mark and Release scheme was its ability to communicate visually the decontamination of sheep, and by association, the decline of radioactive contamination. Across Wales, the number of sheep within contaminated zones, and failing the 1,000 Bq limit, fell each year after 1986. In turn, the scheme made radiation, otherwise invisible to the human sense, visible. Or at least partially visible, as the radiation in the ground was represented only indirectly by the markings on the sheep, while any other residual radiation that did not fall on the uplands would remain invisible.

Olga Kuchinskaya postulates that the imperceptibility of Chernobyl radiation by the human senses means that individuals' experience of it is always highly mediated. This meant that governments were able to shape public visibility of the Chernobyl radiation and were able to prevent the construction of links between radiation and its health effects on people, while focusing the attention on radioactive contamination in sheep. Kuchinskaya believes that this in turn can lead to "the social construction of ignorance."[45] One can see a similar phenomenon in the case of contaminated Welsh sheep, as government policies shaped public perceptions of radioactive contamination, breaking the link between radiation and its impact on human health by focusing attention on the decrease in contamination levels in sheep. As Table 12-1 shows, the number of sheep that were being tested and the percentage of sheep failing those tests was falling each year. The people living in North Wales, then, were able to visualize how these statistics played out before them, as the number of marked sheep dropped in restricted areas as they were moved out to be sold at market. Furthermore, marking the sheep in different colors allowed people to distinguish between those sheep that were irradiated in LDAs and HDAs, and those that were safe under the regulations. Over time, this enabled the government to demonstrate that the amount of radioactive contamination was dropping and that the concern for public health could safely subside. However,

Table 12-1. Sheep Failing Live-Monitoring in Wales

Year	Tested	Failed	Percentage of Failed Tests
1987	175,454	943	0.5
1988	165,382	616	0.3
1989	158,896	233	0.1
1990	137,389	161	0.1

Source: TNA, BD 94/4. "Sheep Failing Live-Monitoring in Wales."

what they were not able to see was the impact radioactive contamination was having on their own health and the wider environment around them.

The Welsh Office had to balance the need to be seen to maintain public safety while ensuring that the scheme to derestrict sheep allowed for sheep to go back to market. On 28 February 28, 1987, sheep in HDAs that had kept their blue paint after the seasonal round of tests were to be derestricted. Farmers were given the choice whether sheep marked with blue paint remained restricted or whether they would sell them to market. The official advice from the government was that if the farmer was in any doubt, then they should keep the sheep under restrictions.[46] However, this was really not a choice for farmers, as keeping them restricted meant keeping them from going to market. So by giving farmers the illusion of this choice, it made it seem as though farmers had confidence in the government's Mark and Release scheme, that radiation levels were indeed coming down and that things were getting better. When people saw sheep returning to the market, it would give the impression that things were getting back to normal.

The risk of nuclear power was being reduced in the public's imagination as they saw its worst impacts mitigated. There was, however, a fundamental gulf between what the government hoped to achieve with the Mark and Release scheme and the experiences of farmers participating within it. For example, it was not guaranteed that after each period of monitoring a noticeable number of sheep would be derestricted. Similarly, the desire to derestrict as many sheep as possible did not happen at every monitoring period. Although this does show that the Welsh Office was keen to reduce the number of restricted sheep, probably to show the decline in radiation and to reduce the number of claims, it is not the full picture of the approach that the Welsh Office took throughout this period.

In telex messages from regional offices in Wales to the central Welsh Office, we can see how civil servants were encouraged to push farmers toward derestricting sheep as quickly as possible.[47] This allowed MAFF and the Welsh Office to visually articulate how the radioactive contamination in sheep was decreasing, and that lamb was safe to consume again. It also crucially communicated to farmers that the damage caused by the radioactive fallout was not permanent and that the damage caused to their livelihoods would not last forever. Finally, it also reduced the financial costs to the government, given that the fewer sheep that were restricted, the fewer compensation claims there would be. This is another reason why this very visible method of communication was valuable to the government. Even though it made other areas, things and people irradiated by Chernobyl "invisible," visibly regulating this aspect of local communities allowed the government to be seen as doing something.

Shifting the Narrative: Financial Difficulties, Compensation, and the Invisibility of Radiation?

It was not until July 24, 1986, that a compensation policy that covered the loss of income experienced by farmers came into force.[48] There was a compensation policy in place before then, enshrined by the Nuclear Installation Act (1965) and the Wildlife and Countryside Act (1981). However, the National Farmers Union (NFU) complained in a press statement that the compensation policy introduced in 1986 was insufficient, and that there was still a lot of uncertainty around the duration of the restrictions.[49] Keith Best's personal papers include numerous letters that he received from farmers on the extent of the financial damages caused by restrictions. For example, one farmer claimed to have lost a total of £1,646, as the restrictions entered their fourth week, while another more extreme case saw a farmer, who would have normally sold 200 lambs at £40 each, lose £8,000.[50]

Additional pressure on the government came from the FUW. In a report evaluating the government's response to the Chernobyl Incident, the FUW published a report in 1988 that recommended that the government stop the development of civil nuclear power in the UK and instead invest in alternate energy sources and energy conservation.[51] As Martin's article on the politicization of Chernobyl mentions, there was already a certain anti-nuclear Welsh pride that had been established by the mid-1980s. This was evidenced by the existence of a decentralized Green Party in Wales, CND Cymru, as well as the anti-nuclear views of several Welsh Labour politicians, and by a left-leaning, environmentalist surge in Plaid Cymru.[52] Afterall, the FUW was not the only organization in Wales to prefer alternative energy sources and energy conservation; Plaid Cymru, the Welsh nationalist party, adopted this exact policy in their 1983 general election manifesto.[53]

At a time when the government and nuclear industry wanted to expand nuclear power in Wales, compensation thus had two purposes. First, it sought to cauterize the perceived political damage caused by the incident. This included the Anglesey Conservative Association writing to Nicholas Edwards and senior Conservative politician Norman Tebbit, to express concern about the potential "adverse political effects" to Best and the constituency.[54] Second, it also sought to quantify the economic costs of the risk of radioactive contamination, to divert attention away from the cause of the risk, nuclear power. The policy would end up costing the UK taxpayer up to £3 million to farmers in Wales.[55] However, the FUW report was published in 1988, which indicates that while the narrative was shifting to socioeconomic concerns, nuclear power remained very much visible to farmers unions who traced their problems back to it.

The NFU presented a more positive perspective on the compensation arrangements, stating that "the government got it right for nearly everybody," and that it had worked effectively for those willing to engage with them.[56] However, there were other moments when farmers criticized the government's approach dependent on whether they were still facing restrictions. This reopened the debate on compensatory measures, which, when compounded by apparently "unacceptable delays" in meeting claims, led to further agitation from farmers, culminating in a mass picket outside the Welsh Office in London in late September 1987 to show their continuing dissatisfaction at the government's response.[57] What this suggests is that the financial response to farmers was inconsistent, which led to the mixed response from farming unions.

Keith Best, Conservative MP for Ynys Môn, played a particularly key role in pressuring the government to extend its compensation policy. In a letter to the secretary of state for Wales, Nicholas Edwards, Best detailed what the feeling was among local farmers:

> The situation is extremely serious and is the worst crisis yet to hit the sheep farmers. . . . As far as Anglesey is concerned it could not have happened in a worse month when, because Anglesey sends its lambs to market at this time. . . . Lambs become too fat, lose subsidy and become unsaleable . . . quite beyond any reluctance of the consumer to buy them when the ban is finally lifted. Welsh Office claimed that lambs would not be ready for market, yet the lambs are ready for market now and are rapidly putting on weight. Such statements have caused grave misgiving among my farmers as to whether Welsh Office officials have a true picture of the gravity of the situation on Anglesey.[58]

This letter illuminates the position that Best, the farmers, and the government were in. They felt that this was a serious situation and that farmers were not solvent enough to sustain financial losses. The timing also made matters worse, with restrictions occurring at a peak season for the selling of lamb. This led to significant political pressures on government, giving it the impression that it could not allow a major industry in the North Wales region to be irreparably damaged. This letter also shows the divergence between the view of the government and the view of the farmers over whether the lambs were ready for market. Here we can see evidence that supports Wynne's arguments made in relation to Cumbria, that the government did not consider the farmers' own knowledge about the location their sheep would graze in.[59] This also caused distrust between the farmers and the Welsh Office, which farmers believed was seeking to cover up the true position that farmers were facing.

Farmers became quite cynical about compensation, with some suggesting that the government was holding back because North Wales enjoyed a higher standard of living than the rest of the country and so the government was trying to bring it in line with the national average.[60] They appeared to believe that people would be willing to buy the products once the restrictions were lifted. Still, there was some concern surrounding the future of consumer confidence in Welsh lamb. Best went on to say further in the letter how

> there will remain the need to rehabilitate Welsh lamb in the eyes of the consumer. . . . I believe that there is a responsibility on the government to put the public mind at rest as to the safety of consuming Welsh lamb in the future.[61]

Best believed that people were initially willing to buy lamb, but because of the prolonged restrictions, consumer confidence in Welsh lamb continued to drop. Therefore, Best believed that because the government was responsible for the restrictions, it should be the one to ensure the rehabilitation of Welsh lamb so that it might be competitively sold at market. The recognition here was that even when restrictions were lifted, the damage would be long term. First, this helps explain why the updated compensation policy to cover this was introduced. Second, although the public may not have fully grasped the differences between different kinds of radioactive isotopes, their half-lives, or their potential impact on the human or ovine body, they knew of their existence and that some will persist for a long time. This suggests that the updated compensation policy was first and foremost implemented to ensure the economic viability of the region, but it was also a means by which to focus the conversation on the economic hardships faced by farmers, rather than the origins of that economic hardship.

Conclusion

The British government initially believed that the Mark and Release scheme, and subsequent compensation, was a short-term policy aimed at alleviating concern over the spread of radioactive contamination. However, this was far from the case. The policy remained for twenty-six years after the accident, with 344 farms in total remaining under restrictions for the duration of this period, and with high levels of cesium still being recorded up to 2011. Nevertheless, the restrictions aimed at maintaining confidence in Welsh lamb were now seen to be no longer proportionate, with a "very low risk" to the consumer if restrictions were to be removed.[62]

This essay has shown how government intervention shaped the narratives about health risks and radiation in the United Kingdom, away from concerns about nuclear power and toward toward socioeconomic concerns caused by the restrictions placed on the movement and slaughter of sheep. Chernobyl brought home to the UK the dangers of nuclear energy in a tangible way that nuclear weapons, thankfully, never had. Fears over the invisible radiation from Chernobyl fueled apprehension over radiation, and the government worried that the intensified scrutiny would hinder the plans for future nuclear plants, or worse, force the closure of existing nuclear plants including Wylfa and Trawsfynydd. However, what created the most public anxiety was the apparent danger to public health caused by the radioactive contamination in the uplands that sheep grazed on. This led the government to introduce the Mark and Release policy to make the invisible danger's reduction over time visible, as well as the compensation policy to make up for the financial losses incurred by those impacted farmers. These policies had some success in diverting public concerns away from nuclear power and instead focused attention on the socioeconomic impacts of the Chernobyl incident. The socioeconomic impacts of the Chernobyl incident on North Wales were something that they could quantifiably measure and see for themselves and understand. They could not do the same with any potentially negative health impacts that could take decades to present themselves.

Notes

1 Official measurements of cesium-137 as high as 1785 Bq/m2 in Wales. In parts of Scotland, officials measuring cesium deposition from rainfall recorded readings as high as 20,000 Bq/m2. See William. A. Kerr and S. Mooney, "A System Disrupted—The Grazing Economy of North Wales in the Wake of Chernobyl," *Agriculture Systems* 28, no. 1 (1988): 13–27, 20; Brian Wynne, "Sheepfarming after Chernobyl: A Case Study in Communicating Scientific Information," *Environment: Science and Policy for Sustainable Development* 31, no. 2 (1989): 10–39, 13.

2 North Wales is a recognized region in Wales, sometimes simply referred to as Gwynedd. Its principal areas are the Isle of Anglesey, Conwy County, Denbighshire, Flintshire, Gwynedd, Wrexham County, and Powys.

3 Jon Agar, *Science Policy under Thatcher* (London: UCL Press, 2019), 158.

4 Jonathan Hogg, "Cultures of Nuclear Resistance in 1980s Liverpool," *Urban History* 42, no. 4 (2015): 1–19; Mark Phythian, "CND's Cold War," *Contemporary British History* 15, no. 3 (2001): 133–156.

5 Christoph Laucht and M. Johnes, "Resist and Survive: Welsh Protests and the British Nuclear State in the 1980s," *Contemporary British History* 33, no. 2 (2019): 226–245.

6 Daniel Salisbury, *Secrecy, Public Relations and the British Nuclear Debate: How the UK Government Learned to Talk about the Bomb, 1970–83* (London: Routledge, 2020).

7 This is often evident; see, e.g., Frank Barnaby and Douglas Holdstock, *The British Nuclear Weapons Programme, 1952–2002* (London: Routledge, 2003), but also central to sociocultural works, e.g., John Baylis and Kristan Stoddart, *The British Nuclear Experience: The Roles of Beliefs, Culture and Identity* (Oxford: Oxford University Press, 2014).

8 Christopher R. Hill, "Nations of Peace: Nuclear Disarmament and the Making of National Identity in Scotland and Wales," *Twentieth Century British History* 27, no. 1 (2016): 26–50; Johnes and Laucht, "Resist and Survive"; and John Baylis, *Wales and the Bomb* (Cardiff: University of Wales Press, 2019). Historical emphasis on the civil nuclear industry in Wales is virtually nonexistent. For a comprehensive overview of historical writing on modern Wales see M. Johnes, "For Class and Nation: Dominant Trends in the Historiography of Twentieth-Century Wales," *History Compass* 8, no. 11 (2010): 1257–1274.

9 Sean Martin, "Politicising Chernobyl: Wales and Nuclear Power during the 1980s." *Transactions of the Royal Historical Society* 29 (2019): 273–292.

10 Key works include a 2012 special edition of the *British Journal for the History of Science (BJHS)* esp. Jonathan Hogg and Christoph Laucht, "Introduction: British Nuclear Culture," *BJHS* 45, no. 4 (2012): 479–493; Hogg, *British Nuclear.*

11 Olga Kuchinskaya, *The Politics of Invisibility: Public Knowledge about Radiation Health Effects after Chernobyl* (MIT Press, 2014); Wynne, "Sheepfarming after Chernobyl."

12 Wynne, "Sheepfarming after Chernobyl"; Brian Wynne, "May the Sheep Safely Graze? A Reflexive View of the Expert–Lay Knowledge Divide," in *Risk, Environment and Modernity: Towards a New Ecology*, ed. Scott Lash, Bronislaw Szerszynski, and Brian Wynne (SAGE, 1998), 44–83. There is a chapter on sheep farming in Cumbria in Harry Collins and T. Pinch, *The Golem at Large: What You Should Know about Technology* (Cambridge: Cambridge University Press, 2009). Collins and Pinch argue that suspicions of scientists stem from the over-confident statements that eroded the view that science was "all about certainty and science being a political conspiracy."

13 The notion that fallout from nuclear power or indeed weapons would not respect national borders was not new. As seen in John G. Fuller, *The Day We Bombed Utah* (Penguin, 1985).

14 Ursula K. Heise, *Sense of Place and Sense of Planet: The Environmental Imagination of the Global* (Oxford University Press, 2008), 177. For a transnational approach to nuclear culture see Laucht, "Transnational Professional Activism and the Prevention of Nuclear War in Britain," *Journal of Social History* 52, no. 2 (2017): 1–29.

15 Fuller, *The Day We Bombed Utah.* Fuller's investigation into the case of sheep dying in Utah and their potential link to radiation from nuclear weapons tests showcases a different outcome for sheep farmers whereby they did not receive compensation as the American atomic authority refuted claims that the tests and radiation killed the sheep.

16 TNA, BD 119/32. "Chernobyl Incident, Memorandum, September 26, 1986."

17 *The Guardian*, May 3, 1986, 1.

18 *Liverpool Daily Post Welsh Edition,* May 5, 1986, 5.

19 TNA, BD 119/32. "Memo from Dr. J. A. V Pritchard, Scientific Adviser Welsh Office on Medical Physics North Wales: Post Chernobyl, August 1986."

20 TNA, POWE 74/452. "Correspondence with Friends of the Earth Ltd, Fallout over Chernobyl: A review of the official monitoring programme in the UK. May 27, 1987."

21 Ibid., 31.

22 TNA, PREM 19/3656. "Chernobyl—Soviet Nuclear Accident." TNA, MAF 691/4. "Sheep Compensation Scheme: Chernobyl Compensation Policy."

23 John Dunster, interviewed by Paul Maurice, "John Dunster on Chernobyl Fallout" [Radio Broadcast], LBC, May 6, 1986. The NRPB would later produce several information leaflets on the issue of radiation protection and monitoring.

24 TNA, PREM 19/3656.

25 John Dunster, interviewed by Paul Maurice, "John Dunster on Chernobyl Fallout" [Radio Broadcast], LBC, May 6, 1986.

26 HC Parliamentary Debates, June 20, 1986, 99, cols. 1333–1338.
27 TNA, BD 85/73. "Welsh Office press statement, May 7, 1986." Initially the government only advised against drinking rainwater and only began monitoring in Wales on May 16, 1986.
28 TNA, PREM 19/3656. "Chernobyl—Soviet Nuclear Accident."
29 TNA, BD 85/45/2.
30 TNA, MAF 298/190/1.
31 Ibid., NFU pamphlet, no date given.
32 The National Library for Wales, Agriculture Keith Best 2. "July 1, 1986, Letter to Nicholas Edwards MP regarding the local agriculture economy."
33 TNA, BD 85/53. "Chernobyl nuclear incident, Soviet Union: telex messages received from agriculture regional offices."
34 TNA, BD 119/32. "Chernobyl incident: monitoring of radiation levels in agricultural produce in North Wales; risks to public health—Welsh Office News, September 26, 1986."
35 National Library of Wales (NLW), KB 21. Chernobyl Radioactivity Lamb, G. Moss Jones, letter to Keith Best, July 7, 1986.
36 TNA, BD 119/32.
37 TNA, MAF 298/190/1. "Chernobyl nuclear accident: response of the Ministry of Agriculture, Fisheries and Food; radioactivity monitoring, sheep restrictions."
38 *London Review of Books* 41, no. 1 (2019): 3.
39 Sheep in all Low Deposit Areas were derestricted on the September 29, 1986. Sheep in High Deposit Areas did not begin to be derestricted until February 27, 1987. After that the rotational use of paints came into effect—so blue to green, green to apricot, and then apricot back to blue.
40 TNA, BD 85/72. Chernobyl nuclear incident, Soviet Union: ministerial correspondence. Letter from H. R. Bollington to Secretary of State providing advice on how to reply to a letter from Dafydd Wigley MP, July 23, 1986.
41 Martin, "Politicising Chernobyl."
42 NLW, WGP 4. Minutes, June 7, 1986.
43 L. A. Reynolds and E. M. Tansey, eds., *Foot and Mouth Disease: The 1967 Outbreak and Its Aftermath* (London: Welcome Trust, 2003), 3, 74. In the transcript of the discussion of an expert audience it was mentioned that "the Government has said, through their Ministers this year, that regulations were relaxed too soon in the 1967–68 outbreak." This was disputed by others in attendance. They also believed that restrictions lasted "far longer than we would consider a veterinary necessity."
44 TNA, BD 85/53. "Chernobyl nuclear incident, Soviet Union: telex messages received from agriculture regional offices."
45 Kuchinskaya, *The Politics of Invisibility*, 2.
46 TNA, BD 85/59.
47 TNA, BD 85/59.
48 Farmers could also sell sheep at a loss to farmers in lowland areas. Collins and Pinch, *The Golem at Large*, 119; Farmers Union for Wales (FUW), *Chernobyl—The Government Reaction* (FUW, 1988).
49 NLW, "Chernobyl Radioactivity Lamb Keith Best 21," NFU press statement, July 7, 1986.
50 NLW, Keith Best 21. "Chernobyl Radioactivity Lamb."
51 This is in addition to the pressure from the NFU, the FUW claimed and claims to this day to be "the only agricultural union that exclusively represents the farmers of Wales." While in other matters the FUW believed that the NFU policies were contrary to the interests of Welsh farmers in the response to Chernobyl they were aligned in their view.
52 Martin, "Politicising Chernobyl," 290.

53 Plaid Cymru, *Plaid Cymru—The Only Alternative: General Election Manifesto 1983: Agenda Wales, Peace, Jobs and Justice* (Plaid Cymru, 1983).
54 NLW, KB 21. M. H. Norris to Nicholas Edwards, July 1, 1986.
55 HC Parliamentary Debates, November 6, 1987, 122, col. 872.
56 House of Commons Agriculture Committee, Second Report, *Chernobyl: The Government's Reaction*, vols. 1–3 (1988).
57 *The Guardian*, September 29, 1987, 3.
58 NLW, Keith Best Papers, Agriculture Keith Best, 2; July 1, 1986 Letter to Nicholas Edwards MP regarding the local agriculture economy.
59 Wynne, "May the Sheep Safely Graze?"
60 NLW, Agriculture Keith Best, 2.
61 NLW, Agriculture Keith Best, 2; July 1, 1986 Letter to Nicholas Edwards MP regarding the local agriculture economy.
62 BBC News, "Chernobyl Sheep Controls Lifted in Wales and Cumbria," March 22, 2021, https://www.bbc.co.uk/news/uk-wales-17472698 (accessed August 15, 2021).

Diplomatic Fallout

Nuclear Power and Cold War Diplomacy from Three Mile Island to Chernobyl

WILLIAM M. KNOBLAUCH

> Had a call that a Soviet Nuc. Sub. [sic] is on fire several hundred miles No. of Bermuda. Russian ships are on hand—no danger of nuclear accident. Message was from Gorbachev. Seems Chernobyl had an effect.
>
> —Ronald Reagan, Diary, October 5, 1986[1]

By "Chernobyl," President Ronald Reagan was of course referring to the 1986 nuclear disaster near Pripyat, Ukraine, then the worst nuclear accident in human history. However, it was more than altruism that led Soviet General Secretary Mikhail Gorbachev into communicating this recent disaster so quickly with his Cold War adversary. Global political pressure had been mounting since Chernobyl, but much of this pointed rhetoric has remained understudied, often overshadowed by the public upheavals on both sides of the Atlantic over the 1980s arms race. While nuclear weapon development, testing, and deployment were clear examples of saber rattling throughout the Cold War, both sides also used public critiques of nuclear power, and especially nuclear power accidents, as another rhetorical weapon. In short, much attention has been paid to the role that nuclear weapons played in Cold War diplomacy, especially during the Reagan administration.[2] Less attention, however, has been paid to the role of nuclear power in diplomacy.[3]

This essay acts as one corrective to this often overlooked aspect of Cold War relations. It does so by framing the rhetoric and actions of the superpowers in the wake of two nuclear accidents: the 1979 Three Mile Island accident in Pennsylvania and the 1986 Chernobyl disaster. Each event led a Cold War adversary to propagate that these accidents were due to ideological flaws in the others' system. For both the United States and the Soviet Union, nuclear power failures became symbols of not just technological shortcomings, but moral failings. In 1979 and 1986, respectively, the superpowers used nuclear power tragedies as part of an ongoing contest of ideas and ideologies. Assigning blame over nuclear accidents, therefore, became yet another rhetorical weapon in the Cold War.

In studying the role of nuclear power in Cold War diplomacy, I have referenced presidential and news archives, historical scholarship on the Chernobyl and Three Mile Island incidents, and Reagan administration officials' memoirs, including interviews. These final sources prove to be the most difficult to discern objectively. Consider former National Security Council member Jack Matlock's book, *Reagan and Gorbachev*, which asserts that while many administration officials wanted to use Chernobyl to "berate" the Soviet Union, under Matlock the NSC restrained American policymakers from using the tragedy for Cold War purposes.[4] But like many Reagan administration officials' memoirs, distant memories can conflict with the archives. Although Matlock's work remains fairly objective, time can distort the past. The question then, is: What did the historical actors—Carter, Reagan, Gorbachev, Matlock, and their subordinates—think and say *at the time* of these disasters? Examining numerous archival documents reveals that in 1979 and 1986, diplomatic exchanges over nuclear tragedies not only took on a tense Cold War dynamic, but resulted in actual policies that contributed to the Cold War's conclusion. By the end of the 1980s, nuclear power rhetoric led to real and lasting impacts on global security. That process began with the nuclear reactor accident at Three Mile Island.

Three Mile Island

On March 28, 1979, in Harrisburg, Pennsylvania, Unit Two of the Three Mile Island nuclear power plant began to overheat. A broken gauge led engineers to leave open one of the valves used to control the inflow of cooling water. As temperatures rose, so did radioactivity. In response, engineers leaked radiation into the air above Harrisburg and the nearby Susquehanna River, but alarmingly a hydrogen bubble formed atop the core's container. Thankfully, an explosion never occurred. However, because the accident released radioactivity into the plant's surroundings, TMI reawakened debates over the potentials and pitfalls of nuclear power. Politicians had promoted nuclear power as a source of salvation during the 1970s, a decade plagued by an oil embargo and energy shortages, a period in which blackouts darkened New York City and smog hung prominently over Los Angeles. Noting growing public concerns over energy and pollution, Richard Nixon had campaigned in 1972 with promises of achieving American energy independence by decade's end. Adding to the public's interest in nuclear power was the timely release of *The China Syndrome,* a film that seemed to eerily mimic events from TMI. Released just weeks before the incident, *The China Syndrome* turned TMI into an even greater political issue.[5]

TMI quickly became a politicized Cold War issue. The day after the accident, the Associated Press reassured the public that despite the fervor, TMI was, "far from the worst nuclear accident on record," citing CIA reports that in 1958 a Soviet accident near Chelyabinsk had spread radioactivity some 1,000 square miles.[6] In Hanover, West Germany, TMI's wake led an already planned antinuclear protest to feature chants of "We All Live in Pennsylvania!" while in Copenhagen a newspaper referred to the Harrisburg disaster as a "blessing in disguise" for nuclear power's critics. But it was the Soviet Union that was the most critical, reporting from the official news agency TASS that, "thousands of panic-stricken people" in Pennsylvania were fleeing for their lives. Another Soviet press release stated that TMI was the natural "consequence of American capitalistic 'energy monopolies' hunting for profits at all costs." The state-run news program *Today in the World* even proclaimed that TMI "had evoked profound concern and continues to alarm the American public."[7]

Comparisons between the two superpowers' nuclear programs arose almost immediately. Communist newspaper *Pravda* bragged that its Leningrad atomic plant, unlike TMI, was safe.[8] Some days later, the party boasted that it had "perfected safety systems at its nuclear generating plants to prevent any radiation leaks like those at the Three Mile Island reactor." In an English-language broadcast aimed at the West, Deputy Chairman of the Soviet Committee on Peaceful Use of Atomic Energy Igor Morokhov boasted that "Soviet safety norms simply rule out any escape of radiation."[9] In a harbinger of what would occur in 1986 at Chernobyl, US experts lashed back, stating that the Soviets "didn't even bother to build containment structures around their pressurized-water reactors" and concluded that "a Pennsylvania-type accident in the Soviet Union would spew high-level radiation all over the countryside."[10] Such premonitions fell on deaf ears. "The Soviet plans are made and they will not be changed," assured Soviet minister of power and electrification Pyotr S. Neporozhny, who promised to build more 1,000-megawatt plutonium reactors—even after admitting to past Soviet nuclear accidents.[11]

Some Soviet scientists were less sanguine. Nikolai Dollezhal, a chief designer of Soviet nuclear reactors, co-authored "Nuclear Energy: Achievements and Problems," an article that warned that the Soviets had only been able to construct such cost-effective nuclear reactors because they spent less on safety measures; US engineers spent "seven to eight times" the amount on safety. The authors also pleaded that new reactors not be built in the European-adjacent section of the Soviet Union, but farther east, to mitigate climate-changing effects and to place reactors closer to uranium supplies. In response to these commonsense concerns and suggestions, Soviet academics pushed back, falsely claiming that reactors being built in the Eastern Bloc, including Chernobyl, remained safe.[12]

The Soviet's public response to TMI threatened to complicate ongoing arms negotiations. The accident occurred during the tenuous talks for a second Strategic Arms Limitation Treaty (or SALT II) between Soviet General Secretary Leonid Brezhnev and US President Jimmy Carter, the latter seeking Senate approval for treaty ratification. But dissent against SALT II was strong, especially from an emerging cadre of "Neoconservatives" such as Henry "Scoop" Jackson. Writing for Canada's *Globe and Mail,* David Lancashire noted that "Harrisburg may have beneficial side effects. Backers of a second Strategic Arms Limitation Treaty with Moscow feel the nuclear accident will swing public opinion enough to ensure its passage by the Senate, the only remaining obstacle."[13] It was the December 1979 Soviet invasion of Afghanistan, however, that dealt SALT II a fatal blow. Carter removed the treaty from Senate consideration, halted technology and grain exports to the Soviets, and intensified his Cold War rhetoric. By 1981, the American electorate—unhappy with domestic economic challenges and an ongoing hostage crisis in Iran—replaced Carter with Ronald Reagan.

What were Reagan's thoughts on nuclear power? As he left the California governor's office in 1975, and clearly eyeing up a run for the presidency, he made plain his support of "environmentally protected" nuclear power plants. He recognized that there was opposition to the creation of new plants, but argued that by the end of the 1980s, "45 percent" of the state's "electric power must be generated by nuclear power." He pushed for more immediate reforms, because nuclear plants could "reduce our oil needs by 24 million barrels a year and help protect against future blackouts and brownouts." Reagan concluded, "Construction of these and other nuclear units must go forward. Time is running out."[14] Of course, Reagan aimed to persuade potential voters as he took on incumbent Gerald Ford in 1976—a presidential primary campaign he only narrowly lost.

In the immediate wake of the TMI incident, Reagan remained supportive of nuclear power. Just three years previous, he had lambasted anti-nuclear power protesters of the proposed Seabrook, New Hampshire, plant site as "Luddites" and "pseudo-environmentalists." "Nuclear plants can't become bombs," Reagan argued.[15] Now, after TMI and in his successful campaign against Carter, Reagan exclaimed, "Some catastrophe! No one was killed, no one was injured, and there will not be a single additional death from cancer among the two million people living within a fifty-mile radius of the plant."

Carter, a former nuclear submarine engineer who visited TMI in person, had somehow remained in Reagan's shadow. Reagan, a former actor, noted that the only real "fallout" from TMI was a "beneficial kind to one small group of Americans—the cast, crew and investors of the movie *The China Syndrome*."[16] While Reagan continued to support nuclear power as he entered the Oval Office, some five years later a new nuclear catastrophe would again test superpower relations—this time, with a new Soviet premier at the helm.

Reagan and Gorbachev

Mikhail Gorbachev's ascent to Soviet General Secretary in 1985 suggested a turning point in superpower relations. Gorbachev's leadership differed from that of his previous three predecessors—Leonid Brezhnev, Yuri Andropov, and Konstantin Chernenko, who all died in fairly quick succession—as he favored expanding the Soviet economy over maintaining military strength. Reflective of this change, early in his administration Gorbachev made two pronouncements. The first was for *glasnost,* or the need for public and international transparency to improve the Soviet image abroad. The second, *perestroika,* focused on economic restructuring to improve Soviet society; or, as he put it so succinctly, "the elimination of everything that interferes with development," meaning weapons. Gorbachev recognized the drain that the arms race was putting on the Soviet economy, not to mention the ongoing quagmire in Afghanistan, and he sought to change things.[17]

Gorbachev's initiatives became tangible when he offered an olive branch to Reagan regarding the Nuclear Non-Proliferation Treaty (NPT). On July 30, 1985, *Pravda* reported that Gorbachev had volunteered a unilateral moratorium on nuclear testing, a move the National Security Council (NSC) assessed as a "public relations effort timed" to coincide with the ten-year anniversary of the Helsinki Accords, the fortieth anniversary of the US bombing of Hiroshima, and an upcoming NPT review conference. Gorbachev's proclamation put the US on the defensive, with NSC communiqués dispatched to all ambassadors on ways to respond. The Soviets made good on this promise too, not detonating another nuclear device until 1987, while US tests continued.[18] In mid-1985, then, the United States was looking like the more hawkish Cold War superpower.

By late 1985, nuclear weapons, and not nuclear power, remained a central point of contention in Cold War diplomacy. As historian Serhii Plokhy has assessed, "Gorbachev paid much more attention to nuclear arms than to nuclear energy." The same could be said for Reagan, although the two men differed in their approaches. Early in his administration, Gorbachev pleaded that his comrades rethink the ongoing arms race as it "threatened to destroy life on earth many times over." When Gorbachev extended his offer of nuclear abolition, Reagan did not bite. He simply could not shake his faith that his Strategic Defense Initiative (SDI, or "Star Wars") posed no threat to the Soviets; he refused to abandon SDI, which would remain a sticking point for years to come. Although claims that "Star Wars" funding contributed to bankrupting the Soviet Union are overblown, the program did concern Soviet leaders who recognized that they lagged behind US technology. Gorbachev envisioned reallocating Soviet resources away from the arms race and toward economic modernization, and the threat of SDI was, at least on some level, a hindrance to achieving that end.[19]

Gorbachev's *perestroika* goals may have inadvertently contributed to the Chernobyl disaster. His calls to cut not just military spending, but to produce more electricity for industrial and commercial output, led the Politburo to increase nuclear power plants' production quotas. In February 1986, party leaders demanded that more plants to be constructed over the next five years—despite KGB warnings that mishaps in construction made plants like Chernobyl potentially unsafe. But dictates from the top, stressing the need for more output, overrode any safety precautions. Publicly, the Soviets had long boasted about the success of their plants. The February 1986 issue of *Soviet Life* (an English-language magazine) even relayed that the Chernobyl station was so safe that its cooling pond had become a successful fish hatchery. In the wake of such propaganda, voices of dissent were either downplayed or ignored.[20]

With 1986 underway, the Soviet Union seemed to be gaining global approval (outside the United States) for its reductions in the arms race. However, that year two technological malfunctions set the stage for another round of diplomatic salvos. The first occurred far above US soil. On January 28, 1986, the NASA shuttle *Challenger* launched to much fanfare. After a well-publicized search, the crew included a civilian, Christa McAuliffe, who would become the first school teacher in space. However, midflight the *Challenger* exploded. There were no survivors. That night, President Reagan was scheduled to deliver another State of the Union address; instead, he delivered one of his best-remembered speeches. Written in part by Peggy Noonan, it praised the lost NASA pilots and closed with the following, poetic line: "We will never forget them, nor the last time we saw them, this morning, as they prepared for their journey and waved goodbye and slipped the surly bonds of Earth to touch the face of God." Less quoted, however, is an earlier section of the speech, one that—even in the face of human tragedy—was a thinly veiled jab at the Soviet Union. In expressing admiration for America's space program, Reagan boasted: "We don't hide our space program. We don't keep secrets and cover things up. We do it all up front and in public. That's the way freedom is, and we wouldn't change it for a minute."[21]

Reagan's boasts of American freedom, bravery, and scientific progress became more poignant later that year when a second disaster came to expose the flaws in Soviet technology. It came in the form of a nuclear reactor meltdown near the Ukrainian city of Pripyat, although it became better known simply as its local regional name: Chernobyl. Much as TMI had afforded the Soviets an opportunity to criticize American nuclear power, Chernobyl would now allow American policymakers an even greater opportunity to counterattack.

Chernobyl

On April 26, 1986, at 1:23:40 a.m., the first of two explosions rocked the Chernobyl nuclear plant; these blasts were so massive that they upended the reactor's two-million-pound upper shield and sent highly radioactive debris half a mile into the night sky. The cause of the explosion was an ill-fated safety test which, in the span of four seconds, took the reactor from 7 percent of capacity to almost 100 percent maximum power. The now exposed reactor sent pieces of hot radioactive graphite atop the asphalt roof of nearby building Number Four. Radioactive steam spewed from the exposed core, contaminating the nearby river (and fish hatchery) that had acted as the plant's cooling source. All of this damage because, as American scientists had previously warned, the Chernobyl plant had no concrete and steel containment structure to cover it.[22]

According to historian Richard Rhodes, the root cause of the accident at Chernobyl was design. The plant housed a "High Power Channel Reactor," or an RBMK reactor, which utilized a large block of graphite into which two types of uranium fuel rods can be inserted, hundreds at a time when necessary. With water cooling the rods, steam powered the turbines to make power. RBMK reactors became the preferred design for plants across the Warsaw Pact as they could produce 1,000 megawatts of electricity per unit, outperforming previous designs. As Plokhy notes, "By 1982, more than half of the electrical power produced in Soviet nuclear plants came from" RBMK reactors. For years, Chernobyl was a supremely efficient power plant, regularly meeting or exceeding the demands of a state-planned economy that relied on such power for industrial output. But now, new demands for output meant that it was being pushed beyond its limits.[23]

The Soviets had long boasted of their safe and efficient nuclear program, but such propaganda was misleading. Even the name for the Soviet nuclear reactor program, "Ministry of Medium Machine Buildings," was purposefully deceiving. In fact, the KGB had attempted to conceal at least thirteen nuclear power accidents spanning the Soviet Union. Some had occurred at the Beloyarsk Plant just east of the Urals, while in "1975, the core of an RBMK reactor at the Leningrad plant partly melted down, [and] in 1982, a rupture of the central fuel assembly of Chernobyl Reactor Number One released radioactivity over" Pripyat. Additionally, in 1985 "a steam relief valve" burst at the Balakovo nuclear reactor near the Volga river.[24] In each case, the KGB attempted to downplay the extent of the damage, doing so not just to the West, but to Soviet citizens. Chernobyl proved no different, with the KGB again pursuing "active measures" for a cover-up. Deploying some 15,000 personnel, they engaged in acts of subterfuge, at one point even swapping a French scientist's soil sample with a less-irradiated one. They also harassed *Newsweek*'s Moscow correspondent Steven Strasser, albeit to little effect. However failed in these attempts, the KGB was correct in assuming that Chernobyl would cause the Soviet Union publicity headaches in the days and weeks that followed.[25]

Chernobyl was both dramatic and tragic. The rushed and secretive efforts to contain the radioactive fire that raged within the core—by heroic firefighters, airdrops of sand and boron from helicopter pilots, and the efforts of miners underground to combat a "China Syndrome"—so has the hallmarks of a drama that it has inspired popular books as well as an acclaimed HBO series.[26] These cinematic likenesses were not lost on Reagan, who was doubly intrigued by Chernobyl for its allusion to biblical prophecy. Loosely translated as "Wormwood," Chernobyl is mentioned in the Book of Revelation: "a great star, blazing like a torch, fell from the sky on a third of the rivers and on the springs of water—the name of the star is Wormwood. A third of the waters turned bitter, and many people died from the waters that had become bitter." It was an imperfect comparison, but for Reagan the coincidence was powerful. Perhaps emboldened by this potential prophecy, Reagan reverted to the rhetorical tricks of his first term, in which he (like the evangelical preacher Billy Graham) equated nuclear war, and now nuclear power, with biblical Armageddon.[27]

However dramatic, Chernobyl held diplomatic ramifications as well. The reactor revealed the stresses that Gorbachev's *perestroika* initiatives, with their calls for more and more electricity, had put on Soviet electrical infrastructure. Also, attempts to obfuscate information and downplay the accident brought into question *glasnost,* which Gorbachev had promoted just a few months previous. Thus, Chernobyl opened the Soviet Union to renewed attacks by US policymakers who questioned not just how genuine, or attainable, Gorbachev's goals were, but the feasibility of the entire Soviet system itself.

The Reagan Administration Responds to Chernobyl

American policymakers learned of the Chernobyl accident not from Gorbachev or TASS, but from Scandinavian scientists. Shortly after the accident,

Finnish researchers hundreds of miles away detected six times the normal levels of radioactivity, while Swedish counterparts publicly guessed that the radioactivity was leaking from a civilian plant, and not from some military test. These insights led TASS to comment obliquely that, "Measures are being undertaken to eliminate the consequences of the accident; aid is being given to those who have suffered injury."[28] That was April 28. One day later, TASS admitted to "a certain leak of radioactive substances." Such assertions, however vague, showed that Chernobyl was putting *glasnost* to the test; this was the first time Soviet news agencies admitted to suffering an atomic disaster, and UPI called the announcement "unprecedented."[29]

The Reagan administration quickly went on the diplomatic offensive. On April 29, arms control administrator Kenneth Adelman called the purported Soviet death toll of only two people "frankly preposterous." This statement was not simply conjecture; that day the White House Situation Room received a briefing from a Soviet source who estimated 2,000 dead already.[30] The same day, White House spokesman Larry Speaks "called on the Soviet Union to minimize the danger to other countries by providing full information about the incident, and repeated a US offer of technical help in containing fire and radiation."[31] He was speaking on Reagan's behalf, who via communiqués with Jack Matlock at the NSC, knew full well that the Soviets were already requesting assistance abroad—just not from the United States. Matlock's memo that day confirmed that the "Soviet embassy in Stockholm has approached Swedish nuclear power officials for advice on how to put out a fire at a nuclear power plant. [The] Swedes reportedly told the Soviets they do not have the type of power plant in question, and advised the Soviets to take the matter up with Britain." He concluded without a doubt that "a fire is indeed raging inside the Chernobyl nuclear plant."[32]

Also that day, the CIA suggested using Chernobyl as a way to challenge the authenticity of *glasnost*. They noted that "the Chernobyl disaster already has put Moscow on the defensive," that they were "seeking to deflect criticism of its reactor program by unfavorable comparisons to nuclear accidents in the West," especially by "stressing the 'terrible consequences' of accidents involving military weapons or weapons production, whose abolition Gorbachev demanded earlier this year." The CIA's recommendation: Use Chernobyl to expose *glasnost* as false.

> The [Soviet] reaction contrasts sharply with Gorbachev's recent efforts to provide more timely and accurate coverage of developments rather than be preempted by foreign reporting and "rumormongering." This policy [of *glasnost*] has failed its first important test. The regime's

> credibility again has been seriously damaged abroad and the incident has undermined Gorbachev's efforts to persuade Soviet citizens they can count on media to inform them.[33]

In a then-classified memo titled "Implications of the Chernobyl Disaster," the CIA predicted how the accident might hinder the Soviet regime—and ways to capitalize on it. Because "a major concern of Soviet citizens [is] how well their system looked out for their safety . . . it will be of great importance to the regime to influence the way this is perceived, either by effective action or, as is the usual Soviet practice, by manipulating information." The agency also suggested that Chernobyl "could exacerbate ethnic and class resentments because . . . it seems likely that lower classes and Baltic and Ukrainian populations will suffer disproportionately." The US, then, should anticipate that if there is "widespread death, illness, and dislocation, this event will be a severe psychological blow to the Gorbachev regime." In any case, "the system under Gorbachev's new leadership will be put to a politically and psychologically important test," especially if it "manifest[ed] the usual sloth, carelessness, evasions, and outright lies" the US had come to expect.[34]

On April 29, Senator Alfonse D'Amato (R-NY) sent the president a letter demanding verification on Chernobyl, asking that Gorbachev "open Soviet nuclear plants to safety inspections by" the International Atomic Energy Agency (IAEA), with assurances that the US comply as well. He went on to say:

> In light of the inferior safety standards for nuclear power plants in the Soviet Union, there is no assurance that such a disaster could not occur again. Since nuclear radiation knows no geographic bounds, I am deeply concerned for the safety of citizens living throughout the Soviet Union, in neighboring countries, and, in fact, worldwide. I believe it was extremely irresponsible for the Soviet Union to fail to provide complete information to the world so that those countries which might be affected could take precautionary measures. Those nations which were directly downwind had a right to know immediately about any accident that would increase radiation levels in their area.[35]

On May 5, the response D'Amato received from Assistant to the President William Ball likely downplayed the impact his letter had on the president, because in just a week's time Reagan sought similar action.[36]

If the Soviets continued to see Chernobyl as an internal matter, others in the NSC agreed with D'Amato and assessed the disaster as having global implications. Jack Matlock was one of them. He had entered the NSC seeking common ground with the Soviets, which he assumed could be found on the grounds of nuclear nonproliferation, even hinting in one classified document that the way to achieve common goals was to use the go-between of the IAEA.[37] But by the end of April 1986, the situation had changed drastically. NSC dispatches kept coming in with new information about the disaster. After confirming that Chernobyl's radioactivity leak crossed national borders, he suggested to Reagan that international action would be required. Later that day, Matlock wired Reagan to ask if the US should offer their assistance; the president's response was a resounding "yes."[38] The Soviets refused. Playing Cold War politics, they first asked Sweden and then Great Britain for assistance instead.

If Gorbachev's silence signaled that he continued to treat Chernobyl as an internal affair, voices from the US made clear the disaster's international impact. Pushing the issue, on April 30, Press Secretary Larry Speaks confirmed that Reagan had ordered that a US task force help the Soviets contain Chernobyl—a not so subtle indication that US policymakers had little faith in the Soviet response. On May 1, Secretary of State George Shultz boasted that the United States already has "a much fuller picture than what the Soviets are presenting to us or, for that matter, to their own people" and expressed shock that the Soviets had "firmly rejected offers of technical and medical help from the United States."[39]

Reagan may have publicly shown restraint, but privately he shared the opinion of advisors who urged for public comparisons between the superpowers' nuclear programs. Speaks reassured the public first, noting that there were only two "graphite reactors" in the US, neither of which was susceptible to a Chernobyl-like meltdown. That same day, Shultz provided his own critique: "It's a great contrast to the way information emerges on something of that kind, let's say, in the United States as compared with the Soviet Union . . . because there would be a tremendous volume of information available" had the event occurred in the US.[40] By May 3, Gorbachev still remained silent, leading Speaks to tell reporters that both Reagan and Vice President Bush "expressed serious concern with the lack of information that the Soviet Government is providing to the public and to the world and to its own citizens." If Gorbachev had hoped *glasnost* would improve the Soviet Union's image, Chernobyl proved to be a major hindrance to that end.[41]

Perhaps recognizing this opportunity to impose oversight on the Soviet Union, Reagan acted quickly. The president was traveling abroad for the twelfth annual G-7 Summit in Tokyo, and although trade was the focus, he remained fixated on Chernobyl. Senior officials encouraged Reagan not to go soft on the Soviets, with one exclaiming that Chernobyl had given the world "a different and less favorable picture" of Gorbachev, whose silence in the wake of the disaster made the case for "realism and verification" in future dealings.[42] This was perhaps because as early as April 30, the State Department recognized that although it was "customary international law [to] notify other states of the possibility of transboundary effects" such as radiation, they had "not identified any specific treaty binding upon the Soviet Union which imposes this requirement."[43] Clearly, the State Department insinuated that the Soviets needed to be held to standards similar to those of Western nuclear powers. It seems that Reagan agreed.

On May 4, in a radio address direct from Tokyo, Reagan made a suggestion that mimicked D'Amato's recommendation and addressed the State Department's concerns:

> The contrast between the leaders of free nations meeting at the summit to deal openly with common concerns and the Soviet Government, with its secrecy and stubborn refusal to inform the international community of the common danger from this disaster, is stark and clear. The Soviets' handling of this incident manifests a disregard for the legitimate concerns of people everywhere. *A nuclear accident that results in contaminating a number of countries with radioactive material is not simply an internal matter. The Soviets owe the world an explanation.* A full accounting of what happened at Chernobyl and what is happening now is the least the world community has a right to expect.[44]

Before leaving the summit, Reagan suggested to the G-7 leaders that they create "a convention providing for prompt notification of nuclear accidents with significant transboundary effects." Others agreed, and worked quickly to reach consensus. That night, Reagan wrote in his diary: "The meeting was long and arduous over . . . our statement of taking Russia to task for being secretive about the nuclear accident when there was a radioactive threat to other countries."[45] Reagan, the renowned deregulator, had spearheaded a movement for international radiological oversight.

A full ten days after the G-7 meeting, Gorbachev finally responded. Appearing "haggard" in a May 14 television appearance, he admitted to the crisis that had befallen the Soviet Union, going on to claim, falsely, that fewer than 300 people had been hospitalized (the actual number hovered around 10,000). Then Gorbachev turned accusatory. He warned that "certain NATO countries, especially the USA" were engaged in "a wanton anti-Soviet campaign . . . a veritable mountain of lies—most brazen and malicious." They were trying "to defame the Soviet Union and its foreign policy, to lessen the impact of Soviet proposals on the termination of nuclear tests and on the elimination of nuclear weapons." Gorbachev then attempted to lump together atomic power and nuclear weapons, reiterating that the "stockpiled nuclear arsenals" were more of a threat than Chernobyl.[46] Ultimately, faced with international pressure, Gorbachev proposed "a special representative international conference on the safe development of nuclear energy to be" hosted by the IAEA in Vienna. The organization quickly accepted.[47]

Well before the IAEA's newly scheduled conference, the US Committee for Energy Awareness was working to dispel fears of nuclear power at home. They ran an ad in the May 12 and 20 issues of the *Wall Street Journal*, the *New York Times*, *USA Today*, and the *Washington Post*, titled "Why what happened at Chernobyl didn't happen at Three Mile Island." Just below a half-page graphic of a US nuclear reactor, a caption explained that the image showed "the major containment barriers . . . of a US nuclear plant's fuel core [and that] many Soviet plants lack such containment structures." Further text assured Americans that their reactors used "defense in depth," employed "multiple protective barriers to *contain* the effects of an accident," and that readers should not worry but continue to enjoy "safe, reliable operations."[48] Additionally, public talking points written up by conservative pundit and advisor Lyn Nofziger also stressed that "had the Soviet Union followed the western example, the consequences of Chernobyl would not have occurred. Refer to Three Mile Island and lack of health consequences."[49]

Other talking points made clearer the comparisons between Chernobyl's casualties and TMI's lack of casualties, and then used that metric to compare ideological differences between US and Soviet ways of life. In a June 5 memo, Energy Secretary John S. Herrington stressed that although Americans had every right to be concerned about a Chernobyl-style accident here, it simply was not going to happen. Proclamations like these were growing increasingly necessary, as anti-nuclear groups were quick to point out that Hanford's N reactor, like Chernobyl's, utilized a graphite core—hence the energy secretary's quick response to dispelling comparisons, which would undermine his rhetoric.[50] In his final comparison between US and Soviet reactors, he concluded that "these differences are no accident. The difference between Soviet and American nuclear safety philosophy for nuclear reactors is as vast as the differences between a closed society and an open one."

The US's special delegation to Vienna August 25–29 was given many of the same comparative talking points. Reagan's Deputy Energy Secretary William F. Martin distributed a memo about Chernobyl to figures such as Communications Director Pat Buchanan, which emphasized the need for preparedness, as the conference would "attract international attention." Primary points reiterated that a Chernobyl-type accident could not happen in the US, that TMI was not nearly as dangerous as Chernobyl, and most important, that US policymakers' efforts at Vienna should be focused on achieving Reagan's goal to strengthen "existing international mechanisms for peaceful uses of nuclear energy" such as "working with IAEA and NEA to developing . . . international initiatives" for safety.[51] These were the initiatives that forced Gorbachev into transparency. It was diplomatic and rhetorical pressure in the wake of Chernobyl that opened the way for international oversight, the very approach the President Reagan had first suggested to the leaders of the G-7.

Conclusion

In 1979 and 1986, both superpowers attacked their opponents over nuclear power disasters. In the process, both governments had to walk a fine line between promoting the necessity of nuclear power and citing its dangers if left in the wrong hands. Both the United States and the Soviet Union made a similar distinction: it was not that all nuclear power was dangerous—just the other superpower's nuclear power. For decades, organizations such as the IAEA and the US Department of Energy had promoted safe nuclear power. In part, both had reassured citizens that the difference in the potency of nuclear power rods for power and materials used for weapons differed greatly. However, as Valery Legasov, the Soviet nuclear chemist who reported to the Politburo on Chernobyl, would later admit, the RBMK Soviet design reactor blurred these distinctions: It could not just produce nuclear power, but weapons-ready plutonium as well. Not that it was doing so at the time of the disaster, but in a 1980s Soviet Union under near-constant threat of Reagan's arms buildup and a proposed SDI program, having the option appealed to Soviet policymakers. This may have led them to continue to rely on RBMK reactors like the one at Chernobyl despite scientific warnings.[52]

How did Gorbachev feel about Chernobyl? As William Taubman assessed in his biography of Gorbachev, "The chain reaction at Chernobyl and the Kremlin's nonreaction to it marked a turning point for Gorbachev and the Soviet regime." He continues:

> Atomic energy had been perhaps the regime's highest priority, especially for its military, but for civilian uses as well. Its nuclear prowess was proof when so much else was going wrong, that the system still worked. The top scientists and engineers in charge of atomic development were virtually beyond criticism, even by the Politburo. That even the untouchable

> nuclear realm turned out to be rotten suggested that the whole system was, too. For the flaws revealed at Chernobyl and afterward were characteristic of the system as a whole: rampant incompetence, cover-ups at all levels, and self-destructive secrecy at the top.[53]

The Reagan administration agreed with this assessment. For much of the Cold War, hawkish policymakers were quick to judge a Soviet system they saw as inefficient, amoral, and secretive—but few publicly questioned the abilities of their nuclear programs. That changed after the explosion near Pripyat publicly exposed the flaws in this once-respected aspect of the communist system.

Looking back, both Reagan and Shultz agreed that Chernobyl had a dramatic effect on Gorbachev. Reagan believed that it showed Gorbachev the dangers of nuclear weapons, while Shultz called it a "turning point" that made a "dramatic impact on Gorbachev." Chernobyl made clear that the Soviets could no longer fully control media messages in a new "information age," one in which "lies are almost impossible to sustain."[54] Gorbachev was no less sanguine. In his memoirs, he reflected that "the accident at the Chernobyl nuclear power station was graphic evidence, not only of how obsolete our technology was, but also of the failure of the old system." Still, he maintained that if "something was not done in a timely manner, it was mainly because of a lack of information." He also chastised nuclear scientists at a June 3, 1986, Politburo meeting, stating that "for thirty years you scientists, specialists and ministers have been telling us that everything was safe . . . but now we have ended up with a fiasco." He had not, however, forgotten the attacks from the West, writing with hindsight that "some foreign propaganda centres, in a stream of accusations, showed that they were less interested in the tragedy itself than in using it to discredit our new policy." He admitted, however, that Chernobyl was indeed "a difficult test for *glasnost*."[55]

Chernobyl should take a more prominent place among the primary influences on the Cold War's end, in part because it provided the Reagan administration with another way to critique the Soviet system. Specifically, it was Reagan's request to Congress in the immediate aftermath of Chernobyl to "consent to ratification" a new Convention on Early Notification of a Nuclear Accident (CENNA), as well as the Convention on Assistance in the Case of a Nuclear Accident or Radiological Emergency. These initiatives increased global political pressure on Gorbachev and the Soviet Union, and were essentially the same conventions Reagan introduced on September 26, 1986, in Vienna. The IAEA's adoption of CENNA was a direct result of the momentum that followed the Chernobyl tragedy. By decade's end, the United States Senate had approved both the CENNA and the Convention on Assistance (on

March 23, 1987, and September 7, 1988, respectively).[56] These were landmark moments not just for those pursuing global nuclear power transparency and safety, but arguably in influencing the Cold War's end.

Notes

1 David Brinkley, ed., *The Reagan Diaries* (New York: HarperCollins, 2007), 443.
2 Frances FitzGerald, *Way Out There in the Blue* (New York: Simon & Schuster, 2000); David E. Hoffman, *The Dead Hand: The Untold Story of the Cold War Arms Race and Its Dangerous Legacy* (New York: Doubleday, 2009); John Newhouse, *War and Peace in the Nuclear Age* (New York: Alfred A. Knopf, 1989); Paul Lettow, *Ronald Reagan and His Quest to Abolish Nuclear Weapons* (New York: Random House, 2005).
3 On nuclear power, see Jacob Darwin Hamblin, *Poison in the Well: Radioactive Waste in the Oceans at the Dawn of the Nuclear Age* (New Brunswick, NJ: Rutgers University Press, 2008); Jacob Darwin Hamblin, *The Wretched Atom: America's Global Gamble with Peaceful Nuclear Technology* (New York: Oxford University Press, 2021); Natasha Zaretsky, *Radiation Nation: Three Mile Island and the Political Transformation of the 1970s* (New York: Columbia University Press, 2018).
4 Jack Matlock, *Reagan and Gorbachev: How the Cold War Ended* (New York: Random House, 2004), 188–189.
5 On TMI, see Zaretsky, *Radiation Nation*, 1–2; on TMI's influence on reawakening the anti-nuclear movement, see William Knoblauch, *Nuclear Freeze in a Cold War* (Amherst: University of Massachusetts Press, 2017), 12–15; on Nixon's "Project Independence" see *Foreign Relations of the United States*, 1969–1976, vol. 36, "Energy Crisis, 1969–1974," in *Office of the Historian*, https://history.state.gov/historicaldocuments/frus1969-76v36/d237 (accessed July 6, 2020).
6 Floyd Norris, *The Courier News* (Bridgewater, NJ), March 29, 1979, p. 10.
7 Associated Press, March 31, 1979, AM cycle.
8 Associated Press, April 2, 1979, Monday, AM cycle.
9 *Washington Post*, April 5, 1979, First Section, A14.
10 David Pauly, Ronald Henkoff, John J. Lindsay, Mary Hager, and Sylvester Monroe, "Nuclear Power on the Ropes," *Newsweek*, April 16, 1979, p. 41.
11 Jim Adams, Associated Press, April 23, 1979, AM cycle; on admitting past accidents, see "Other Atomic Energy News: Soviet A-Accidents Disclosed," *Facts on File World News Digest*, May 4, 1979.
12 Serhii Plokhy, *Chernobyl: The History of a Nuclear Catastrophe* (New York: Basic Books, 2018), 49–50.
13 David Lancashire, "Near-Disaster Fails to Blunt World's Nuclear Ambitions: Warning Noted, but It's 'Press On,'" *Globe and Mail*, April 14, 1979.
14 Ronald Reagan, State of the State Address, January 9, 1974. Transcript, p. 4, https://www.reaganfoundation.org/ronald-reagan/reagan-quotes-speeches/state-of-the-state/ (accessed September 4, 2020),
15 Rick Perlstein, *Reaganland* (New York: Simon & Schuster, 2020), 527.
16 Ibid., 539.
17 Richard Rhodes, *Arsenals of Folly: The Making of the Nuclear Arms Race* (New York: Alfred A. Knopf, 2007), 4; Adam Higginbotham, *Midnight in Chernobyl* (New York: Simon and Schuster, 2019), 212, 246.
18 The US even extended an offer to allow Soviet observers to watch a US nuclear test. See Memo, "To All Diplomatic Posts Immediately," July 21, 1985, US-USSR, p. 30, folder

"Soviet Union—Nuclear," Box 29, Jack Matlock Files, Ronald Reagan Library (hereafter "RRL").

19 Plokhy, 15–16; Higginbotham, 19.

20 Plokhy, 52–53.

21 Ronald Reagan, "Address to the Nation on the Explosion of the Space Shuttle Challenger," January 28, 1986, https://www.reaganlibrary.gov/research/speeches/12886b (accessed September 3, 2020); Reagan struck similar language from the draft of his June 9 remarks. See "Presidential Remarks" in folder "Presidential Commission on the Space Shuttle Challenger Accident," White House Office of Speechwriting: Speech Drafts, 1981–1989, Box 6 of 8, RRL.

22 Kate Brown, *Manual for Survival: An Environmental History of the Chernobyl Disaster* (New York: W. W. Norton, 2019), 15–16; Rhodes, *Arsenals of Folly*, 5–6; Plokhy, *Chernobyl*, 77–86; Higginbotham, *Midnight in Chernobyl*, 84–89.

23 Rhodes translates RBMK as "reactor, high-power, boiling, channel type" in *Arsenals of Folly*, 3; supporting Rhodes's translation is Plokhy, *Chernobyl*, 18–19.

24 Rhodes, *Arsenals of Folly*, 7.

25 The organization's treatment of Strasser involved a failed forgery attempt—a fake letter sent to a US Representative stating intentions to publicly shame the US for using Chernobyl as a propaganda tool—which was later released by the FBI to expose the KGB's methods. "The Deadly Fallout of Disinformation," *Belfast Telegraph*, July 11, 2020, Edition 1, National Edition, p. 29.

26 Vladimir Gubaryev, *Sarcophagus: A Tragedy* (New York: Penguin Books, 1987); Higginbotham, *Midnight in Chernobyl.*

27 The name "Chernobyl" dates back at least to the twelfth century C.E., when Kyivan people settled between the Carpathian Mountains and the Volda River. Noting the surplus of *Artemisia vulgaris,* or common wormwood, a blackish shrub native to the region, they settled on Chornobyl (or Chernobyl), a variant on the Ukrainian word for black: "Chornyi." See Plokhy, *Chernobyl*, 27.

28 "News Summary," *The MacNeil/Lehrer News Hour,* April 28, 1986 (transcript); United Press International, April 28, 1986, PM cycle.

29 G. Luther Whitington, United Press International, April 28, 1986, AM cycle.

30 On reports that there were only two casualties, see Higginbotham, *Midnight in Chernobyl*, 177; on Adelman's response, see page 9 of File: USSR: Nuclear Accident—Chernobyl April 29, 1986 (1/9), Box 29, Jack Matlock Files, RRL.

31 Norman Black and Jill Lawrence, "Meltdown Occurred, Fire May Still Be Burning, Intelligence Sources Say," Associated Press, April 29, Tuesday, AM Cycle.

32 Series II: USSR Subject File, Soviet Union— Nuclear Accident, Box 29, Jack Matlock Files, RRL.

33 Memo, Central Intelligence Agency Directorate of Intelligence, April 29, 1986, pp. 33–34, Folder: USSR Nuclear Accident (1/9), box 29, Jack Matlock Files, RRL.

34 "Implications of the Chernobyl Disaster," April 29, 1986, pp. 57–60, Folder: USSR Nuclear Accident (1/9), box 29, Jack Matlock Files, RRL.

35 Letter, William Ball to Alfonso D'Amato, May 5, 1986, ID#412202, DI001 Accidents, WHORM: Subject File, RRL.

36 Ibid.

37 "National Security Council Message Center," Folder "Soviet Union—Nuclear," box 29, Jack Matlock Files, RRL.

38 The NSC documents offered assistance in the form of "Atmospheric Release Advisory Capability (ARAC)" to predict radioactive material dispersion; Arial measuring systems (AMS), a helicopter-borne radiological measurement system to trace contamination radiological assistance teams "including health physicists and equipment to measure radioactive

contamination in water, air and soil"; "technical assistance in assessing the environmental effects of the radioactive materials released"; "medical personnel experienced in the diagnosis and treatment of radiation exposure in people"; and finally "technical expertise and assistance in radiological decontamination and recovery from a nuclear reactor accident." That same day, Assistant Secretary Ridgway met with Soviet charge Sokolov and made clear that the U.S. was "prepared to make available to the Soviet Union humanitarian and technical assistance in dealing with the accident." See page 32, Folder USSR: Nuclear Accident—Chernobyl, (1/9), Box 29, Jack Matlock Files, RRL.

39 Statement by Principal Deputy Press Secretary Speakes on the Soviet Nuclear Reactor Accident at Chernobyl, May 3, 1986, RRL, https://www.reaganlibrary.gov/research/speeches/50386a (accessed August 8, 2020).

40 "Statement by Principal Deputy Press Secretary Speakes on the Soviet Nuclear Reactor Accident at Chernobyl," RRL, https://www.reaganlibrary.gov/research/speeches/43086b (accessed August 8, 2020);https://www.reaganlibrary.gov/research/speeches/50186b; David Hoffman, "Reagan Angered by Soviet Delay, Moscow Rejects US Aid Offer," *Washington Post,* May 2, 1986, A27.

41 Statement by Principal Deputy Press Secretary Speakes on the Soviet Nuclear Reactor Accident at Chernobyl, May 3, 1986, RRL.

42 Higginbotham, *Midnight in Chernobyl,* 200; Lou Cannon, "Chernobyl in Spotlight at Summit Soviet Secrecy, Terrorism High on Reagan Agenda," *Washington Post,* May 4, 1986, A01.

43 Memo, Jonathan Schwartz to Mr. Devine, April 30, 1986, page 7 of USSR Nuclear Accident – Chernobyl (2/9), Box 29, Jack Matlock Papers, RRL.

44 Italics mine. See Higginbotham, *Midnight in Chernobyl,* 201; see also Jack Nelson, "Reagan Criticized Disaster Secrecy," May 4, 1986, *Los Angeles Times,* https://www.latimes.com/archives/la-xpm-1986-05-04-mn-3685-story.html (accessed September 5, 2020).

45 Brinkley, *The Reagan Diaries,* 409.

46 Rhodes, *Arsenals of Folly,* 23; Higginbotham, *Midnight in Chernobyl,* 200.

47 Letter, Helmut Zilk to Ronald Reagan, May 15, 1986, ID#407773, DI001 Disasters, WHORM: Subject File, RRL. (Note that Zilk's letter is incorrectly dated May 15, 1985.)

48 Letter, Gary Knight to John A. Svahn, May 20, 1986, (advertisement attached), ID#415695PD, DI001, WHORM: Subject File, RRL.

49 Memo, John M. Poindexter to Dennis Thomas, July 10, 1986, ID#424379, AT folder 401000-429999, WHORM: Subject File, RRL.

50 Note to John Tuck from Henry M. Gandy, July 15, 1986, ID#411277, DI001 to DI002, WHORM: Subject File, RRL.

51 Memo is attached to Letter, William F. Martin to Mitch Daniels, August 22, 1986, ID#401754, DI001 Accidents, WHORM: Subject File, RRL.

52 Brown, *Manual for Survival,* 22, 54.

53 William Taubman, *Gorbachev: His Life and Times* (New York: W. W. Norton, 2017), 241.

54 George Shultz, *Turmoil and Triumph* (New York: Charles Scribner's Sons, 1993), 724.

55 Mikhail Gorbachev, *Memoirs* (New York: Doubleday, 1996), 189–192.

56 Memo, Colin Powell to President Reagan, September 14, 1988, ID#57687955, DI 001—Accidents, WHORM: Subject File, RRL.

Nuclear Weapons, Ionizing Radiation, and the Principle of Unnecessary Suffering

JAROSLAV KRASNY

The year 2020 was one of reckoning. A global pandemic of novel coronavirus (COVID-19) resulted in hundreds of thousands of deaths and forced people worldwide into isolation. The virus caught states by surprise, unprepared to deal with an infection on such a large scale. 2020 was also a year marking the anniversary of yet another globally significant event. It had been 75 years since nuclear bombs exploded above the Japanese cities of Hiroshima and Nagasaki, when tens of thousands, including women and children, lost their lives instantly.[1] In just a few seconds, people lost family members and neighbors, and whole communities disappeared. Most of the densely populated inner city and its social structures vanished—obliterated in mere seconds; no firefighters, no hospitals, no one to help. In the following weeks, thousands kept dying from horrible burns and high doses of ionizing radiation—an invisible disease plaguing the survivors.

The long-term effects of ionizing radiation haunt the survivors, their children, and grandchildren to this day. In his study, Robert Jay Lifton vividly describes this crippling intergenerational anxiety: *"Nor are the fears of hibakusha limited to their own bodies; they extend to future generations. Survivors are aware of the general controversy about genetic effects of the atomic bomb—a very serious emotional concern anywhere."* [2] As the number of original *hibakusha*[3] declines each year, it is crucial to remember their experiences and reflect on the tragic long-lasting effects of a nuclear explosion to keep the message of Hiroshima and Nagasaki survivors very much alive. Unlike the virus that is plaguing the world in 2020 and beyond, nuclear weapons are entirely under human control. Since 1945, during the Cold War, multiple bilateral and multilateral efforts have been made to curb the risk posed by nuclear weapons, strengthen the nuclear arms control regime, and move forward toward comprehensive disarmament.

Scholars and practitioners alike have pointed out that the reality of mutually assured destruction with massive retaliation, including high-yield nuclear or even thermonuclear weapons, is no longer applicable to the current warfare.[4] A majority of current armed conflicts are categorized as non-international, low-intensity, urban conflicts.[5] Today, when geographically limited or regional conflicts are dominant, such an effective deterrence is possible only with limited yield and number of strikes as opposed to a massive launch of hundreds of high-yield missiles resulting in the obliteration of the Northern Hemisphere, if not the globe. In its 2018 Nuclear Posture Review (NPR), the United States re-introduced its flexible deterrence policy, which includes tailored nuclear response even to a non-nuclear attack.[6] Flexible deterrence requires new types of nuclear weapons with "tailored" yield to respond to a variety of security threats. In 2020, the Russian Federation issued its own nuclear doctrine, making similar statements by including the possibility of responding to a non-nuclear attack with nuclear weapons.[7] Hence, multiple countries or nuclear "haves" are upgrading nuclear arsenals and re-thinking their nuclear strategy.[8]

The United States defines its strategy as *tailored deterrence* toward a range of perceived nuclear and even non-nuclear threats coming from China, Russia, North Korea, and Iran. According to the 2018 NPR, this tailored deterrence requires *flexible capabilities* that include a combination of nuclear weapons of various yields, arguably from very low to very high.[9] Thus, the Cold War threat of mutually assured destruction (MAD) employing massive retaliation with high-yield nuclear weapons might today be even undermining effective deterrence since such a strike is not only highly improbable but an ineffective response to a modern, in many cases urban, battlefield situation. Therefore, low-yield nuclear weapons and their limited, tactical use, according to the 2018 NPR, are described as strengthening the nuclear deterrent since they are perceived as being of rather limited destructive power and as such more "usable" on today's battlefield. The stationing of low-yield warheads of approximately 5 kilotons (kt), that is, an explosive power of 5,000 tons of TNT, on the *USS Tennessee* submarine and upgrading the B61-12 with an option of yield modification of up to 50 kilotons are both actions reflecting these policies.[10] Others have argued that flexible deterrence with the introduction of low-yield, theater nuclear weapons significantly lowers the threshold for their use. This perception is reflected in the *Bulletin of Atomic Scientists* Doomsday Clock, currently set at 90 seconds to midnight.[11]

As the nature of warfare has been changing, so are the principles and rules of the law of war.[12] Precision strikes, "smart" guided missiles, and international oversight and pressure to respect the law are all characteristics of modern warfare. These changes have resulted in a lower number of battle deaths and collateral damage.[13] The first two World Wars saw an unprecedented number of casualties. The tragedy of these wars led to major improvements in the international system, such as the creation of the United Nations (UN) with the purpose of maintaining peace and security, prohibition on the lawless use of force and aggression, and revision of the Geneva Conventions in 1949. The revised Geneva Conventions added a much-needed wider scope of protection for non-combatants (civilians) and combatants who are no longer participating in hostilities, i.e., are "*hors de combat.*"[14] Additional Protocol I (API)[15] and Additional Protocol II (APII)[16] to the Geneva Conventions of 1949 were added after the horrors of the Korean and Vietnam Wars in 1977, significantly expanding and updating the Hague Convention of 1907 traditionally regulating the conduct of hostilities. Reprisals against civilians, use of chemical weapons, strategic bombardment of whole cities without any distinction, or for that matter strikes against non-military targets for psychological effect—these *warfare methods* would today constitute what is called a "*grave breach*" under International Humanitarian Law, or IHL.[17] All such grave breaches are war crimes invoking individual criminal responsibility. Furthermore, a number of legal scholars agree that the nuclear bombings of the cities of Hiroshima and Nagasaki would today constitute such a grave breach of IHL by making the civilian population the object of the attack, causing excessive loss of life, injury, or damage. Such an act would be considered a war crime, despite its arguable great military advantage.[18]

The fact, however, is that nuclear weapons are not all the same. As they differ significantly in their yield, so does their method of deployment. The legal question posed by nuclear weapons concerning the law of armed conflict is, therefore, twofold. Are nuclear weapons inherently illegal as a whole weapons category, in a fashion similar to chemical weapons, i.e., that their mere possession is or should be prohibited? Or is it a specific method of their deployment that could violate the law of armed conflict? It is unfortunate that nuclear weapon states, including those under another's "nuclear umbrella," are showing limited interest in the recently adopted Treaty on the Prohibition of Nuclear Weapons (TPNW). The TPNW, while far from being perfect and including a number of legally unclear issues, is so far the only treaty prohibiting the use and possession of nuclear weapons and as such, worthy of working toward its improvement.[19] Therefore, without an effective agreement between nuclear states, nuclear weapons use would have to comply with the customary principles of the law of armed conflict, since customary international law is a form of long-established international practice applicable to all states, unlike a treaty that binds only states parties to it.

The policies expressed by the United States have reiterated that any possible future use of nuclear weapons would comply with the US government's interpretation of the law of armed conflict. *But could it?* Could a state use nuclear weapons and still be in compliance even with the most fundamental and basic customary principles of the law of armed conflict?[20] Could low-yield nuclear weapons be deployed legally against military objectives, as an electromagnetic pulse (EMP) strike or on high seas against warships or submarines, i.e., places where there are no civilians?

Such use might be reconcilable with principles of the law, such as the distinction between combatant and a civilian (non-combatant), by attacking remote military objectives, or proportionality by using very low-yield weapons or even the principle of military necessity by attacking only high-value targets. However, such use would not be in compliance with the prohibition of unnecessary suffering. Weapons cause injuries, and some, such as incendiary weapons, inflict painful, horrendous burns difficult to treat. Although controversial, incendiaries are not prohibited.[21] Nuclear weapons cause burns and trauma through shock wave. Although on a much larger scale than any conventional weapon, these are not particularly different effects from those observed in other weapons.

What is particular to nuclear weapons is ionizing radiation causing excruciating death to those surviving the initial blast but receiving high doses. An invisible, incurable plague torments the survivors of nuclear explosion. Victims of nuclear blast who would survive long enough would not know whether their health would betray them tomorrow, next week, month, or year. They might endure a life of diseases, deteriorating health conditions, continuous mostly ineffective treatment, constant health checks and psychosomatic problems. Never-ending suffering. The effects of ionizing radiation on the human body must play an essential role when discussing the legality of the use of nuclear weapons in an armed conflict and their incompatibility with the prohibition of unnecessary suffering. These effects caused by invisible particles penetrating the body like microscopic bullets are not yet fully comprehended even by the legal, scientific, and medical community. The suffering caused by an invisible unknown is what differentiates nuclear weapons from any other weapon, making them unlawful tools of war.

Health Effects of Ionizing Radiation in Hiroshima and Nagasaki

All explosives cause extensive injuries that are often complex to treat. These include burns, trauma to the lungs caused by the blast, and injuries by bomb fragments, flying glass, or other foreign objects blown by the shock wave. A nuclear explosion differs from the conventional one not only by the sheer amount of energy released but by producing ionizing radiation in the process.

Japanese scientists initiated studies into the health effects of a nuclear explosion almost immediately after the nuclear bombings, despite the imposed censorship by the US Occupation forces on publishing anything concerning the effects of this new weapon. It was not until 1947 that President Truman gave the order to establish in Hiroshima the Atomic Bomb Casualty Commission or ABCC.[22] The stated objective was to understand the acute and long-term effects of a nuclear explosion, especially ionizing radiation, to improve defense against nuclear explosion and provide proper treatment to the possible future victims, primarily military personnel. Therefore, the purpose of the ABCC was not to provide medical treatment to the victims of Hiroshima and Nagasaki bombings, so many survivors and residents initially distrusted and criticized the commission because the *hibakusha* were being treated as mere subjects of scientific interest.[23] In March 1975, the ABCC dissolved and transformed into the Radiation Effects Research Foundation, the RERF. Today the RERF is a joint US-Japanese organization with research centers in both Hiroshima and Nagasaki.[24]

For the purpose of legal analysis of the suffering caused by the use of nuclear weapons, the Hiroshima and Nagasaki bombings certainly provide indispensable case studies and data. Indeed, compared to today, the social and economic conditions of the citizens of Hiroshima and the state of medical knowledge including treatment in 1945 were very different. Most of the houses were wooden and therefore easily combustible. The lack of medicines and general malnutrition further contributed to increased fatalities. Today, it is known that countermeasures, in the form of a relatively light shielding or positioning of the body, substantially reduced the injuries caused by the shock wave and thermal radiation assuming sufficient distance.[25] Modern medicine can treat many of the sustained physical injuries. However, countermeasures against highly penetrative ionizing radiation are not that simple. Further, the effects of ionizing radiation complicate the healing of sustained wounds.

The Committee for the Compilation of Materials on Damage Caused by the Atomic Bombs in Hiroshima and Nagasaki pointed out four stages of exposure that were included in the Report of Investigation of Atom Bomb Casualties sent to the UN by both cities in 1976.[26] In an early or initial stage heat was the primary cause of death together with trauma caused by collapsed buildings. This included large doses of radiation causing death within hours. The following second or *intermittent* stage, beginning from the third week after the bombing, marks the demarcation between so-called acute and recovery phases. In this stage victims showed signs of acute radiation poisoning following by what appeared to be a "recovery" with the symptoms disappearing, albeit only temporarily. Stage III or late stage from the third to the end of the fourth month, that is, December 1945, is defined by recovery in most cases. Long-term or delayed effects begin to appear five months after the bombing.

Overall, by the end of December 1945, from a population of 327,457 in Hiroshima and 286,702[27] in Nagasaki, 140,000 people perished in Hiroshima and 70,000 in Nagasaki.[28] However, because of the delayed effects of radiation, this number was not final.

There were other ways to assess the severity of injury. When it comes to the physical manifestations of radiation poisoning, the Japanese army medical school observed and subsequently organized radiation injuries into three categories. Stage I (symptoms presenting between August 6 and 17), Stage II (August 18 to early September), and Stage III (early to late September).[29] From the medical standpoint, Stage I is an acute phase. Acute radiation syndrome (ARS) occurs after a dose of 1 Gray (Gy) and above.[30] This means, that after receiving a dose of one gray there are observable effects on the human body. The symptoms of such acute doses include vomiting, diarrhea, nausea, hemorrhage from the mouth, rectum or urethra, convulsions, and delirium. Fatal cases terminating after forty days (10 percent of all fatalities) would not necessarily succumb to radiation poisoning since many other factors inhibited their recovery, including lack of medicine, poor nutrition, and infections.[31] Through all the following stages, many of the above symptoms, including epilation and purpura, kept reappearing, thus further tormenting the survivors. It appeared as if an invisible disease was "plaguing" the *hibakusha,* claiming the lives of those who developed some of the symptoms. As these effects were not yet fully understood, the appearance of the symptoms was considered by many as "a mark of death."

For a better numerical perspective, Table 14-1 shows the estimated doses according to the distance from the hypocenter at the time of bombing. It is apparent that the threshold dose of 1 gray for observable acute radiation poisoning was significantly surpassed at a distance less than 1 km or 0.6 miles.

Table 14-1. Non-Shielded Dose by Distance from the Hypocenter (DS02)

Distance from the Hypocenter (m)	Gamma Rays (Gy)	Neutron radiation (Gy)	Total (x 10 for neutron)
1,000	4.22	0.260	6.82
1,200	1.81	0.067	2.48
1,500	0.527	0.0090	0.617
1,800	0.165	0.0013	0.178
2,000	0.076	0.0004	0.080
2,500	0.013	0.0000	0.013

Source: Robert Young and George Kerr, Reassessment of the Atomic Bomb Radiation Dosimetry for Hiroshima and Nagasaki. Dosimetry System 2002. DS02. Volume 1 (Hiroshima: RERF, 2005).

Regarding secondary radiation, that is, not stemming from the bomb itself but through irradiated soil and materials, one hour after the explosion, exposure for five hours would result in a dose of approximately 0.2 Gy while the very next day, those who stayed for 8 hours received a half of that, i.e., 0.1 Gy.[32] While these doses may at first glance appear negligible, there is not a conclusive study as to their effects on human health. On July 14, 2021, the victims of the so-called black-rain that fell on Hiroshima following the bombing claimed victory at a Hiroshima High Court granting them medical benefits as *hibakusha* due to this type of secondary radiation exposure.[33]

William Schull points out that according to the estimates of the Joint Commission for the Investigation of the Effects of the Atomic Bombing in Japan, over half of all the casualties were due to burns compared to 30 percent due to ionizing radiation and 18 percent attributed to blast or shock wave.[34] This estimate shall be differentiated from the estimate of the RERF that divides the energy output of the explosion between 50 percent blast, 35 percent heat-wave, and 15 percent radiation.[35] While energy distribution is 15 percent for ionizing radiation, overall percentage of fatalities due to ionizing radiation is 30 percent, making it the second major effect and cause of death. This means that ionizing radiation is one of the major causes and not secondary and unintended as some might argue.[36] As suggested by the Hiroshima Radiation Effects Research Foundation estimates, heat and blast were initially the major cause of death, and although burns and trauma are not particular to nuclear weapons, those who survived, as described above, would develop delayed radiation symptoms depending on their absorbed dose.

It is because of these radiation effects that the suffering of victims is not limited in time. Even if victims survive the tremendous heat and shock wave, there is a subsequent period where those receiving doses of 4 Gy and above die,[37] and those receiving 3 Gy have a 50:50 chance of survival.[38] This torment comes at times, when, for instance combatants, would in no way be engaging in combat anymore but still would not be spared. It is as if on a battlefield, a soldier comes upon a wounded enemy combatant who is severely injured and thus, not fighting anymore, but executes him anyway. Moreover, as exposure to ionizing radiation increases risks of solid cancers and especially of leukemia, those affected would live a life of anxiety and constant fear.

It is indeed complicated to distinguish whether a malignant tumor originates with a physical predisposition, lifestyle, or exposure to ionizing radiation. The RERF confirms Schull's observation[39] that the late radiation effects are often not radiation specific; in other words, it is challenging to distinguish radiation-induced tumors from those occurring due to other reasons not related to an exposure.[40] As explained earlier, threshold doses of about 1–4 Gy show clear radiation-dose response, while for doses around 10–100m Gy, it is incredibly challenging to obtain the same dose-response data. RERF argues there is no scientific data or basis for concluding that internal exposure, for instance, due to inhaling irradiated particles, is more dangerous than external exposure. Both contribute to cancer development. However, the radioactive particle location in the body would be of importance, reflecting the differing response of internal organs to radioactive particles.[41]

On the contrary, among the well-documented major late effects is a significantly increased risk of leukemia. Of four subtypes, that is, "acute myelogenous leukemia, acute lymphatic leukemia, chronic myelogenous leukemia, and chronic lymphatic leukemia, all were detected amongst the survivors, except chronic lymphatic leukemia."[42] The risk starts about two years after the bombing and reaches its peak five to ten years after the exposure.[43] However, the risk does not disappear entirely. Especially for those exposed before adulthood, there is a high probability that the risk may persist throughout their lives. The Relative Risk[44] for leukemia at 1 Gy is 4.92. In other words, a person exposed to a dose of one gray is almost four times more likely to develop leukemia than a non-exposed person. With a risk that high, it is almost a certainty. The relative risk for solid cancers at 1 Gy is 1.29, meaning that the risk of cancer incidence is 0.29 times higher than unexposed. Indeed, when compared to the relative risk of leukemia, this number might appear somewhat low, but is such risk acceptable? Such risk is persistent throughout the lifetime and with an increasing tendency with age. See Table 14-2 with site-specific cancers.

Radiation exposure also carries with itself an independent risk for hepatocellular carcinoma, and it may further multiply the risk associated with Hepatitis C infection in a subset of cases.[45] The RERF study also suggests that radiation exposure increased the risks of heart disease, stroke, benign thyroid tumor, hyperparathyroidism, hypertension, kidney disease, liver disease, and

Table 14-2. Relative Risk concerning Leukemia and Malignant Tumors

Leukemia	4.92
Bladder	2.02
Breast	2.00
Colorectal	1.56
Lung	1.46
Liver	1.12
All malignant tumors	1.29

Source: Y. Shimizu, H. Kato, and William J. Schull, "Studies of the Mortality of A-bomb Survivors. 9. Mortality, 1950–1985: Part 2. Cancer Mortality Based on the Recently Revised Doses (DS86)," Radiation Research 121, no. 2 (1990): 120–141.

cataract.[46] However, some of these diseases are also due to rapid westernization occurring after the war, including dietary changes. Some effects are observed among those exposed in utero during the gestational period. These include structural abnormalities in exposure at 8–15 weeks after conception and decreased academic performance and IQ scores.[47] However, overall, RERF's study of the second generation is inconclusive without evidence of an increased risk for the children of A-bomb survivors, including no genetic effects.[48] Nevertheless, inconclusive does not mean the risk is not present. For many survivors, the mere possibility of such risk of defects for descendants is a severe social and psychological trauma and life-long anxiety.

The risk of developing solid cancer or leukemia is not linear, and its increase depends on multiple factors including distance, shielding, or predisposition. The question posed, however, asks: Is the above calculated risk to the victims acceptable? Although simplified, for instance, if we communicate to a survivor of a nuclear explosion the mere fact that they have a risk of 47 percent of developing cancer in the next twenty years and almost 100 percent for leukemia, would that be tolerable? As experiences of Hiroshima and Nagasaki revealed, the survivors continue to "live" in a constant fear for their health, with anxiety caused by an overall "unknown." This unknown of long-term effects of ionizing radiation would not be different today as these effects are still not yet fully comprehended. The life-span of the survivors is demonstrably shortened, and quality of life is significantly decreased. Lifelong health checkups, treatment, and anxiety related to the lingering question of how one's condition deteriorates next month, week, or year. Such injury can be defined as superfluous and suffering as unnecessary. Prohibition of inflicting unnecessary suffering is one of the very fundamental principles of the law of war.[49]

Nuclear Weapons and the Law of War

Rules applicable during the conduct of hostilities have been in one way or another evolving since ancient times: rules of conflict between Sparta and Athens were recorded by Thucydides in the fifth century BCE.[50] Various philosophical or religious concepts incorporate or, at some point, introduce rules regulating the conduct of war and use of weapons.[51] The Law of Armed Conflict (LOAC), also known as the Law of War or International Humanitarian Law (IHL), besides its humanitarian aspect under *Geneva law* named after the Geneva Conventions, places limits on means and methods of warfare, that is, conduct of hostilities—usually referred to as the law of the Hague or *Hague law*, named after the two Hague Conventions of 1899 and 1907.[52] It is important to remember that LOAC is applicable only in situations of armed conflict (international and non-international) and not in all cases of violence, for example, sporadic internal disturbances.[53]

The codification of these rules in the modern sense can be traced in most cases to the second half of the nineteenth century.[54] In 1859, a Swiss businessman and humanitarian, Henry Dunant, called for the codification of rules protecting those who are not engaging in combat after witnessing the horrors of the battle of Solferino during the war for Italian independence. Stunned by the tremendous suffering endured by the combatants, Dunant's helplessness is vividly described in his book *A Memory of Solferino* (1862). Dunant founded the International Committee of the Red Cross (ICRC) in 1863,[55] and his humanitarian efforts in the ICRC resulted in the adoption of the First Geneva Convention on the Amelioration of the Conditions of the Wounded in the Field in 1864.[56] Thus, the ICRC headquartered in Geneva, Switzerland, remains the "guardian" of the Geneva Conventions. An international conference in 1906 further expanded the agreement on the protection of soldiers who are not in combat.[57] Following the horrendous experiences of World War I, including the extensive use of chemical weapons, in 1925 the so-called Geneva gas protocol was adopted with the purpose of prohibiting poisonous and asphyxiating gases and bacteriological warfare.[58] The Hague Conventions of 1866 with revisions in 1907 codified customary rules of conduct of hostilities on the land, sea, and partly already in the air, for example, by prohibiting launching projectiles from balloons.[59]

Whereas Hague law limits the conduct of hostilities, Geneva law is concerned with protecting persons and their humane treatment. The introduction of new weapon technologies onto the battlefield has often resulted in massive casualties as demonstrated by WWI and WWII.[60] Therefore, the Geneva Conventions were significantly expanded and revised again in 1949. Subsequently, both bodies of law—Hague and Geneva—later merged into the Additional Protocols I and II[61] of 1977, and the division between *Hague* and *Geneva law* is no longer necessary, except for academic discussion. Hence, the whole purpose of the law of armed conflict is to protect civilians (noncombatants) and alleviate the suffering of combatants by limiting the effects of armed conflict while reducing unnecessary suffering, loss, and damage. Law of war is not concerned with the nature or origin of the conflict or attribution of guilt to the parties of a conflict, and as such it is rather pragmatic; that is, it does not prevent states from gaining military advantage, while at the same time it aims at alleviating the horrors of war. Moreover, conduct of hostilities is mainly regulated by customs, which is a general practice accepted by states as law—i.e., *opinio juris*— and as such is applicable to all states, unlike a treaty, convention, or any agreement that applies only to those agreeing to it.[62]

The Geneva Conventions of 1949 have received universal acceptance, and its two Additional Protocols were accepted with near universal recognition after the horrors of the Vietnam War. Additional Protocol I reflects a large number of customary rules applicable to all states. These fundamental principles of customary rules include distinction, necessity, proportionality, precautions, and prohibition of unnecessary suffering. Therefore, "the means and methods of warfare are not unlimited," as reaffirmed in Additional Protocol I, Art. 35(1). Besides these five fundamental principles, in 2005, the ICRC, after conducting a large-scale research project, identified and published a total of 161 customary rules applicable in international and non-international armed conflicts.[63]

This study represents a respected source on customary law of armed conflict by the majority of states. The use of nuclear weapons is most conceivable in an international armed conflict. While the Treaty on the Prohibition of Nuclear Weapons that would prevent such a use lacks necessary level of support from nuclear weapons states, the Geneva Conventions of 1949 with their universal acceptance, customary rules included in Additional Protocol I,[64] and customary law in general that include the principle of humanity as reflected in the so-called Martens Clause, do apply. The Martens Clause for the first time refers to "humanity" as a principle in the laws of war.

The clause appeared in the Preamble of the 1899 Hague Convention. It stated: "Until a more complete code of the laws of war is issued, the High Contracting Parties think it right to declare that in cases not included in the Regulations adopted by them, populations and belligerents remain under the protection and empire of the principles of international law, as they result from the usages established between civilized nations, from the laws of humanity, and the requirements of the public conscience" (Hague Convention II, Laws and Customs of War on Land). Additional Protocol I, Art. 1(2) further reaffirms the rule. In its Advisory Opinion on the legality of the use or the threat of use of nuclear weapons from 1996, the International Court of Justice (ICJ) referred to the clause *as an effective means addressing the evolution of military technology*, however, the statement lacks further clarification of its practical application in weapons prohibition.[65]

It is unfortunate that the 1996 ICJ Advisory Opinion did not offer any more clarity into the legal issues behind the use of nuclear weapons. In its conclusion, however, the Court stated that the use of nuclear weapons would be contrary to the principles of humanitarian law, without providing any explanation or jurisprudence on how such use would be contradicting such law.[66] However, in its somewhat perplexing concluding statement, the ICJ stated that "it cannot decide whether such use would be illegal in extreme circumstances where survival of the state is at stake."[67] What a *survival of state under extreme circumstances* means from a legal perspective is far from clear. Further, if we accept that nuclear weapons are in fact incompatible with IHL, shall we further accept that a "survival of a state" allows us to disregard the rules and principles of the law as the conclusion suggests? The law of war and particularly its most fundamental, core principles and rules such as the distinction between civilians and combatants or prohibition of unnecessary suffering do not change based on the situation. The applicability of the law of war remains the same whether the survival of a state is at stake or not.

A large number of states include customary principles identified by the ICRC in their military manuals on the law of armed conflict.[68] As pointed out above, the most fundamental principles include proportionality, distinction, precautions, necessity, and avoiding unnecessary suffering. States thus accept these principles as the most fundamental. Military necessity is, of course, challenging to define objectively and is often determined on a case-by-case basis. However, it is a principle that the US Department of Defense Manual of the Law of War refers to:

> The principle that justifies the use of all measures needed to defeat the enemy as quickly and efficiently as possible and are not prohibited by the law of war.[69]

Comparably, the UK Ministry of Defence Manual of the Law of Armed Conflict interprets the principle as follows:

> permits the use of only that degree and kind of force not otherwise prohibited by the law of armed conflict to achieve legitimate purpose . . . complete or partial submission of the enemy at the earliest possible moment with the minimum expenditure of lives and resources.[70]

The UK Manual seems more restrictive; yet both definitions acknowledge that the means and methods are not unlimited.

Another fundamental principle restricting means and methods of warfare is the distinction principle. This principle requires parties to the conflict to distinguish between military objectives and civilian objects and between combatants and non-combatants. To deliberately target non-combatants is a grave violation of IHL and a war crime under the Rome Statute.[71] Related to distinction is the precautionary principle where all feasible precautions must be taken by the warring parties to spare the civilian population and civilian objects. This includes, inter alia, properly verifying the target. Specific rules also apply to a defending party, for instance, not to locate military objectives near civilian objects, that is, objects that serve no military purpose. The most straightforward example of such a limitation in war are hospitals, which shall not be used for military purposes and are protected.[72]

On the other hand, another fundamental rule of war as identified by the ICRC study is proportionality. This calls for an attack to be proportionate to an anticipated military advantage. In other words, incidental loss of life and destruction, the so-called collateral damage shall be proportionate to expected advantage stemming from reaching a concrete military objective.[73] Loss of civilians' lives shall be minimized, if not avoided. Proportionality is in fact an important fundamental rule going hand in hand with the distinction.

It might appear as if these two rules counter each other. Distinction is made between combatants and non-combatants with the purpose of protecting non-combatants from an attack and harm, but proportionality allows non-combatants to be harmed as collateral damage. In fact, "intent" plays a key role. It is prohibited to *intentionally target* civilians; that is, killing civilians cannot become a military objective. Unintentional, unavoidable, or proportionate collateral damage, that is a damage outweighed by the value of the military target, is lawful.[74] It is this exact argument that many proponents of nuclear weapons employ. That the use of nuclear weapons would be in extreme cases where the value of the target outweighs all the unavoidable collateral damage. Even if we stretch our imagination and admit such a "proportionate" use is in fact realistic, customary prohibition of unnecessary suffering is hard to refute.

The principle of the prohibition of unnecessary suffering—that is, avoiding inflicting suffering for the sake of suffering without any military necessity—first appeared in the so-called Lieber Code of 1863[75] and later was codified in an international agreement, the St. Petersburg Declaration.[76] The rule prohibits suffering that has no military purpose. Accordingly, the only legitimate purpose of military combat is to employ only the force necessary to neutralize an enemy, in the words of humanitarian law, to put an enemy combatant "*hors de combat.*" This would be exceeded by, for instance, employing projectiles with poison or exploding projectiles that would unnecessarily increase the suffering, complicate medical treatment, and/or render death inevitable. It is important to remember that the prohibition of unnecessary suffering applies only to combatants because civilians (non-combatants) enjoy the ultimate protection under distinction. The law rationally cannot legitimize infliction of unnecessary or necessary suffering on non-combatants. The principle was then restated and codified in the 1899 Hague Conventions as a prohibition of weapons and methods "*of a nature*" to cause unnecessary suffering or superfluous injury[77] and later in the revised Hague Conventions of 1907 in a slightly different form, as weapons *"calculated"* to cause superfluous injury or unnecessary suffering.[78] The opposing nuance of intent or what lawyers define in criminal law as *"mens rea"* (guilty mind, i.e., criminal intent) in both wordings is fairly apparent. Some defending the justified use of nuclear weapons in war often argue that ionizing radiation is not an *intended* effect of nuclear weapons and as such shall not be considered in discussing the legality of these weapons. In other words, they are denying there is any intent to cause unnecessary suffering. However, the final, internationally accepted version of codification of unnecessary suffering is included in the Additional Protocol I of 1977, Art. 35(2) in the following form:

> It is prohibited to employ weapons, projectiles, and material and methods of warfare of a nature to cause superfluous injury or unnecessary suffering.

Hence, an intent of those developing the weapon or using it is irrelevant if the nature of the weapon is such as to cause superfluous injury or unnecessary suffering. The ICRC study on customary rules identified the principle under the rule number 70 in the same wording confirming its customary status and as such applicable to all, even states who are not parties to the Additional Protocol I.[79]

According to Marco Sassoli, the use of the words "suffering" and "injury" seems to indicate a more extensive application.[80] It is safe to say that psychological suffering would more likely accompany any superfluous injury. How to qualitatively and quantitively define and assess "superfluous" injury or "unnecessary" suffering is, of course, extremely difficult given the subjective element in perception of pain and suffering. In 1997 through the so-called SIrUS Project (SIrUS as an abbreviation of "Superfluous Injury or Unnecessary Suffering) initiated by the ICRC, a group of medical experts attempted to medically define superfluous injury and unnecessary suffering.[81] This project attempted to give normative substance to the principle and clarify its application.[82] The SIrUS Project included a specific proposal on how to determine unnecessary suffering and superfluous injury empirically. The proposal specified four criteria:

> Criterion 1: specific disease, specific abnormal physiological state, specific abnormal psychological state, specific and permanent disability or specific disfigurement.
>
> Criterion 2: field mortality of more than 25 percent or a hospital mortality of more than 5 percent.

> Criterion 3: Grade 3 wound as measured by the Red Cross wound classification (such as the location of the wound, nature of the wound, including depth and size, need for transfusion or complex medical operation . . .)
>
> Criterion 4: effects for which there is no well recognized and proven treatment.
>
> (the SIrUS Project, 1997)

According to some critics, the SIrUS Project and its proposal on defining superfluous injury and unnecessary suffering applied only a one-dimensional, health effects–based approach without considering military necessity.[83] For this particular reason, states disputed the study. The ICRC subsequently ended the project and withdrew the whole proposal in 2001.

Applications of the principle to incendiary weapons such as napalm, or flamethrowers, blinding lasers, white phosphorous, and, of course, nuclear weapons are often contested. It is contested since design purpose is what plays a crucial role. A laser that is designed to cause permanent blindness is prohibited under the Convention on Certain Conventional Weapons, Protocol IV; however, if such an effect is caused by a "misuse" of a laser used for instance for navigational purposes, this would not render the weapon itself in violation of IHL as causing unnecessary suffering.[84] The same argument goes for incendiaries which shall not be used as anti-personnel weapons against an enemy in the open without any cover.[85] Despite this, the international community made substantial achievements in banning certain weapons through various treaties. However, if a country decides not to be a part of a specific treaty, it is not bound by the agreement. Hence, where there is no specific treaty, customary rules and principles apply to any weapons use.

Weapons Prohibition and the Principle of Unnecessary Suffering

Among these first attempts to prohibit means of warfare is the short-lived attempt to ban the crossbow by the second Lateran Council in 1139.[86] Efforts to prohibit specific weapons with more or less success were made throughout history. Beginning with the St. Petersburg declaration of 1868, representing the first international treaty to prohibit specific weapons while referring to unnecessary suffering, subsequent agreements prohibiting or limiting development, use, production, stockpiling, or transfer of weapons gradually followed.

Certainly, with all the arms limitation agreements, not all states are a party to an agreement, and others have reservations about certain provisions.[87]

Although the ban on cluster munitions, incendiaries, or for that matter nuclear weapons has not received universal adherence, some agreements do. Chemical and biological weapons conventions would be such an example: a Chemical Weapons Convention signed in 1993 was preceded by the Biological Weapons Convention signed in 1972. While, without a treaty, customary principles applicable to chemical or biological weapons, including the principle of distinction, would apply—since gas or viruses cannot be effectively controlled and thus are naturally indiscriminate—the principle of unnecessary suffering applies as well. Especially after witnessing the horrible effects of chemical weapons during the Iran-Iraq war (1980–1988), the international community imposed an absolute ban on these weapons and their use as a method of warfare. Many veterans from the war who were exposed to chemical agents suffer from malignant tumors, deteriorating respiratory diseases, or severe eye damage.[88]

As is the case of poison gas, weapons that are indiscriminate by nature, including indiscriminate attacks, are prohibited.[89] Inherently indiscriminate weapons are such that cannot be targeted, or their targeting system is rudimentary, such as the German V2 rocket or the first generation of SCUD missiles.[90] However, advocates of nuclear weapons might argue that this is not the case with nuclear warheads since they can be targeted at a specific and delimited location in compliance with the law of war. Such a case includes an attack on a remote military base or military installation built in mountains by a low-yield missile. The law of armed conflict must protect civilians without hindering state capability for defense, and, therefore, military objectives and combatants are legitimate targets. As pointed out above in regard to collateral damage, under the law of armed conflict, not every attack resulting in civilians' deaths is unlawful.[91]

The unnecessary suffering principle, as stated before, applies only to combatants, since targeting civilians is strictly prohibited and the applicability of this principle to civilians would constitute a legal paradox. As William Boothby points out, it is the aggravation of suffering without military utility that is prohibited.[92] What Boothby further points out is the distinction between effects-based and design-purpose approaches.[93] An effects-based approach refers to calling for a ban on weapons that exhibit particular effects that might be unlawful. This, according to Boothby, would be too wide an interpretation.[94] The effects-based approach alone would undoubtedly be an insufficient argument for a prohibition of a specific weapon. Therefore, the application of the unnecessary suffering principle has to include an argument regarding military necessity. The design-purpose approach then follows a premise that it is the intended use of the weapon that matters, as depicted by the blinding laser example discussed earlier. While blinding lasers are generally prohibited, if a laser that is designed to navigate a military aircraft causes permanent blindness by accident, then this would not violate the principle of the prohibition of unnecessary suffering, for it is not the intended design and purpose of such a laser to blind the enemy.

Furthermore, unnecessary suffering is of minor importance in naval warfare, where the intended target is a platform—e.g., warships or submarines—and not personnel per se.[95] The International Committee of the Red Cross, though, sees this as flawed logic. The argument that the principle does not apply on the high seas is illogical. The same argument could thus be made when attacking a military base by merely stating that the objectives are military vehicles such as tanks, artillery, or armed personnel carriers, and not combatants.[96] The argument that attacks on the high seas does not invoke unnecessary suffering is thus flawed because warships are usually occupied by sailors, unlike unmanned surface vehicles.

Nuclear Weapons and the Principle of the Prohibition of Unnecessary Suffering

The legality of nuclear weapons is often discussed regarding their deployment against remote military bases as an electromagnetic pulse (EMP) strike or in naval warfare against military vessels or submarines. Proportionality is said to be assured through yield modifications, that is, selecting a low-yield device against remote targets or with minimal civilian objects in proximity. Precautions could include a warning in advance. Although it is often argued that such use with a low-yield nuclear weapon would be perfectly legal under the law of war, the unnecessary suffering principle would not agree with such a premise. An EMP strike might, in general, be victimless, however, as stated above, it is the standard and intended purpose and use of the weapon that shall be considered and not every single possible use. Since nuclear weapons are intended to explode and destroy mainly through shock wave, the fact that these weapons can be utilized in an EMP strike is irrelevant.

Any use of nuclear weapons would not reconcile with the SIrUS principle precisely because of the effects of ionizing radiation on combatants. Although the effects of shock wave or thermal radiation, even though incomparably of a far graver nature than conventional weapons, could be reduced and arguably are not specific to nuclear weaponry, ionizing radiation would uselessly aggravate the suffering of affected combatants and is the second major cause of death, as explained above. An extremely high risk of a deadly leukemia uselessly increases the suffering and makes death inevitable in many cases.

Superfluous injury or unnecessary suffering is a comparative rule that makes it difficult to imagine circumstances where such use could be deemed necessary, and without violating this fundamental principle. As Simon O'Connor points out, the fact that people exposed to radiation from nuclear weapons develop cancer or other diseases long after the attack serves no military advantage.[97] That is, a combatant would succumb to the effects of a deadly, invisible weapon—ionizing radiation—long after he or she ceased to be a combatant, as a civilian when his or her right to life and health shall be protected as stipulated in Article 6 of the ICCPR[98] and Article 12 of ICECSR.[99]

Conclusion

Today, when some states have access to enough funding to create options in weaponry, it is difficult to argue for situations that would require the use of even low-yield nuclear weapons and where such use would be necessary and in compliance with humanitarian law. The SIrUS Project expressed that the mere fact certain weapons inflict horrible injuries is not sufficient to deem the weapons in violation of the principle of the prohibition of unnecessary suffering. The point was that even though some immediate effects of weapons might be gruesome (e.g., napalm, white phosphorous), these effects in themselves do not create a basis for their ban. An *inhumane* weapon, a notion often employed by activists or NGOs, is indeed a problematic concept since *a humane* weapon is an oxymoron.

Delayed effects of ionizing radiation, the persistent crippling anxiety and fear for one's health can, without a doubt, be considered superfluous and unnecessary.[100] The risk of leukemia is significant and unnecessary; so is life with a constant threat of solid incurable cancers. It is as if the weapon is still attacking the combatant long after they ceased to be a combatant. It is a life under a constant and invisible attack causing harm that the medical community still does not fully understand. Therefore, it is of the utmost importance to study, comprehend, and emphasize the effects of ionizing radiation on the human body when discussing legal issues behind the use of nuclear weapons.

At the beginning of the nuclear era, these weapons were considered by some military and political strategists as merely extremely powerful conventional weapons. Given the overall rhetoric of the Nuclear Posture Review and activities in nuclear weapons development, this "thinking" seems to be coming back. We can no longer consider nuclear weapons merely political or rhetorical instruments. The reality shows that these weapons are what they are: war tools with specific plans for their battlefield deployment. Proponents can argue for their compatibility with the Law of Armed Conflict, including fundamental principles under specific circumstances—e.g., in outer space, on the high seas, or in remote areas. Yet these experts often forget the overall purpose of the law of armed conflict is not to inhibit state security, but to limit suffering, including that of combatants. Quite to the contrary, these laws strengthen a state's security if one believes no state desires entering the spiral of reprisals toward a total war like World War II ever again. The ultimate ban on chemical and biological weapons serves as testimony. The fact that there are violations does not mean that the law is ineffective, just as the unfortunate fact that murders happen does not necessarily mean that criminal law is ineffective. The health effects of ionizing radiation are difficult to reconcile with military necessity. As pointed out above, besides long-term effects, acute high doses resulting in vomiting, epilation, cataracts, purpura, and hemorrhaging from different body parts leading to an excruciating death can hardly be perceived acceptable, just as excruciating death caused by chemical nerve agents or weaponized pathogens is unacceptable.

In WWI, chemical weapons were originally justified as "humane." Nowadays, they are seen as taboo. Hardly anyone would argue for the necessity and acceptability of injuries caused by nerve agents such as Sarin even though arguments for its utility against combatants hidden underground or in cave labyrinths could be made. No one would dare, though, to make such statements. Delayed and long-term effects represented by a considerably increased risk, that is, an almost certainty of developing leukemia and other malignant tumors, serve as a practical example of how these weapons keep on damaging the human body, often resulting in an excruciating death caused by cancer.

Hiroshima and Nagasaki serve as a reminder that even seventy-eight years since the explosions, the suffering of those affected continues. The processes in which ionizing radiation damages the human body are far from being fully understood. Still, the research continues into the second and third generation. The children and grandchildren of *hibakusha* live in constant worry about whether they will remain healthy in months to come. They are under the constant threat of something invisible and not understood well enough. Nuclear weapons are not ordinary weapons with just a more powerful explosion.

Any use, including low yield, would cause *unnecessary suffering* of the affected combatants, with harm extending way past their life-span to their descendants. Means and methods of warfare are not unlimited.[101] Unlike COVID-19 that is plaguing our world today, these weapons are fully under our control: we choose whether we want this threat to exist or not. The COVID crisis has demonstrated how quickly a health care system can collapse even in the most developed countries. Influx into hospitals of those severely burned and injured by a nuclear explosion would be unprecedented, of course, if there would be any hospitals left. It would include a lifelong possibility of developing fatal diseases as a result of an explosion that incorporates a covert, invisible attack, that is, an attack utilizing ionizing radiation.

Notes

1 On the day of the bombing many schoolchildren were outside, including near the hypocenter working on fire prevention by tearing down houses. The estimated number of instant deaths varies greatly, with the lowest figure of 70,000 to the highest of 140,000 people. See Alex Wellerstein, "Counting the Dead at Hiroshima and Nagasaki," *Bulletin of the Atomic Scientists,* August 4, 2020, https://thebulletin.org/2020/08/counting-the-dead-at-hiroshima-and-nagasaki/ (accessed August 5, 2021).

2 Robert Jay Lifton, *Death in Life: Survivors of Hiroshima* (Chapel Hill: University of North Carolina Press, 1991), 105.

3 The word "hibakusha" (被爆者) refers to "a person affected by an explosion" and originally referred to the survivors of the 1945 bombings. Therefore, outside Japan, the word is also used for those affected by nuclear tests, including residents in the proximity of the nuclear *polygon* (a testing venue for Soviet nuclear weapons) in Kazakhstan. Akira Matsumura, *Japanese Language Dictionary,* vol. 3, 4th ed. (Tokyo: Sanseido, 2019), 2315.

4 Jeffrey A. Larsen and Kerry M. Katchner. *On Limited Nuclear War in the 21st Century* (Stanford, CA: Stanford University Press, 2014), 145.

5 For conflict classification see RULAC, "Rule of Law in Armed Conflict Project," Geneva Academy, https://www.rulac.org/ (accessed August 3, 2021).

6 US Department of Defense, "Nuclear Posture Review," https://dod.defense.gov/News/SpecialReports/2018NuclearPostureReview.aspx (accessed August 5, 2022).

7 President of Russia, "Ukaz Prezidenta Rossiskoi Federatsii Ob Osnovakh Gosudarstvennoi Politiki Rossiskoi Federatsii V Oblasti Yadernovo Sderzhivania [Decree of the President of the Russian Federation on the Foundations of the State Policy of the Russian Federation in the Area of Nuclear Deterrence], June 2, 2020, http://static.kremlin.ru/media/events/files/ru/IluTKhAiabLzOBjIfBSvu4q3bcl7AXd7.pdf [in Russian] (accessed August 1, 2020).

8 China is expanding and intertwining its nuclear arsenal with conventional systems and capacities. This poses a danger of mistaking a conventional attack as nuclear, see Gerald C. Brown, "Understanding the Risks and Realities of China's Nuclear Forces," *Arms Control Today* 51 (June 2021): 5. For the United Kingdom's nuclear buildup see Tom Plant and Matthew Harries, "Going Ballistic: The UK's Proposed Nuclear Build-up," Royal United Services Institute, March 16, 2021, https://rusi.org/explore-our-research/publications/commentary/going-ballistic-uk%E2%80%99s-proposed-nuclear-build (accessed August 5, 2021). For the DPRK's nuclear upgrades since 2020 see Reuters in New York, "North Korea Upgraded Nuclear Missile Programme in 2020, Says UN Diplomat," *The Guardian,* February 8, 2021, https://www.theguardian.com/world/2021/feb/08/north-korea-developed-nuclear-missile-program-during-2020 (accessed August 6, 2021).

9 In the 1996 Doctrine for Joint Theater Nuclear Operations nuclear yields are defined as very low (less than 1 kt), low (1 to 10 kt), medium (over 10 to 50 kt), high (50 to 500 kt), and very high (over 500 kt). This is not recognized internationally as an official categorization of nuclear yields; however, it serves as a reliable quantitative indicator. US Air Force, "Doctrine for Joint Theater Nuclear Operations," February 9, 1996, https://www.nukestrat.com/us/jcs/JCS_JP3-12-1_96.pdf (accessed August 6, 2021).

10 Hans M. Kristensen and Matt Korda. ""United States Nuclear Forces, 2020," *Bulletin of the Atomic Scientists* 76, no. 1 (2020): 46–60, https://doi.org/10.1080/00963402.2019.1701286.

11 See *Bulletin of the Atomic Scientists*, "Doomsday Clock," https://thebulletin.org/doomsday-clock/ (accessed May 6, 2023). 90 seconds is the closest threat estimate to nuclear Armageddon yet, the farthest away was 17 minutes in 1991. Even during the Cuban missile crisis, the Doomsday Clock was set at 7 minutes to midnight.

12 *Jus in bello* or law of war, law of armed conflict, or international humanitarian law are more or less synonymous terms. The United States military uses the term Law of War in its manual. The United Kingdom employs the term Law of Armed Conflict in its manual. Humanitarian organizations such as the International Committee of the Red Cross and Red Crescent (ICRC) prefer the term international humanitarian law.

13 See Max Roser, Joe Hasell, Bastian Herre, and Bobbie Macdonald, "War and Peace after 1945," Our World in Data, https://ourworldindata.org/war-and-peace (accessed April 12, 2022).

14 Literally "*out of combat,*" a French expression used in the field of the law of armed conflict denoting combatants who are unable to continue engaging in combat due to wounds, sickness, or being captured, lit. "out of combat." See Geoffrey Corn, Victor Hansen, Richard Jackson, M. Christopher Jenks, Eric Talbot Jensen, and James A. Schoettler, *The Law of Armed Conflict: An Operational Approach* (New York: Wolters Kluwer, 2012), 78.

15 Protocol Additional to the Geneva Conventions of August 12, 1949, and relating to the Protection of Victims of International Armed Conflicts (Protocol I) (adopted June 8, 1977, entered into force December 7, 1978) 1125 UNTS 3.

16 Protocol Additional to the Geneva Conventions of August 12, 1949, and relating to the Protection of Victims of Non-International Armed Conflicts (Protocol II) (adopted June 8, 1977, entered into force December 7, 1978) 1125 UNTS 609.

17 Reprisal is a forcible measure that would ordinarily be an excess of the law, however, as such is legitimate to stop enemy's violations of IHL. Such counter-measures are of course limited and highly controversial. For use of nuclear weapons in reprisals see Marco Sassoli, *International Humanitarian Law: Rules, Controversies, and Solutions to Problems Arising in Warfare* (Cheltenham: Edward Elgar, 2019), 397. Grave Breaches are defined in all four Geneva Conventions (GC) of 1949 (GCI Art. 50, GCII Art. 51, GCIII Art. 130, GCIV Art. 147) and the 1977 Additional Protocol I to Geneva Conventions of 1949 (API Art. 11 and Art. 85).

18 Katherine E. McKinney, Scott D. Sagan, and Allen S. Weiner. "Why the Atomic Bombing of Hiroshima Would Be Illegal Today," *Bulletin of the Atomic Scientists* 76, no. 4 (2020): 157–165, https://www.tandfonline.com/doi/full/10.1080/00963402.2020.1778344.

19 The TPNW entered into force on January 22, 2021, and currently has 92 signatory States, 68 of which have formally ratified. The treaty and its provisions are binding only for ratifying States. Signatory States which did not ratify are not bound by the treaty; however, signatories under treaty law as codified in the Vienna Convention on the Law of Treaties (1969, Art. 18) cannot engage in activities that would defeat the object and purpose of the treaty. None of the nuclear States, including States under a nuclear umbrella, signed the document; that includes Japan due to the US-Japan Security alliance. The treaty is therefore not binding for these nuclear non-signatory States.

20 These fundamental principles are distinction, proportionality, necessity, humane treatment, and prohibition of unnecessary suffering. See Andrew Clapham and Paola Gaeta, eds., *The Oxford Handbook of International Law in Armed Conflict* (Oxford: Oxford University Press, 2014).

21 The Convention on Certain Conventional Weapons or CCW is an armed conflict or IHL treaty. Its purpose is not disarmament or arms control but regulations on the use of weapons defined in its five protocols. Use of incendiary weapons is regulated in Protocol III.

22 William Schull, *Effects of Atomic Radiation: A Half-Century of Studies from Hiroshima and Nagasaki* (New York: Wiley-Liss, 1995), 19.

23 Gerald F. O'Malley, "The Grave Is Wide: The Hibakusha of Hiroshima and Nagasaki and the Legacy of the Atomic Bomb Casualty Commission and the Radiation Effects Research Foundation," *Clinical Toxicology (15563650)* 54, no. 6 (2016): 526–530. doi:10.3109/15563650.2016.1173217.

24 See Radiation Effects Research Foundation, "Objective and History," https://www.rerf.or.jp/en/about/establish_e/ (accessed August 5, 2021).

25 Los Alamos Scientific Laboratory and Norris E. Bradbury, *The Effects of Atomic Weapons.* (Whitefish, MT: Literary Licensing, 2013), 200–201.

26 Committee for the Compilation of Materials on Damage Caused by the Atomic Bombs in Hiroshima and Nagasaki, *Hiroshima and Nagasaki: The Physical, Medical and Social Effects of the Atomic Bombings* (Tokyo: Iwanami Shoten, 1981), 114.

27 Committee, *Hiroshima and Nagasaki,* 348.

28 Committee, *Hiroshima and Nagasaki,* 113.

29 Committee, *Hiroshima and Nagasaki,* 130.

30 Tsutomu Shimura, Ichiro Yamaguchi, Hiroshi Terada, and Naoki Kunugita, "Lessons Learned from Radiation Biology: Health Effects of Low Levels of Exposure to Ionizing Radiation On Humans Regarding Fukushima Accident," *Journal of the National Institute of Public Health* 67 (2018): 1, 115–122, https://www.niph.go.jp/journal/data/67-1/201867010013.pdf (accessed August 5, 2021).

31 Shimura et al., "Lessons," 146.

32 Robert W. Young and George D. Kerr, "Reassessment of the Atomic Bomb Radiation Dosimetry for Hiroshima and Nagasaki: Dosimetry System 2002: DS02: Volume 2," International Nuclear Information System (2005), 148, https://inis.iaea.org/search/searchsinglerecord.aspx?recordsFor=SingleRecord&RN=39019932 (accessed May 6, 2023).

33 The Asahi Shimbun, "Editorial: Pressing Need to Recognize 'Black Rain' Victims as Hibakusha," December 2, 2021, https://www.asahi.com/ajw/articles/14493395 (accessed February 13, 2022),

34 Schull, *Effects,* 12.

35 RERF, *A Brief Description* (Hiroshima: RERF, 2016), 2.

36 See International Court of Justice, "Letter Dated June 16, 1995, from the Legal Adviser to the Foreign and Commonwealth Office of the United Kingdom of Great Britain and Northern Ireland, together with Written Comments of the United Kingdom," June 16, 1995, https://www.icj-cij.org/public/files/case-related/95/8802.pdf, 48 (accessed May 6, 2023).

37 Schull, *Effects,* 126.

38 RERF, *A Brief Description,* 4.

39 Schull, *Effects,* 162–167.

40 Interview with the RERF staff, Jeffrey Hart, Hiroshima, November 29, 2019 [in person].

41 RERF, *A Brief Description,* 15.

42 RERF, *A Brief Description,* 19–20.

43 Itsuzo Shigematsu, Mitoshi Akiyama, Hideo Sasaki, Chikato Ito, and Nanao Kamada, eds., *Genbaku hoshasen no jintai eikyo* [Effects of A-bomb radiation on the human body] (Tokyo: Bunkodo, 1995), 17–20.

44 "Relative Risk is the ratio of a particular health effect in an exposed group to that in a comparable unexposed group. A relative risk of 1 implies that exposure has no effect on risk." See Radiation Effects Research Foundation, "Relative Risk (RR)," https://www.rerf.or.jp/en/glossary/riskrela-en/ (accessed August 4, 2021).

45 Visit and interview with the staff of the Radiation Effects Research Foundation in Hiroshima, December 2, 2019.

46 RERF, *A Brief Description.*

47 Schull, *Effects*.
48 RERF, *A Brief Description*, 37.
49 *Legality of the Threat or Use of Nuclear Weapons (Advisory Opinion)* [1996] ICJ Rep 226, para. 95.
50 John Lazenby, *The Peloponnesian War: A Military Study* (London: Routledge, 2004), 116.
51 Such as Book of Manu prohibiting "hooked" spikes. See Stuart Casey-Maslen and Tobias Vestner, *A Guide to Disarmament Law* (New York: Routledge, 2019), 4.
52 See Stuart Casey-Maslen and Steven Haines, *Hague Law Interpreted: The Conduct of Hostilities under the Law of Armed Conflict* (New York: Hart, 2020).
53 Additional Protocol II to the Geneva Conventions of 1949, Art. 1 (2).
54 William H. Boothby, *Weapons and the Law of Armed Conflict* (New York: Oxford University Press, 2016), 10.
55 David Forsythe, *The Humanitarians* (Cambridge: Cambridge University Press, 2005).
56 Convention for the Amelioration of the Condition of the Wounded in Armies in the Field (signed August 22, 1864, entered into force June 22, 1865) (1864) 129 CTS 361.
57 Convention for the Amelioration of the Condition of the Wounded in Armies in the Field (1906) (signed July 6, 1906, entered into force August 9, 1907) (1907) 1 AJIL Supp 201.
58 Protocol for the Prohibition of the Use in War of Asphyxiating, Poisonous or other Gases, and of Bacteriological Methods of Warfare (signed June 17, 1925, entered into force February 8, 1928) 94 LNTS 65.
59 Casey-Maslen and Haines, *Hague Law*, 247.
60 Paul Scharre, *Army of None: Autonomous Weapons and the Future of War* (New York: W. W. Norton, 2019), 38.
61 Additional Protocol I is concerned with the rules applicable in International Armed Conflict (IAC), while AP II regulates Non-International Armed Conflict or NIAC.
62 See Malcolm Shaw, *International Law* (Cambridge: Cambridge University Press, 2017).
63 See Jean-Marie Henckaerts and Louise Doswald-Beck, eds., *Customary International Humanitarian Law*, vol. 1, *Rules* (Cambridge: Cambridge University Press, 2005).
64 During the negotiations of the Additional Protocol I, State parties expressed common understanding that the protocol would not apply to nuclear weapons; however, some of the rules included in API are customary and therefore apply to all Parties; see Boothby, *Weapons*, 209.
65 Laurence Boisson de Chazournes and Philippe Sands, eds., *International Law, the International Court of Justice and Nuclear Weapons* (Cambridge: Cambridge University Press, 1999).
66 Although in their dissenting opinions several judges dwell onto radiation and its particular effects. See, for instance, *Legality of the Threat or Use of Nuclear Weapons* (Advisory Opinion), (Dissenting Opinion of Judge Koroma), 1996, pp. 569–570, or Dissenting Opinion of Judge Weeramantry, 1996, pp. 458–461. See International Court of Justice, "Advisory Opinions," https://www.icj-cij.org/en/case/95/advisory-opinions (accessed August 5, 2021).
67 Full text available at the International Court of Justice, "Advisory Opinions: Advisory Opinion of July 8, 1996," https://www.icj-cij.org/files/case-related/95/7497.pdf (accessed July 6, 2021).
68 For military manuals, see the International Committee of the Red Cross, "IHL Database: Military Manuals," September 11, 2021, https://ihl-databases.icrc.org/customary-ihl/eng/docs/src_iimima.
69 Michael Newton, ed., *The United States Department of Defense: Law of War Manual Commentary and Critique* (Cambridge: Cambridge University Press, 2018), 116.
70 UK Ministry of Defence, *The Manual of the Law of Armed Conflict* (Oxford: Oxford University Press, 2004), 21–22.

71 Rome Statute of the International Criminal Court (adopted July 17, 1998, entered into force July 1, 2002) 2187 UNTS 90.
72 "Medical units shall be respected and protected at all times and shall not be the object of attack." Additional Protocol I, Art. 12(1).
73 Art. 51(5)(b) of AP I: "an attack which may be expected to cause incidental loss of civilian life . . . excessive in relation to the concrete and direct military advantage . . ."
74 Corn et al., *Law of Armed Conflict,* 124.
75 Instructions for the Government of Armies of the United States in the Field, *US Army General Order No. 100,* April 24, 1863.
76 Declaration Renouncing the Use in Time of War of Explosive Projectiles under 400 Grammes Weight 138 CTS 297 (entered into force December 11, 1868).
77 See the text of the Convention at the International Committee of the Red Cross, "Treaties, States Parties and Commentaries: Convention (II) with Respect to the Laws and Customs of War on Land and its annex: Regulations concerning the Laws and Customs of War on Land. The Hague, July 29,1899," https://ihl-databases.icrc.org/applic/ihl/ihl.nsf/Treaty.xsp?documentId=CD0F6C83F96FB459C12563CD002D66A1&action=openDocument (accessed October 11, 2021).
78 "In addition to the prohibitions provided by special Conventions, it is especially forbidden . . . To employ arms, projectiles, or material calculated to cause unnecessary suffering" Art. 23(e), Convention (IV) respecting the Laws and Customs of War on Land and its annex: Regulations concerning the Laws and Customs of War on Land. The Hague, October 18, 1907. Available at the International Committee of the Red Cross, "Treaties, States Parties and Commentaries," https://ihl-databases.icrc.org/ihl/INTRO/195 (accessed October 11, 2021).
79 Henckaerts et al., *Customary,* 237.
80 Marco Sassoli, *International Humanitarian Law: Rules, Controversies, and Solutions to Problems Arising in Warfare* (Cheltenham: Edward Elgar, 2019), 382.
81 See the SIrUS Project, available at the Library of Congress, "Collection Items," https://www.loc.gov/rr/frd/Military_Law/RC_SIrUS-project.html (accessed August 7, 2021).
82 Ritu Mathur, *Red Cross Interventions in Weapons Control* (London: Lexington Books, 2017), 55.
83 Donna Marie Verchio, "Just Say No! The SIrUS Project: Well-Intentioned, But Unnecessary and Superfluous," *Air Force Law Review* 51 (2001): 183–228.
84 Yoram Dinstein, *The Conduct of Hostilities under the Law of International Armed Conflict* (Cambridge: Cambridge University Press, 2004), 73.
85 Convention on Prohibitions or Restrictions on the Use of Certain Conventional Weapons which may be Deemed to be Excessively Injurious or to have Indiscriminate Effects: Protocol III: Protocol on Prohibitions or Restrictions on the Use of Incendiary Weapons (adopted October 10, 1980, entered into force December 2, 1983) 1342 UNTS 171.
86 Michael Bryant, *A World History of War Crimes: From Antiquity to the Present,* 2nd ed. (New York: Bloomsbury Academic, 2021), 161.
87 For instance, the US made a reservation to Chemical Weapons Convention Article 1(5) prohibiting the use of riot control agents (RCAs) as a method of warfare during armed conflict. Under the Executive Order 11850 issued by the President Ford, the US retained the right to use RCA's in combat under specific circumstances. 1975 Ford Executive Order 11850. Office of the Federal Register. "Executive Order 11850." National Archives. April 8, 1975, https://www.archives.gov/federal-register/codification/executive-order/11850.html (accessed August 22, 2021).
88 Kouki Inai, *Atlas of Mustard Gas Injuries: Building Bridges between Iran and Japan through the Relief of Victims Exposed to Mustard Gas* (Hiroshima: MOCT, 2012).
89 Indiscriminate attacks definition in API, 51(4).

90 Boothby, *Weapons*, 220.
91 William H. Boothby, *The Law of Targeting* (Oxford: Oxford University Press, 2012), 70.
92 Boothby, *Weapons*, 46.
93 Boothby, *Weapons*, 49.
94 Boothby, *Weapons*, 49.
95 Louise Doswald-Beck, ed., *San Remo Manual on International Law Applicable to Armed Conflicts at Sea* (Cambridge: Cambridge University Press, 1995), 118.
96 Personal interview by the author with Magnus Lovold of the ICRC Arms Unit, in Geneva November 2019.
97 Simon O'Connor, "Nuclear Weapons and the Unnecessary Suffering Rule," in *Nuclear Weapons under International Law*, ed. Gro Nystuen, Stuart Casey-Maslen, and Annie G. Bersagel, 128–147. Cambridge: Cambridge University Press, 2014.
98 International Covenant on Civil and Political Rights (adopted December 16, 1966, entered into force March 23, 1976) 999 UNTS 171.
99 International Covenant on Economic, Social and Cultural Rights (adopted December 16, 1966, entered into force January 3, 1976) 993 UNTS 3.
100 For further information regarding anxiety among the survivors, see Robert J. Lifton, *Death in Life: Survivors of Hiroshima* (Chapel Hill: University of North Carolina Press, 1991).
101 Additional Protocol I, Art. 35(1).

Reflections on the *Golden Rule*

HELEN JACCARD

In 1958 four activists sailed a small boat, the *Golden Rule,* toward the Marshall Islands in an attempt to interfere with nuclear weapons tests. I've sailed the very same *Golden Rule,* a project of Veterans For Peace, with many activists who tell me why they care about the boat that sails for a nuclear-free world. Many are veterans, and some are atomic veterans who were exposed to atomic blasts as a form of human experimentation. Many, including some of the *Golden Rule's* crew, have since died from cancer, likely caused by the radiation. But their stories are often unseen, unheard, and the harms unrecognized.

Some are atomic cleanup veterans who, from 1977 to 1980, with no protection, moved 3.96 million cubic feet of plutonium and other radioactive materials into the "Cactus" atomic blast crater and covered it with a layer of concrete—Runit Dome, which has been leaking radioactivity into the ocean in Enewetak Atoll. Almost all are now dead. As Keith Kiefer, commander of the National Association of Atomic Veterans, and a cleanup veteran from Enewetak, explains, "Of the 4,000 plus veterans we know of who were involved with the Enewetak cleanup, there are only about three hundred we can find alive. We are losing about two every three months."[1]

The material in Runit Dome is primarily from the Castle Bravo test, the biggest nuclear bomb the United States ever detonated. A Japanese fishing boat, *Lucky Dragon V,* was in the area, and very soon after the blast the crew became ill with radiation poisoning. I visited the *Lucky Dragon V* museum in 2017 to learn more about the effects of nuclear testing in the Pacific. There were many other fishing boats in the area at the time, and the precious food had to be discarded. The effects of the sixty-seven nuclear tests are still seen today—in addition to cancers and birth defects, the Marshallese cannot eat the fish and are malnourished because of the poor-quality food that is shipped to them by the US government—their only source of food. Their culture may never recover. Many have relocated to the US—where they face discrimination and bullying in schools, instead of the admiration their stories of resiliency deserve.

The *Golden Rule* in San Francisco Bay, 2017. Photograph by Gerry Condon.

The organization Veterans For Peace gives an avenue for the Marshallese to tell the stories of how these tests have affected them—for example, at the 2020 Veterans For Peace National Convention, other webinars, and when the *Golden Rule* visited a community of 800 Marshallese in Dubuque, Iowa, October 9–11, 2022. In addition to exposure from nuclear weapons tests, the Marshallese were forced to drink or be injected with radioactive substances, more human experimentation, so that the US government could study the effects of radiation on the human body. These experiments are part of the secret "Project 4.1" which has only recently been declassified.[2]

Because of the *Golden Rule,* I have learned from other survivors of nuclear history. I give presentations with Tina Cordova, whose entire family for four generations has experienced cancers from the first nuclear explosion, Trinity, on July 16, 1945. Tina talks about "Downwinders," and why the mutagenic radiation has caused those exposed and their descendants to develop cancer.[3]

Another person who makes harms visible in public presentations with me is Klee Benally. His family lived near the disastrous Church Rock uranium tailings dam breach. In 1979, 1,100 tons of radioactive tailings and 94 million gallons of contaminated water spewed eighty miles downstream in the Puerco River on the Navajo Reservation. This was the worst nuclear disaster that most people have never heard about.[4] Klee and I work with Charmaine Whiteface, Speaker of the Lakota Sioux tribe, to advocate for cleanup of 15,000 abandoned uranium mines in fifteen western states, primarily in Indigenous lands. These mines contaminate the air, soil, and water and cause cancer, birth defects, hypertension, and many other health effects. Even the smallest mine takes over a million dollars to remediate—so the total cost will be huge! In the meantime, uranium is again on the US "Critical Minerals" list, meaning that such mining is not likely to come to an end any time soon.

And so it continues . . .

The Golden Rule Project has given me quite a tragic view of the often unseen consequences caused by the nuclear arms and power industry. In 2017, when President Trump and Chairman Kim Jong-un started threatening each other's countries with nuclear annihilation, Veterans For Peace decided to send this little boat, the *Golden Rule,* back into the Pacific to stop the possibility of nuclear war. The plan was to sail to Hawai'i, where the US illegally occupies this never-ceded land, with over 20 percent of the island of O'ahu alone taken over by the US military and with two admitted sites of depleted uranium contamination. Then we would sail on to the Marshall Islands, the 1958 hoped-for destination of the *Golden Rule,* where people mostly don't know that the *Golden Rule* was on the way to bring attention to their plight. From there, we planned to sail to Guam and Okinawa, other US-military occupied islands, and finally to Japan in time for the 75th anniversary of the bombings of Hiroshima and Nagasaki. Alas, due to Covid 19, we were unable to go beyond Hawai'i. We had to cancel the *Golden Rule's* voyage to Japan.

The planned voyage led me to undertake, however, in 2017, a ten-city speaking and listening tour of Japan. I cried at the Hiroshima and Nagasaki Peace Memorials and spoke with five *hibakusha* (survivors of the Hiroshima bomb) and four evacuees from Fukushima. An important part of the trip to Japan was learning about how radiation contamination had affected the lives of the people of Fukushima and Tokyo, which has some hot spots that are still very unsafe for women and girls. Research shows radiation regulations privileging male bodies as the average body underestimated radiation and cancer risks to women and girls.[5] I listened to four women who took their children and fled, voluntarily, from the radiation from the triple meltdown of the Fukushima Daiichi nuclear power plant starting on March 11, 2011.

Their stories were heartbreaking. The men stayed behind due to work or because their parents refused to leave. They visited their wives and children only periodically, taking long, expensive train rides to places all over the Japanese islands. The Japanese government never helped those who were voluntary evacuees, so the financial burden is immense. But the need to protect their children from radiation is a higher priority.

One of the mothers wanted to tell me her story, but only if she were kept anonymous due to her fear of reprisals:

> In Tokyo our health condition deteriorated. We were not sure if it was due to radiation or fatigue or something, but my son and I were alternately hospitalized. My son had at least twice a high fever and virus-related stomach symptoms and it was so serious that he had to go to the hospital.
>
> We tried to wait until our health symptoms calmed down, but it never did.
>
> We know that Fukushima Daiichi has been emitting radiation and there is no future for decommissioning so far. Nothing has changed.
>
> We didn't feel secure, even in Tokyo, which is 250 km away from Fukushima, because we found out that the water supply was contaminated. Knowing that, we wondered, is it okay to bathe in the water, is it okay to brush our teeth? Those kinds of concerns just continued.
>
> We found out that the sandbox in a nearby park had a radiation level way above the safety limits. They took away the contaminated sand and they brought in fresh sand. We had evacuated, but we witnessed that this new place was not safe, either, so that was not an effective evacuation.
>
> So we decided to be away from Eastern Japan and that's how we ended up in Osaka. Since then our health condition improved.
>
> After three years my home in Fukushima received decontamination service, and they measured the radiation level of the soil in our back yard. It was 25,000 Bq/kg, which is enough to receive an evacuation order. It was supposed to be safe for the kids to go play outside. I'm too scared to go back home. This is fucking ridiculous.

It is important to tell these stories so that everyone can understand that it is not just those who live close to a nuclear disaster that suffer—the entire country does. Golden Rule Ambassadors will continue to listen and provide a voice for those affected by radiation.

We visited Motomi Ushiyama, M.D., in Yokohama, who is studying and helping children who have developed thyroid problems as a result of the Fukushima radiation. She gave me a copy of her presentation in English, "Current Status of Childhood Thyroid Cancer." I have some of Dr. Ushiyama's PowerPoint slides and other research in English. She said in her testimony that radiation-caused thyroid cancers amount to crimes against children by the Japanese government.[6]

Here are some of the key points from Dr. Ushiyama's research:

- Usually thyroid cancer is very rare, especially in children.
- Thyroid cancer is usually more common in adult women. It grows very slowly and rarely poses a threat of death.

- However, in Chernobyl, the thyroid cancer that developed in people who were exposed to radiation as young children was a very aggressive, fast-growing, metastatic cancer. Some of the thyroid cancers in children in Fukushima show similar characteristics.
- In Chernobyl, the younger the age of exposure to radiation, the more likely the child was to develop thyroid cancer later in life. Similarly in Fukushima, children who were young at the time of the accident are likely to develop many problems as they grow up. Continued testing will be necessary in the future.
- The survey conducted by Fukushima Prefecture authorities may not have captured all cases of thyroid cancer. A survey by a private organization revealed that they found the number of thyroid cancers in Fukushima children is actually 15 percent higher than the number reported by Fukushima Prefecture. The prefectural government's announcement, "Childhood thyroid cancer in Fukushima is unrelated to radiation exposure from the nuclear accident," is based on inadequate data and may be unreliable.
- Thyroid cancer was found in Fukushima children at a rate more than 60 times higher than normal.

Dr. Ushiyama later explained that the Japanese prefectural and national governments, as well as official UN bodies like the United Nations Scientific Committee on the Effects of Atomic Radiation, UNSCEAR, dismiss this very high rate of thyroid cancer in Fukushima children as due to their improved screening with high-performance ultrasound. However, she and others working with harmed individuals doubt the screening effect alone could be solely responsible. UNSCEAR in a recent 2020/2021 report asserted that the thyroid cancer found in large numbers in Fukushima is unrelated to radiation exposure, stating to the UN General Assembly, "No adverse health effects among Fukushima residents have been documented that are directly attributable to radiation exposure from the Fukushima Daichi Nuclear Power Station accident. The Committee's revised estimates of dose are such that future radiation-associated health effects are unlikely to be discernible."[7]

Many, like Dr. Ushiyama, continue to raise questions about the veracity of such findings by the UN and scientific and government bodies invested in promoting nuclear technologies who may have vested interests in minimizing the Fukushima accident's consequences and harms. Dr. Ushiyama wants researchers to make every possible effort to dispassionately understand radioactive contamination and any health problems that result. She is especially concerned about the ongoing internal exposure inside human bodies in order to better protect children from radiation contamination.

Our *Golden Rule* Japan delegation also saw unexpected harm to livelihoods. We visited a rancher who refused to kill his herd of irradiated cattle. The radiation caused white spots on their hides, and the meat would be considered unsafe to eat, so he will care for his herd for as long as they live. Other ranchers abandoned their cattle to die a horrible death of starvation or lack of water. Most killed their animals and then evacuated.

Now a new concern is that the Japanese government intends to dump 1.4 million tons of radioactive water into the sea in 2023. There are 62 radioactive isotopes in the current water storage tanks. Water used to cool the melted cores accumulates at a rate of about 140 tons per day. Fishers are opposed to the release—who would eat contaminated fish? So the Japanese government said that they will buy the fish—huh? What will they do with it? The people we met in Japan want the public to see these harms. They do not want to see any communities experience what they have survived. These legacies of harm are vast and enduring.

Before joining the Golden Rule Project, my partner, Gerry Condon, and I learned firsthand from the people of Sardegna, Italy, about the effects of radiation on health. We visited the island and learned about the high rates of contamination, cancer, and birth defects as a result of depleted uranium weapons and chemicals used by several huge military bases on the island. In one small community, 25 percent of babies were born with defects one year, the highest percentage in the world. Throughout the area are also serious malformations in animals: two-headed lambs, calves with deformed legs, a pig with one huge grotesque eye, and other horrors.[8] This same pattern happened in Fallujah, Iraq. Analysis of pre-2003 data compared to now showed that "the rate of congenital heart defects was 95 per 1,000 births—13 times the rate found in Europe."[9]

Activist Linda Modica took us to visit Jonesborough, Tennessee, where depleted uranium weapons are manufactured by Aerojet Ordnance. When Aerojet decommissioned its settling ponds, it spread the contaminated pond sludge onto nearby fields—with the state's and landowners' permission—but the farmers noticed that their cows were getting sick. Aerojet was required to take back their "fertilizer." Times changed and farms were subdivided for modest homes whose land was later found to be contaminated. DU from Aerojet was detected through mass spectrometry conducted by forensic chemist Dr. Michael Ketterer. The water and sediment of Little Limestone Creek—into which Aerojet discharges its waste and which flows into the Nolichucky River near there—is also contaminated with Aerojet's radioactive waste.

Upstream in Erwin, where highly enriched uranium naval reactor fuel is made by Nuclear Fuel Services, the Nolichucky takes another hit from that process's toxic cocktail of radionuclides. Owners of a dog who drank from creek water along Erwin's Linear Trail downstream from Nuclear Fuel Services had to bring their pet to the emergency clinic when it suffered convulsions soon thereafter. They later reported to a county commissioner that the vet who euthanized their dog believed that poison caused the convulsions.[10]

Although my education continues, one thing I am certain of—invisible as it is, the visible effects of radiation are increased cases of cancer, thyroid problems, heart disease, birth defects, and premature death. The entire nuclear chain must end—mining, enrichment, energy, and weapons. And all of the contamination needs to be cleaned up. The money that is wasted on nuclear weapons and energy can be used for remediation of contaminated sites and for societal good.

We sail the *Golden Rule* to support the United Nations Treaty on the Prohibition of Nuclear Weapons (TPNW) and endorse other measures to bring us back from the brink of nuclear war. It is important to remember that every step of the way, from mining to production to waste disposal, there are contamination and health consequences. The TPNW supports providing care for those exposed to radiation, something that has been done in a piecemeal and incomplete fashion. All radiation survivors deserve quality medical care—and an apology wouldn't hurt!

Each new story brings more into view. It is becoming clear that we must learn from affected communities and individuals about what surviving radiation contamination is like. When I retired in 2010, I never thought retirement from computer programming would lead me into sailing, activism, and research about militarism and radioactivity. Eleven years into "retirement" here I am, working harder and happier, sailing and living a life filled with purpose toward a nuclear-free future.

Let's end the whole nuclear era and leave a safe place for the next seven generations.

Notes

1 Keith Kiefer, email and telephone conversation with Linda M. Richards, June 12, 2022.

2 The 4.1 "Study of Response of Human Beings Exposed to Significant Beta and Gamma Radiation Due to Fallout from High Yield Weapons," with research by Beverly Deepe Keever, "Suffering, Secrecy, Exile: Bravo 50 Years Later," *Honolulu Weekly,* February 25, 2004, is documented in the House Congressional Record, March 17, 2004, p. H1338-42, https://www.congress.gov/108/crec/2004/03/17/CREC-2004-03-17-pt1-PgH1135.pdf.

3 Learn more at the Tularosa Basin Downwinders Consortium, https://www.trinitydownwinders.com/ (accessed September 3, 2022).

4 To learn more, see Doug Brugge, Esther Yazzie-Lewis, and Timothy H. Benally, eds., *The Navajo People and Uranium Mining* (Albuquerque: University of New Mexico Press, 2007), and Doug Brugge, Jamie L. deLemos, and Cat Bui, "The Sequoyah Corporation Fuels Release and the Church Rock Spill: Unpublicized Nuclear Releases in American Indian Communities," *American Journal of Public Health* 97, no. 9 (September 1, 2007): 1595–1600.

5 Pandora Dewan, "Nuclear Radiation Risk Impacts One Group Far More Than Any Other" *Newsweek.com,* October 10, 2022, https://www.newsweek.com/newsweek-com-nuclear-radiation-risk-impacts-one-group-more-other-1750413. For more information, see the work of the Radiation + Gender Impact Project, https://www.genderandradiation.org/.

6 Dr. Motomi Ushiyama email communications with author and editor, October 2022, with her most recent updated PowerPoint (September 2022), "Nuclear Accidents and Thyroid Cancer, Issues Recognized by a Clinician and What We Can Do—Mysteries of Why Pediatric and AYA Thyroid Cancer in Fukushima Is Considered Unrelated to Radiation Exposure," also shared at the symposium "No More Hibakusha! Let's Talk Together" (Tokyo, November 2019) and the Hong Kong Medical Women's Conference meeting, August 2017, http://www.vfpgoldenruleproject.org/wp-content/uploads/2018/01/hongkong-for-CD-R.pdf.

7 United Nations Scientific Committee on the Effects of Atomic Radiation, "Sources, Effects and Risks of Ionizing Radiation, Report to the General Assembly with Scientific Annexes, Volume II, Scientific Annex B," UNSCEAR 2020/2021, p. 106, https://www.unscear.org/unscear/uploads/documents/publications/UNSCEAR_2020_21_Annex-B.pdf.

8 Helen Jaccard, "Sardinia: Cancer, Contamination, and Militarization in Paradise," https://www.academia.edu/8672890/Sardinia_Cancer_Contamination_Militarization (accessed August 30, 2022).

9 Nafeez Ahmed, "How the World Health Organisation Covered up Iraq's Nuclear Nightmare: Ex-UN, WHO Officials Reveal Political Interference to Suppress Scientific Evidence of Postwar Environmental Health Catastrophe," *The Guardian,* October 13, 2013, https://www.theguardian.com/environment/earth-insight/2013/oct/13/world-health-organisation-iraq-war-depleted-uranium.

10 To learn more, see Linda Cataldo Modica, "Atomic Appalachia: The Sickening Results of Weapons Contamination in East Tennessee," *Peace and Freedom* 79, no. 1 (Winter 2019): 6–7, https://wilpfus.org/documents/peace&freedom/P&F-WinterSpring19.pdf.

PART III

Remembering and Forgetting

Marshallese Downwinders and a Shared Nuclear Legacy of Global Proportions

DESMOND NARAIN DOULATRAM

An enduring injustice continues for the Marshall Islands, a former UN Strategic Trust Nuclear Testing site that served as ground zero.[1] Nuclear issues detailing long-term harm of the nuclear testing program were brought up in the first face-to-face meeting with Joseph Yun, special presidential envoy ambassador and negotiations team leader, regarding the Compact of Free Association negotiation on June 14, 2022 on Kwajalein Atoll of the Marshall Islands.[2] Once code-named "Sand Niggers" by the US Military, the Marshallese people have a real history of grievances.[3] Shortly after Japan's surrender, after the atrocities of the atomic bombs, the nuclear cycle took a radical turn signifying official entry into the Cold War.[4] America's arms race against the Soviet Union provided the perfect opportunity to continue nuclear weapons development in the Marshall Islands through a policy of nuclear deterrence.[5]

Although the UN Strategic Trust granted the United States several requirements in fostering the inhabitants of the Marshall Islands toward self-government and economic self-sufficiency, including protecting their health and natural resources, military strategies took precedence despite the fact that the United States willingly signed the United Nations Charter Trusteeship agreement in 1947. This was a far cry from American exceptionalism and more nearly evidenced American realism. In the words of former Secretary of State Henry Kissinger who served as National Security Adviser, "There are only 90,000 people out there. Who gives a damn?"[6]

On the noon of the Nations Constitution on May 1, 1979, a proud founding father stated to the United States delegation attending his inauguration, "The most difficult long-term problem, both for this government, and your administration, as well as for the future of the people of this country is meeting the impact of lingering effects of nuclear radiation in the Marshall Islands."[7] More than four decades later, the Marshall Islands continues to meet these impacts head on with the creation of its National Nuclear Commission.[8]

Burdened by the inevitable consequences of the infamous nuclear testing period in the Marshall Islands, the Marshall Islands government has sought legal remedies for the social injustices faced by Marshallese nuclear victims.[9]

Petitions and Unwitnessed Effects

The nuclear testing period in the Marshall Islands did not go unchallenged. Described as a "peace and freedom loving people,"[10] the Marshallese petitioned US administrators and military officials for years, but Micronesia's unique postwar status as a strategic trusteeship omitted them from US and international legal remedies.[11] As early as 1953, an original request to cease nuclear testing was presented by Marshallese congresswoman Dorothy Kabua, the first Indigenous inhabitant of the Trust Territory to sit in a UN Trusteeship Meeting.[12] Unfortunately, her request fell on deaf years.

A year later, Kabua's warning seemed to come into fruition in 1954 during the infamous Bravo incident, a US nuclear test in which people were subjected to radioactive fallout.[13] Weeks after the incident, a formal petition was lodged to the United Nations by Marshallese petitioners Dwight Heine and Atlan Anien and customary chieftains Kabua Kabua and Dorothy Kabua.[14] The petition sought to cease the nuclear testing program after fallout victims were identified in the Marshallese atolls of Utrik and Rongelap. Sadly, this petition was defeated by a UN Resolution.[15] Another petition in 1956 presented by Marshallese petitioner Dwight Heine was also defeated, and US officials were so angry that Dwight Heine was suspended from his job. Two UN resolutions in response to the Marshallese people's formal petitions in 1954 and 1956, Trusteeship resolutions 1082 and 1493, remain the only time in which a UN body explicitly authorized specific use of nuclear weapons. The Marshallese people have carried a burden that no people should ever have to bear, exclaimed Marshallese UN Ambassador Amatlain E. Kabua.[16] There was also a petition from Representative Amata Kabua concerning the Pacific Islands Trust Territory relating to human experimentation.[17] The latter reported that Americans had conducted human blood tests of the inhabitants of the region.[18] When taken together with other reports to the Trusteeship Council and Security Council regarding the status of the health care system in the Trust Territory of the Pacific Islands, one expert stated that "the United States clearly understood that the nuclear tests had a negative impact on the physical well-being of the inhabitants."[19] Regardless, the tests continued until 1958.[20]

Sadly, as nuclear weapons abolition leader Tilman Ruff puts it, "These largely secret operations were not subject to usual laws, accountability or standards of protection for people and the environment."[21] In fact, the United

States practiced a policy of isolationism in Micronesia up to the 1960s where only nuclear scientists and anthropologists were allowed into a type of ethnographic zoo to keep strategic operations intact.[22] The United States did not allow anyone in or out without the express permission of the admiral in Guam until 1968.[23] In fact, there were no legitimate foreign investments in the Marshall Islands until 1978.[24] According to the late Marshallese politician Tony DeBrum, the entire future of the Marshall Islands was premised on the United States building these nuclear arsenals and testing them while keeping the world from knowing what they were actually doing.[25] Debrum stated vividly that the Marshall Islands was still figuring out exactly what the US was doing in the Marshall Islands.[26] In other words, a general lack of transparency on the part of the US left the nuclear events unwitnessed. With other avenues closed, the Marshallese turned to anti-nuclear activism focused on environmental law and international law, to bring their causes to a wider audience.

The lingering effects of the nuclear testing program are best explained through a general comparison of the detonated yields of explosives dropped in the Marshall Islands as compared with other nuclear explosions in relation to reparations established through American-sponsored institutions. The obvious disparity is apparent with the Marshall Islands' explosive yield dwarfing other US nuclear detonations particularly those discharged during WWII and at the Nevada nuclear test site.[27] The total yield of the sixty-seven tests conducted in the Marshall Islands was 108 megatons.[28] This is equal to between 6,000 to 7,200 Hiroshima bombs and equivalent to the explosive force of over one hundred million tons of TNT, more than seventy-five times the total yield of all the US nuclear tests in Nevada.[29]

According to the US Centers for Disease Control and Prevention, approximately 6.3 billion curies of radioactive iodine-131 was released in the Marshall Islands.[30] In a striking comparison on US soil, estimates show 150 million curies of iodine-131 were released during the Nevada tests, 40 million curies during the Chernobyl accident, and 739,000 curies in Hanford during its weapons production operations.[31]

Despite the greater explosive yields, higher levels of radioactivity, and wider affected geographic area, US sites received by 1998 a total of $115.7 billion in cleanup costs while the Marshall Islands was only afforded $0.34 billion as of 1999. The total explosive power of the nuclear tests in the Marshall Islands accounts for 80 percent of the yield of all atmospheric tests exploded by America.[32] To put this level of testing at Bikini and Enewetak in perspective, this is the same as "1.6 Hiroshima bombs being detonated every day for the 12 years of testing," an angry surviving spouse of a nuclear victim named Giff Johnson pointed out.[33]

Even as one fully dissects the Bravo incident of 1954, the inequality is evident. Before the first Hiroshima-sized atomic tests at Bikini in 1946, the American military evacuated Enewetak, Rongelap, and Wotho residents as a safety precaution.[34] Yet during the Bravo test of March 1, 1954, which was predicted to be 250 times as strong as Hiroshima but was revealed to be 1,000 times stronger, nobody was evacuated until after direct exposure.[35] This mere fact alone has fostered a sense of distrust toward America among nuclear victims who still hold the view that they were treated like lab rats or guinea pigs. While many claim that the nuclear issue started in 1946 when the Bikinians were duped into believing the "Good for Mankind" argument, late Paramount Chieftain Imata J. Kabua and former First Lady Emlain S. Kabua state that the mistake began when the real Paramount Chief of Bikini Jeimata Kabua was denied his agency and his legitimate right to resettle the Bikinian population into more fruitful lands to provide for their sustenance via a subsistence lifestyle. Journal records of Dorothy Kabua, the wife of Jeimata's son Lejolan Kabua, showcase the Americans going against Jeimata's wishes to resettle the population in the Kabinmeto area triangle of Wotho, Ujae, and Lae in favor of Rongerik, a plan that Jeimata Kabua knew nothing about. As a result, the Bikinians nearly died of starvation in Rongerik atoll where the land was insufficient to provide for their sustenance. In fact, one elderly lady died as a result of this settlement in Rongerik atoll. According to Rongelap activist Abacca Anjain Maddison and United Nations Ambassador Amatlain Elizabeth Kabua, this is the classic divide and conquer strategy that America is known for. This was finally settled in the Bikini Act, however, the damage had already been done, with a clear separation of ties between the Bikinian people and their original chief. It is also stated in High Court archives that Lejolan Kabua and Dorothy Kabua believed that American advisors amplified the separation of ties between Jeimata Kabua and the Bikinians.

While the American personnel had the luxury of being aware of the fallout path of the Bravo test and were thus evacuated a day later on March 2, the people of Ailinginae and Rongelap were evacuated two days later after being directly exposed to the showering radioactive white dust.[36] A day later on March 4, the people of Utrik atoll were also evacuated.[37] Unfortunately, the people of Ailuk who suffered fallout mirroring that of Utrik atoll were not evacuated because it was deemed inconvenient to evacuate 400 people.[38] However, this is but a mere portion of the problem, for one also needs to consider the subsequent events that followed after the nuclear test to get a holistic glimpse of the argument for amends.

Resettlement and Reparations

In 1957, the aforementioned Rongelap people were prematurely returned home to their atoll by the United States. A Brookhaven National Laboratory report admitting to conducting human subject experimentation stated: "Even though the radioactive contamination of Rongelap island is considered perfectly safe for human habitation, the levels of activity are higher than those found in other inhabited locations in the world. The habitation of these people on the island will afford most valuable ecological radiation data on human beings." Despite numerous appeals from the people of Rongelap to the United States government to move out again due to nuclear-related illnesses, the Rongelapese were not taken out of the atoll. Many were later voluntarily evacuated by Greenpeace in 1985 because of evidence of excessive radiation levels unacceptable in other US jurisdictions. Ruff explains that the "soil in Rongelap at that time contained about 430 times the amount of plutonium and other transuranics than the northern hemisphere average."[39]

The people of Bikini underwent a similar fate after President Lyndon B. Johnson publicly announced America's decision to resettle them in 1968.[40] Unlike the Rongelap people, the people of Bikini were not directly exposed to Bravo. After the Atomic Energy Commission released a public statement in 1969 stating that "there's virtually no radiation left and we can find no discernible effect on either plant or animal life," the people of Bikini went back home not knowing that the same fate awaited them that the Rongelapese experienced. By 1978 the people of Bikini were once again evacuated after ingesting high levels of strontium and radioactive cesium in the water and soil, which was considered the highest in the world at the time.[41]

When the Marshall Islands was officially granted independence by the United States in 1986 through the Compact of Free Association, the political settlement payment (known as Section 177) was entirely one-sided with key classified evidence being withheld from Marshallese negotiators.[42] No Marshallese knew the cost of cleanup and the lingering effects of radiation that would take a toll on the environment that a subsistence community depends on for food and water.[43] This is why a Nuclear Claims Tribunal through the Compact of Free Association was set up to address future claims that might arise. Yet the implication is obvious. Withholding classified information was key in keeping the situation contained and the American colonial narrative maintained to avoid responsibilities for a nuclear legacy unforetold, censored, and supported via structural violence using an espousal clause in the Section 177 settlement.[44]

Fortunately, Marshallese negotiators inserted a changed circumstance provision to the infamous 177 settlement, allowing the Marshall Islands to petition the US Congress for additional funding if losses or damages were discovered after the effective date of the agreement. These injuries could not have been reasonably identified at the effective date of the agreement, and such failures provide a legal humanitarian basis to render the agreement manifestly inadequate. The US congressional record is clear in showcasing America's commitment toward international law, particularly in upholding human rights as they crafted the Compact agreement. However, Senator Alan Cranston elaborated doubts on the agreement, stating that the provisions established in the $150 million trust fund denied 5,000 Marshallese, who had already filed claims, a day in court. Senator James McClure, then ranking minority member of the Committee on Energy and Natural Resources responded directly to these concerns, stating that:

> Article IX of the subsidiary contains a changed circumstance clause which would allow the Marshallese to ask Congress for relief if circumstances develop which could not have been foreseen, such as newly identified claimants. As you indicated, there is a continuing moral and humanitarian obligation on the part of the United States to compensate any victims—past, present, or future of the nuclear testing program. For this reason, I fully expect that if new claims develop, Congress should and will provide any assistance required, absent compelling contradictory evidence. . . . There is an enormous burden on Congress to state affirmatively that if future valid claims develop we will do everything possible to compensate adequately all newly-identified victims.[45]

Nuclear and Climate Activism amid American Hegemony

Unfortunately, a lack of sympathy is evident due to the lack of transparency. One needs only search in an American textbook. Though the crimes of the Nazi Holocaust are readily exposed, Project 4.1, the "Study of Response of Human Beings Exposed to Significant Beta and Gamma Radiation Due to Fallout from High Yield Weapons," remains unseen in most accounts. Being taken out of sight and out of mind, coupled with the Marshall Islands' small island power status, has led to a general lack of transparency. American hegemony in the Indo-Pacific region is very real. It was not until 1994 that more documents were declassified by the US revealing greater damage than was previously known to the Marshallese people. The recently declassified 1955

Atomic Energy Commission report showcased a significant number of atolls that were measured and received radiation doses exceeding internationally recognized standards deemed safe.[46] This new report offered a list of newly affected atolls that were not included in the aforementioned 177 political settlement. When the Marshall Islands under President Amata Kabua suggested doing a feasibility study for storing nuclear waste just to draw attention to the issue in the mass media, the nuclear legacy resurfaced on the international scene. Nuclear activism in the Marshall Islands increased, with more attempts to reveal America's nuclear legacy with a fuller narrative.[47]

The year 1996 was an important year for such activism as the Marshall Islands hosted for the very first time the Pacific Islands Forum on Majuro, Marshall Islands.[48] Previously, in 1995, the Marshall Islands had encountered stiff opposition from the Pacific power, Australia, and a few other countries, when it sought to insert the unresolved nuclear issue in the Pacific Island Forum Communiqué, in an effort to remedy past injustices.[49] Prime Minister Paul Keating and his Pacific Islands' Minister Gordon Bilney spoke out and lobbied against it.[50] In the end, after an exchange of sympathetic views and extensive consultations, the Pacific Islands Forum, under the leadership of Papua New Guinea's Prime Minister Sir Julius Chan, adopted the entire text for the very first time as originally submitted by then President Amata Kabua.[51] For the sake of clarity, the entire text is given in full below.

> The Forum again reaffirmed the existence of a special responsibility toward those peoples of the former United Nations Trust Territory administered by the United States, the Marshall Islands, who had been adversely affected as a result of nuclear weapon tests conducted during the period of the Trusteeship. This responsibility included safe resettlement of displaced human populations and the restoration to economic productivity of affected areas.[52]

This responsibility in the text given above was echoed in the 1995 Conference of the Treaty on the Non-Proliferation of Nuclear Weapons, and was again reintroduced in 1996.[53] It seemed that President Amata Kabua's political tactic of causing a media frenzy, by alluding to doing a feasibility study for nuclear waste, had succeeded in drawing attention to a much-neglected nuclear past.[54] The plans for such a study were never approved, despite numerous offers to conduct it. Kabua's former assistant Fred Pedro noted that he refused to sign off on any documents relating to the feasibility study because President Kabua did not give him permission to do so. Unfortunately, President Amata Kabua passed that same year, leaving an unresolved task for nuclear justice.

This untold story was relayed by his former assistant Fred J. Pedro, who is participating in the new Compact negotiations still ongoing as of 2023.[55]

Another attempt at legal remediation came in 2000 when the Nuclear Claims Tribunal through the Marshall Islands Parliament presented its Changed Circumstance Petition to the US Congress stating that new facts in light of new knowledge rendered the original agreement manifestly inadequate.[56] The new facts included unforeseen environmental damages that showed greater radioactivity in Enewetak. One expert stated that the outside of the Runit Dome was just as radioactive as inside it.[57] It should be fairly noted that unforeseen circumstances could also be in the form of ingestion because of subsistence dependency on crops and fish. If the person were to consume only foods grown in the islands for their entire lifetime, then the initial annual radiation dose would be higher, given bioaccumulation consideration and standards. This request for additional compensation to account for unforeseen circumstances also fell on deaf years, and as late as 2023 when this essay was published, it still sits with the US Congress awaiting formal action.[58] The United States through the George W. Bush administration claimed that it had already dispensed $5 billion and that fifty-plus years of money was enough. Five billion dollars is surely a lot of money, but, as Barbara Rose Johnston has argued, if you tear apart that $5 billion, it includes all of the cost of human subject experimentation and costs of a whole host of things that don't compensate in a just way the people of the Marshall Islands.[59] The United States has greatly devalued international obligations to human rights, preferring instead to place price tags on Indigenous Marshallese victims, recording them as less than others. US foreign policy on Marshallese people has set an existing precedent of inhumane treatment, and it has gone largely unseen and ignored, institutionalized under the existing global hierarchy.

In response to inaction, a United Nations special rapporteur visited the Marshall Islands in 2012 to do a formal evaluation of continuing human rights obligations on the part of the United States. The special rapporteur noted that his 2012 visit was the first by a United Nations official in almost sixty-five years.[60] The United Nations Special Rapporteur expressed particular concern about the radioactive dump site on Runit Islands in Enewetak, Marshall Islands, stating that there was evidence of lack of structural integrity. This nuclear waste storage facility (i.e., Runit Dome) is where the US "dumped 35 Olympic swimming pools' worth of atomic soil and debris created by its Cold War nuclear weapons testing program," as the *Los Angeles Times* put it.[61] In keeping with its elements of secrecy to avoid transparency, "The United States did not tell the Marshallese that in 1958, it shipped 130 tons of soil from its atomic testing grounds in Nevada to the Marshall Islands," the newspaper added.[62]

This is in light of the fact that under Title One, Article VI of the Compact of Free Association between the Republic of the Marshall Islands and the United States, the United States government is obligated to apply environmental standards to its activities that are substantively similar to US environmental statutes.[63] Section 161 of that same compact requires the United States government to develop judicially reviewable standards and procedures to regulate its activities. In short, the compact requires the United States, in cooperation with the Marshall Islands, to develop and apply to its activities a set of standards and procedures that will provide environmental protections that are appropriate to the particular environments of the Republic of the Marshall Islands, cognizant of the special political relationship between the United States and the Marshallese people. It states that the United States government must apply the National Environmental Policy Act (NEPA) to its activities as if the Marshall Islands were within the United States of America.

Today, the Runit Dome is "leaching as predicted"[64] amid the effects of sea level rise.[65] According to Ruff, "The United States has generally ignored the landmark recommendations of the United Nations Special Rapporteur."[66] However, given the historical role of the United Nations in designating the Marshall Islands as a strategic Trust Territory, there is still a moral and humanitarian obligation for the international community, through the United Nations, to assist in the care, fair compensation, and effective cleanup owed to the environment, health, and well-being of the Marshallese people, he adds.[67] This remains to be done. Lack of effective remediation by the United States is but one of the most pressing concerns for the nuclear community of the Marshall Islands.

The new changing circumstance in the form of climate change is an equally pressing one to consider, and it adds further environmental concerns to the existing nuclear ones. Flooding in Kili, where the people of Bikini were resettled by US military personnel, is becoming more and more frequent at every king tide. This repeated ocean water flooding over the past years has pushed the Bikini Council to request US government assistance to relocate the population that has lived in exile since the commencement of the US nuclear testing program. As of 2015, the Bikini Council formally approved two resolutions requesting the US Department of the Interior to allow the Resettlement Trust Fund, established in 1982 by US public law, to be used for relocation of Bikinians outside of the Marshall Islands. A formal request was given to Esther Kiaaina, the assistant secretary of the interior for insular areas, by former Bikini Mayor Nishma Jamore. Nishma Jamore explains this decision, stating: "In the future, we may have no option but to relocate. We are preparing for the future. Climate change is real. We are feeling and experiencing it." Former

Bikini Councilwoman Lani Kramer expresses further frustration about the situation. "The Americans said 'we will always take care of you as the children of America' and we believe that if promises are made, they will be kept," says Lani Kramer. After more than seventy years, there has yet to be any action to get the people off this small island, she adds.[68]

Low-lying atolls and islands in the Marshall Islands remain the most vulnerable to climate change impacts, making it increasingly difficult for nuclear victims. With fragile ecosystems and massive coral bleaching due to temperature increases, a loss of livelihood becomes inevitable as the exacerbation of climate change impacts remain unchecked. The Human Rights Office of the High Commissioner has indicated that "climate change impacts severely limit the range of human rights by people throughout the world, including the rights to life, water and sanitation, food, health, housing, self-determination, culture and development."[69] As Pacific islanders, climate change poses the greatest threat to the existence of Marshallese nuclear victims, given their natural ties to their land and waters. By not formally addressing these various changing circumstances, including the formal Changed Circumstance Petition submitted to US Congress in the year 2000, human rights violations continue to plague a large number of Marshallese. These are nuclear victims who sacrificed immensely to the American military cause, and many have paid the ultimate price for what the Americans called peace and security.[70]

Although the Radiation Exposure Compensation Act (1990) has aided some atomic veterans, uranium miners, and nuclear weapons test Downwinders in the continental US, the Marshallese variant in the form of the Nuclear Claims Tribunal has yet to be replenished. The explosive yields in the Marshall Islands was seventy-five times higher than in Nevada and the release of radioactive iodine-131 was more than forty times higher.[71] Despite the higher yield, the inequality is apparent with Marshallese nuclear victims receiving, on average, less compensation than their American counterparts. A 2002 study concluded that there is an obvious disparity and inequality between American Downwinders as compared with Marshallese nuclear victims. The study concluded that on every atoll of the Marshall Islands, the average external radiation dose from US nuclear testing surpassed the average level for Americans living in the six counties nearest the Nevada test. Over $1 billion has been granted to date by the Downwinders' Act on behalf of 24,266 individuals as compared to only $72.9 million dollars for 60,000 Marshallese nuclear victims.[72] Despite their best efforts, the existing failure on the part of the US Congress to provide for these injuries has officially rendered the original 177 Agreement manifestly inadequate. This apparent inequality brings back past experiences into the fore, particularly those of the Bravo victims of Rongelap, who were also denied legal remedy in the 1960s when eighty Rongelapese were paid a mere $950,000 as opposed to the twenty Japanese Lucky Dragon Fishermen who received $2 million in the 1950s.

The only solid evidence of sympathy came in 2006 by the US House of Representatives. In 2006, the US House of Representatives unanimously voted to "commend the people of the Republic of the Marshall Islands for the contributions and sacrifices they made to the United States nuclear testing program."[73] However, words and actions are two different things. Due to the apparent failure to date by the US Congress to deliver just compensation, this commendation rings hollow.

Despite the Paris Agreement becoming international law on November 4, 2016, President Donald Trump and his administration took progress backward by repealing many of President Barack Obama's environmental initiatives. Trump's administration proposed budgetary cuts to many of the institutions that provide relief to the nuclear community of the Marshall Islands. This included budget cuts to the Department of Energy and the Department of the Interior that handle many of the programs for the affected populations mentioned. It did not help that President Trump and his administration were climate change deniers, having pulled the United States from the Paris Agreement, despite the overwhelming evidence of climate change impacts presented even in US military installations such as the Ronald Reagan Test Site located in Kwajalein, Marshall Islands.

The scarred population seeking redress from the US government has been extremely active in the international scene. Grassroots movements represented through the nonprofit organizations REACH-MI (Radiation Exposure Awareness Crusaders for Humanity–Marshall Islands) and Jo-JiKum and student activism through MISA4thePacific are coming into fruition despite a depressing reality. Pope Francis has also called for greater environmental concern, thereby allaying the collective fears of Marshallese nuclear victims suffering from the current state of sea level rise on top of unaddressed outstanding nuclear issues. Yet President Trump delayed action for climate change by cutting EPA regulations and called into question the recent humanitarian and environmental progress achieved thus far, leaving Marshallese nuclear victims in a state of quandary as to what the future may hold.

Conclusion: The Argument, the Position, the Case

To conclude, the apparent current circumstances of Marshallese nuclear victims reveal the justified aggressive political stance of the Marshall Islands in attaining some modicum of justice for its victims. The history of the

Marshallese people and their nuclear victims speaks for itself in relaying this reality. Although expanding environmental laws and international laws have provided greater transparency and avenues to address human rights concerns, words and actions are two different things. The fearless decision to seek disarmament by suing the nuclear powers of the Indo-Pacific region, including its hegemon—the United States—reveals this startling reality that the Marshall Islands is desperate to achieve justice for its people in the name of human rights. Furthermore, the ongoing international discourse established by bilateral and multilateral frameworks reveals on record that these efforts are not completely futile as an increasing conscience has taken root.

Although the state of US politics by the Trump years revealed a depressing reality for Marshallese nuclear victims awaiting just reparations, the international community has provided greater remedy in telling and re-telling their stories. With this in mind, human rights are thus permanently woven and embedded in the ongoing discourses defining international law and environmental law. These expanding and emerging discourses depicting the situation of Marshallese people provide greater understanding of the human consequences of nuclear weapons development as well as unsustainable development, for both are intricately connected through the lens of the environment. Although the nuclear lawsuit initiated by the Marshallese government failed to deliver promising results, as it lost in both the International Court of Justice and in the US Ninth Circuit Court, it nonetheless created an avenue of transparency by re-stirring the pot, creating multiple discourses surrounding the issue of international law, environmental law, and the human rights that come with it. The recent Treaty on the Prohibition of Nuclear Weapons, passed by the UN General Assembly in July 2017, shows that the fight for nuclear justice remains. Article 7 of that treaty reminds global audiences of user state responsibilities to former test regions such as the Marshall Islands. The right of human dignity has been etched clearly in international humanitarian law, revealing the importance of equal treatment of all humans. However, the reality of the experience of Indigenous populations in the Marshall Islands is to the contrary.

The unique situation of the Marshall Islands reveals a depressing historical reality coming into greater transparency in the international scene. With outstanding human rights claims, it is no surprise that the Marshall Islands has been aggressive in the international scene in drawing attention to climate change and nuclear disarmament. Similarly, it is no surprise that the Marshall Islands leads the discussion in nuclear disarmament and climate change action, given the current state of affairs of its nuclear victims.[74] The United States has greatly devalued international obligations to human rights, promoting instead to placing price tags on Indigenous Marshallese victims, recording them as less than others. Despite the best efforts of the Marshall Islands government in presenting its Changed Circumstance Petition to US Congress in 2000 and those of the United Nations in presenting its Special Rapporteur Report in 2012 solidifying lack of the US Congress's response to that Changed Circumstance Petition, institutionalized discrimination within the current international machinery continues unabated.

The right of self-determination is promised by the United Nations Charter. However, the human rights of the Marshallese people have been infringed on, perhaps because the United States is the biggest financial contributor to the UN. The right of self-determination promised by the UN Charter includes the right to be a self-determining people but only if afforded adequate health care to enjoy the right to life as rightfully promised in the Universal Declaration of Human Rights. In this light, US congressional sympathy is thus shown to have been diluted greatly. It has not honored human rights commitments etched in their bilateral treaties of Free Association with the Marshall Islands government, though their governments were founded upon respect for human rights and fundamental freedoms for all.[75] Still, international treaties such as the recent Treaty on the Prohibition of Nuclear Weapons have provided an avenue for open dialogue revealing that the Marshall Islands is still living up to its commitment, with its recently established National Nuclear Commission, to ensure that the voiceless nuclear victims are heard.[76] Ironically, these once brave "sea navigators"[77] are now bravely navigating the anxieties of the United States' New World Order, kindly reminding the world that America is more than its politics.

Notes

1 Desmond N. Doulatram, "An Enduring Injustice," review of *Blown to Hell: America's Deadly Betrayal of the Marshall Islanders,* by Walter Pincus, Arms Control Association (Armscontrol, May 1, 2022), https://www.armscontrol.org/act/2022-05/book-reviews/blown-hell-americas-deadly-betrayal-marshall-islanders.

2 Michael Martina and David Brunnstrom, "Biden Envoy to Visit Marshall Islands as US Concerns Grow about China's Pacific Push," *SWI,* June 8, 2022, https://www.swissinfo.ch/eng/biden-envoy-to-visit-marshall-islands-as-u-s--concerns-grow-about-china-s-pacific-push/47656318.

3 Amata Kabua (First President of the Marshall Islands) in discussion with Ambassador Amatlain E. Kabua, July 13, 1996.

4 Tsuyoshi Hasegawa, "The Atomic Bombs and the Soviet Invasion: What Drove Japan's Decision to Surrender," *Asia-Pacific Journal: Japan Focus* 5, no. 8 (2007): 1–30.

5 Hilda C. Heine and Julianne M. Walsh, *Etto nan Raan Kein: A Marshall Islands History* (Honolulu: Bess Press, 2012), 289.

6 Tilman A. Ruff, "The Humanitarian Impact and Implications of Nuclear Test Explosions in the Pacific Region," *International Review of the Red Cross* 97, no. 899 (2015): 777.

7 Amata Kabua, "Formal Address of RMI President on Constitution Day," Speech at Court House, Majuro, Marshall Islands, March 1, 1979. http://www.pacificdigitallibrary.org/cgi-bin/pdl?e=d-000off-pdl--00-2--0--010-TE--4-------0-1l--10en-50---20-about-may+1%2c+1979+inauguration--00-3-1-00bySR-0-0-000utfZz-8-00&a=d&cl=search&d-=HASH01b3d73be08790dd2ba72bbc.13

8 See National Nuclear Commission Act: https://rmiparliament.org/cms/images/LEGISLATION/PRINCIPAL/2017/2017-0034/NationalNuclearCommissionAct2017_1.pdf

9 Desmond Narain Doulatram, "Evidence of Inconsideration and Inequality towards a Pacific Island State Serving US Military Interests in the Asia Pacific Region: Marshall Islands Provide Greater Transparency through Moral Angle Despite American Hegemony," *Journal of Humanities and Cultural Studies R&D* 3, no. 2 (2018): 1–90.

10 Dorothy Kabua, "Micronesian Representative Addresses Trusteeship Council," *Department of State Bulletin* 29, no. 732 (1953): 151.

11 Sheri Englund, "Historian Chronicles Islanders' Fight for Environmental Justice," *Cornell Chronicle*, last modified April 12, 2017, https://news.cornell.edu/stories/2017/04/historian-chronicles-islanders-fight-environmental-justice.

12 Heine and Walsh, *Etto nan Raan Kein*, 305.

13 Desmond Narain Doulatram, "A Marshallese Tale of Modernity in the Pacific Rim: Bridging the Modern and the Traditional," *Journal of Humanities and Cultural Studies R&D* 3, no. 3 (2018): 1–66.

14 Nic Maclellan. "*The Survivors*—Lemeyo Abon and Rinok Riklon," in *Grappling with the Bomb: Britain's Pacific H-Bomb Tests*, ed. Nic Maclellan (Canberra: Australian National University Press, 2017), 44.

15 Hilda Heine, "Formal Address of RMI President on Nuclear Victims Remembrance Day," Speech at Delap Park, Majuro, Marshall Islands, March 1, 2016.

16 Amatlain E. Kabua, "Official Statement of Marshall Islands UN Permanent Representative on Vote of Draft Treaty on the Prohibition of Nuclear Weapons," Speech at United Nations conference to negotiate a legally binding instrument to prohibit nuclear weapons, New York, New York, July 5, 2017.

17 Petition from Representative Amata Kabua Concerning the Pacific Islands, T.PET.10/30/Add.1, November 3, 1959. In United Nations Trusteeship Council, Petitions, T/PET.10/1-T/PET.10/150, June 1950–June 1979.

18 Aaron Steven Wilson, "West of the West?": The Territory of Hawai'i , the American West, and American Colonialism in the Twentieth Century" (PhD diss., University of Nebraska-Lincoln, 2008), 180.

19 Ibid.

20 Ibid.

21 Tilman A. Ruff, "The Humanitarian Impact and Implications of Nuclear Test Explosions in the Pacific Region," *International Review of the Red Cross* 97, no. 899 (2015): 777.

22 Carl L. Heine, *Micronesia at the Crossroads: A Reappraisal of the Micronesian Political Dilemma* (Canberra: Australia National University Press, 1974), 20–21.

23 Tony Debrum, "Acting Locally and Globally to Pressure the Capitals: What the US Owes the Marshall Islands," St. John's University, New York City, September 20, 2014. Statement of Guest Speaker Minister Tony Debrum. https://www.youtube.com/watch?v=aPPyQPfFTBs&t=3s

24 Ibid.

25 Ibid.

26 Ibid.

27 Harvard Law Student Advocates for Human Rights, *Keeping the Promise: An Evaluation of Continuing US Obligations Arising Out of the US Nuclear Testing Program in the Marshall Islands* (Cambridge, MA: Harvard Law School Student Advocates for Human Rights, 2006), 4.
28 Ibid
29 Ibid.
30 Ibid.
31 Ibid.
32 Giff Johnson, *Nuclear Past: Unclear Future* (Micronitor: Majuro, Republic of the Marshall Islands, 2009), 4.
33 Ibid.
34 See Bikini Atoll Act of 1994, http://rmiparliament.org/cms/images/LEGISLATION/PRINCIPAL/1994/1994-0086/CustomaryLawBikiniAtollAct1994_1.pdf
35 Johnson, *Nuclear Past: Unclear Future*, 4.
36 Harvard Law Student Advocates for Human Rights, *Keeping the Promise*, 4.
37 Ibid., 4–5.
38 Johnson, *Nuclear Past: Unclear Future*, 3–4.
39 Ruff, "The Humanitarian Impact and Implications of Nuclear Test Explosions," 796.
40 Doulatram, "Evidence of Inconsideration and Inequality," 1–90.
41 Ibid.
42 Ibid.
43 Ibid.
44 Ibid.
45 Ibid.
46 There has been recorded political action addressing this, particularly through Bill S.1756. A substitute version of Bill S.1756 would have expanded the health care program with an increase in funding to include atolls not recognized in the original 177 settlement, such as Ailuk, Mejit, Likiep, Wotje, Wotho, and Ujelang atoll, but this bill was never introduced and the original bill died in the 110th Congress in December 2008. Another bill in 2010, Bill S.2941 that was nearly identical to Bill S.1756, also died in the 111th Congress in December 2010. Another bill, Bill S.342, was introduced in February 2011 to provide $4.5 annually to ten atolls previously mentioned in Bill S.1756 and Bill S.2941. It was officially referred to the Committee on Energy and Natural Resources, but no further action was taken.
47 Doulatram, "Evidence of Inconsideration and Inequality."
48 Ibid.
49 Ibid.
50 Ibid.
51 Ibid.
52 Ibid.
53 1995 NPT/CONF.1995/MC.III/1
54 Doulatram, "Evidence of Inconsideration and Inequality."
55 There is also a Nuclear Steering Committee in the Compact Negotiation Team, https://www.facebook.com/PresidentOfficeRMI/posts/2607918549481207.
56 Doulatram, "Evidence of Inconsideration and Inequality."
57 Mark Willacy, "A Poison in Our Island," *ABC NEWS*, May 3, 2019, https://www.abc.net.au/news/2017-11-27/the-dome-runit-island-nuclear-test-leaking-due-to-climate-change/9161442?nw=0
58 Doulatram, "Evidence of Inconsideration and Inequality,"

59 Barbara Rose Johnston, "We Are All Downwind: The Environmental Health History and Continuing Consequences of Nuclear Militarism in the Marshall Islands," Public lecture by Center for Political Ecology environmental anthropologist Barbara Rose Johnston, Recorded in Majuro, RMI, March 2, 2017, as part of the Marshall Islands Nuclear Legacies Conference: Charting a Path Towards Justice, video editing by Brian Cowden, *Center for Political Ecology*, https://www.youtube.com/watch?v=LewmoMFh3l8&t=127s

60 The UN Special Rapporteur concluded: "The Nuclear testing resulted in both immediate and continuing effects of the human rights of the Marshallese. . . . Radiation from the testing resulted in fatalities and in acute and long-term health complications. The effects of radiation have been exacerbated by near-irreversible environmental contamination, leading to the loss of livelihoods and lands. . . . Many people continue to experience indefinite displacement."

61 Susanna Rust, "US Won't Clean Up Marshall Islands Nuclear Waste Dome But Wants It Free of Anti-US Graffiti," *Los Angeles Times*, November 14, 2019, https://www.latimes.com/environment/story/2019-11-14/marshall-islands-runit-nuclear-waste-dome-site-graffiti.

62 Susanna Rust, "How the US Betrayed the Marshall Islands, Kindling the Next Nuclear Disaster," *Los Angeles Times*, November 10, 2019, https://www.latimes.com/projects/marshall-islands-nuclear-testing-sea-level-rise/.

63 Doulatram, "Evidence of Inconsideration and Inequality."

64 "Interview: President Amata Kabua," *Pacific Magazine*, March–April 1983.

65 The House Committee on Armed Services introduced H.R.2500 in June 2019, which included a requirement for the secretary of energy to produce a detailed report on the status of the Runit Dome 180 days after the enactment of said act, which was titled "National Authorization Act for Fiscal Year 2020."

66 Ruff, "The Humanitarian Impact and Implications of Nuclear Test Explosions," 798.

67 Ibid.

68 Doulatram, "Evidence of Inconsideration and Inequality."

69 Ibid.

70 Ibid.

71 Ibid.

72 Ibid.

73 Ibid.

74 "World Court Rejects Marshall Islands Suit Seeking Nuclear Disarmament," *Reuters*, last modified October 16, 2016, http://www.reuters.com/article/us-nuclear-marshallislands-court-idUSKCN1250U0.

75 See Compact of Free Association Preamble.

76 See National Nuclear Commission Facebook Page: https://www.facebook.com/pg/RMINNC/about/?ref=page_internal

77 Lejolan Kabua and Amata Kabua, "Marshall Islands," In *Micronesian Navigation, Island Empires and Traditional Concepts of Ownership of the Sea*, by Masao Nakayam and Frederick L. Ramp (Saipan, Mariana Islands: Congress of Micronesia, Trust Territory of the Pacific Islands, 1974), 84–85.

Our Action Now, Our Future, Our Resilience

DESMOND NARAIN DOULATRAM

"Our Actions Now" Should be more than about "Wows"!
PICs[1] need to redefine their ways
This is what I have to say
It's more than "Dark Anthropology"[2]
Seeking only an Apology
Like those on "Runit Dome" seeking only apologists Believing in the "The Way" of the Conformists Absorbed by the global narrative of Cowards
What are we moving Towards?
"Anthropology of the Good?"[3]
To Change the Mood?
Neglecting the truth is very contradictory!
This future of ours is imaginary!
Our Actions are LAME!
Seeking only that of FAME!
Our Resilience!
Is Merely Diplomatic Silence!
Speak Up!
Act Up!
Our Future!
Is Mature!
Our Resilience!
Is not Silence!
Speak UP!
Act UP!
Our Future!
Needs to Mature!
To Be Resilient!
Does not mean to be Silent!
Your Future!

Becomes when you Mature!
Time is not on our Side
Climate Change is Near
So why do we conformingly Abide
When we will lose what we hold Dear!
Speak UP!
Act UP!
Our Future!
Is Mature!
We are Resilient!
We are no longer Silent!
This is Our Future!
We are NOW Mature!

Notes

1 PIC is an abbreviation for Pacific Island Country.

2 "Dark Anthropology," according to Sherry B. Ortner, is "anthropology that focuses on the harsh dimensions of social life (power, domination, inequality, and oppression), as well as on the subjective experience of these dimensions in the form of depression and hopelessness." See Sherry B. Ortner, "Dark Anthropology and its Others," *HAU: Journal of Ethnographic Theory* 6:1 (2016), 47-73.

3 "Anthropology of the Good," according to Sherry B. Ortner, is explicitly or implicitly a reaction to the dark turn of the "Dark Anthropology" mentioned above under the rubric of "anthropologies of the good," including studies of "the good life" and "happiness," as well as studies of morality and ethics.

History Uncontained at the B Reactor

JEFFREY C. SANDERS

When I first visited the B Reactor near Hanford, Washington, in 2009, I was a new professor in charge of a group of eager students from my Columbia Plateau history class. As the day approached for the long-planned journey, I wondered if I had done the right thing by orchestrating this field trip to a place where time is measured in half-lives. Since then, I've returned repeatedly, like a moth to a flame, shepherding students, middle school teachers, and even colleagues to the old reactor.

Before Congress designated the reactor as part of the new Manhattan Project National Historical Park in 2015, visitors in the know like us had to sign up a year in advance for the limited number of spots offered each short season. The sense of scarcity, the advance planning, and for me, getting up so early, heightened my anticipatory jitters. Weeks before, I had dutifully emailed my students the link to the Department of Energy's (DOE) list of dos and don'ts: *Children under 17 not allowed. No opened-toed shoes. Picture ID required. Incendiary devices prohibited.* The journey included a 90-minute drive, down from the calm, rolling wheat-filled hills of the Palouse. Then we rode a bus out of Richland, Washington, for another 45 minutes, traveling along the sloping base of a big beige mountain. On the way we half-watched the Disney film *Our Friend the Atom* about the wonders of the nuclear industry. Mostly we absorbed the austere landscape zipping by our windows. By then the students had plenty of time to consider their choice to come with me to this once secret and still restricted death-dealing place.[1]

As we neared the backside of the Hanford site, my group lowered their voices, maybe out of respect for history's heaviness or maybe just out of their sense for protocol. The bus pulled off the highway at a simple ranch gate guarding a strip of road a mile from the reactor. The stick across the path seemed hardly up to the task of keeping anything in or out. The bus crunched and swayed its way over the gravel to a fenced yard in front of the windowless reactor building. An American flag waved to us out front. We held our water bottles tightly and ducked out under the late September sun, crossing to the refuge of the vaguely Soviet-era building. Inside the old B Reactor we scanned a claustrophobic government green entryway and huddled close to our docent, an officious veteran of the industry. Two DOE representatives in identical blue polo shirts stood nearby like overseers enforcing a party line. This was back when the DOE, industry retirees, and local boosters had a tight grip on the presentation that was only emerging, later to become a repertoire of stories and set pieces I would hear many times.

We stood in the passageway like soldiers waiting for orders, anticipating lectures that could explain this half mausoleum, half museum in the middle of nowhere.

Then just as our tour leader began to introduce himself, a large bat drunkenly flapped out of the reactor room and into the hallway above our heads. We gasped in unison, then laughed uncomfortably, a release of nervous energy. The bat seemed both dramatically out of place and perfectly appropriate. It matched the unease we all were feeling, thinking, but not saying.

Bats are often associated with death and doom, vectors of invisible diseases and blood-borne pathogens, pollution. Both bird and mammal, they straddle species, travel at night, and upend the status quo. They hang from their feet in caves and, as it turns out, the dark recesses of old nuclear reactors. Like us, they leave behind prodigious piles of waste.[2]

In the years since, when I think of the novel materials America cooked up in that gray forge in Central Washington, I always recall the moment with the bat in the corridor. On its way past us, and through the open doors, the creature punctured our sense of distance from history. Most of all it revealed the fact of permeability, and our own physical vulnerability, even as the stories we were about to hear held fast to an illusion of control and safety. No matter how it tries to suggest otherwise, the B Reactor remains an example of failed containment. I don't think this is such a bad thing.[3]

The Bats Out of Town

With this essay I am interested in the opposite of containment. I prefer proliferation rather than control in this case and will probably go too far. Keeping the bat in mind, I've been pondering the public history potential of the B Reactor informed by the insights of environmental history. Environmental historians are known for planting their intellectual stakes deep into local soil. Here it might seem simple enough: Hanford and the B Reactor is a story of a local environment defiled. But environmental history is much more than that. It's also about people, buildings, and food, to paraphrase David Byrne. It's more than a catalog of environmental misdeeds.

In a 1992 essay, "Kennecott Journey: The Paths Out of Town," historian William Cronon suggests that the first question we might ask is, "How does this place cycle?" In other words, how do seasons, water, soil, and climate give shape and limits to what happened here? Cronon used the defunct copper mining ghost town of Kennecott, Alaska (a National Park too), to illustrate his approach for making sense of an inert environmental and social ruin. In the same essay, though, he also suggests moving beyond any fixation on the local. No matter how off the beaten track an old mining town or an old reactor may seem, he argues, historians ought to follow the paths of trade and exchange that animate places, that brought new people and animals and materials, that created new social and ecological relationships, then moved out of town to remake the world again. His recipe, with some additions, is especially suited to the B Reactor and the land where it sits.[4]

The reactor has the potential to dramatize how industrial materials —mid-century wonders, like pesticides or synthetic hormone disruptors or radio-nuclides—produced in out-of-view places like Kennecott or Hanford, moved into ecosystems, and then into different human and animal bodies. These stories are rarely triumphant. When we begin to follow by-products—products, really—as environmental historians such as Nancy Langston and the late Linda Nash have, it also reminds us that the personal, historical, and the ecological are always braided together. These approaches to history emphasize movement as much as the containment of things; supply chains as much as local ecosystems; and moments like World War II as well as much longer and deeper time scales. The legacies are also borne unevenly and unequally, depending on race, class, gender, or location.[5]

Like a journalist told to follow the money, we might similarly follow substances that cross thresholds and keep on moving. Materials themselves may have a certain agency that we will want to acknowledge too. Historians and scientists now believe that everything from the microbiome in our guts to the cows we raise or the metals we smelt have made us, and our ecological niches, as much as we've made them. Historian Timothy LeCain calls this "thing power," or the "creative dynamism of both biotic and abiotic matter, particularly in their interactions with humans." Chemicals, metals, microbes, animals, and soils, then, are the "matter of culture," the "things" that "help to create humans in all their dimensions."[6]

Together these approaches show that distant places are connected, our bodies are implicated in the story, and no matter how much control we think we assert over the material world, or what we call nature, we are ultimately tangled with it in ways we don't yet understand.

This may all sound a bit too high-minded, especially when applied to the B Reactor, a National Park no less, a relic whose story seems clean-cut. Open-ended narratives are not very satisfying or affirming either. Stories with long half-lives are especially unsatisfying. The events that surround the B Reactor remain unresolved. They get under our skin. We want clear beginnings, middles, and ends. But perhaps this is another important legacy of the atomic age. It opened a can of material worms that can't be easily closed up again. Instead, we might learn to better explain the can, the worms, and what may come next.

The B Reactor's main job was to create plutonium. According to the Washington Department of Ecology website, "More than 100,000 tons of uranium were processed to make about 75 tons of plutonium" at Hanford, and "virtually everything that wasn't plutonium remains on site." This included "the millions of gallons of chemicals used to extract pounds of plutonium from tons of irradiated uranium." By-products of the fission process—radionuclides, or the chemicals used in processing irradiated rods in separation plants—were really the main products here. Every step in the process of making plutonium—producing it, moving it around the country, detonating bombs at test sites in Nevada or the Pacific, storing weapons or wastes—involved releasing radiation but especially new radioisotopes, things that had never existed before in nature.[7]

Physicists borrowed the word fission, an analogy from biology, to describe the process that created these new things. In that discipline the term describes animal cells splitting to produce new organisms, new creatures, sometimes monsters. In the case of nuclear fission, each time nuclear engineers bombarded and split uranium atoms in their reactors, or when army engineers detonated bombs with plutonium cores at distant hinterlands, they also gave birth to novel forms of elements, unstable isotopes like cesium-137, iodine-129, and strontium-90—the seeds of new fears as well as new hopes after the war. The "primary contaminants of concern" at the site, according to the Washington Department of Ecology, include strontium-90 and iodine-129, among several others.[8]

At Hanford, these millions of gallons of pollution burrowed into the local ground. But they also followed the paths out of town—through air, water, and flesh. This story, however, is oddly downplayed the most at the reactor, despite the fact that the primary mission at Hanford and so much of the former Manhattan Project lands today is about "cleanup." Indeed, just miles away from the B Reactor itself scientists and engineers at the Hanford site continue to grapple daily with much longer time scales than the events of World War II or the Cold War. Like archivists, they endeavor to package waste and other by-products safely for a deep future, vitrifying, storing, and marking by-products containing radionuclides that will continue to be lethal for thousands of years. It waits to be reckoned with at a later date. The B Reactor is significant for its role in this history as well.

But how do you commemorate this history, making it visible? Telling a story about pollution at a National Park just sounds wrong, even if the B Reactor sits mere steps away from the largest superfund site in the United States. How do you dwell on this reality of pollution in a place that is so preoccupied with moving beyond it?

I think dwelling on environmental health, supply chains, and the products that the reactor produced can only reinforce the B Reactor's significance to visitors. Many who visit are already familiar with similar local examples from their own experience. After all, most all of us now live within a few miles of a superfund site and the legacies of this era.[9]

I say we follow the bat.

In the corridor we felt apprehensions, our understandable fears and misunderstandings about nuclear energy, its real and imagined risks, risks that the DOE would like to put to rest here but can't. The B Reactor is one of the few places in the country where the general public might actually engage with this complicated history and the thicket of questions it raises. The reactor offers a rare opportunity to make it all more visible. It may sound odd but the story of these by-products—this pollution—may even have the potential to connect us.

B Reactor Stories

As the B Reactor has become more available to the general public, folded into the Manhattan Project National Park and therefore the messiness of democracy, it is first worth reflecting on the stories that the structure explicitly and implicitly tells, avoids, or might yet show us. The B Reactor's very leakiness especially undermines our desire for containment or closure, leaving doors open. To think of it in this way is to let the metaphorical bats out and invite more in that might explain the building's productive life, its different contexts, and the reach and consequences it created between 1945 and 1968, and long after.

With historical buildings we obviously choose certain moments to fix significance, constraining some narratives and contexts in favor of others. A building appears to be a static thing too, even if it is always part of the flow of time. Structures deemed "historical" are most susceptible to this way of seeing. By choosing wartime production as the frame for this building, the B Reactor seems especially frozen in time. The avoidance of a more expansive time frame is partly a result of a script written and funded by the DOE and their mission to clean up the site or decommission it. It's a worthy mission for engineers, but not one for historians. The story, however, is slowly being pried from the department's hands by the next generation of historians and the National Park Service. We know that the significance of the place will stretch and change over time as more people like them get their hands on it.[10]

The ever-growing numbers of visitors to the old reactor arrive by way of route 240. You could easily miss the building on your way to Yakima. Nearby, not far from the banks of the Columbia River, are more reactors and related production facilities built after 1945 to supply fuel for the Cold War arms race. They are "cocooned" now, sealed up with corrugated metal like TV dinners in the back of an abandoned mid-century freezer. Wrapped up tight by the DOE, they wait for residual radiation to safely decay. As the Hanford "cleanup" has progressed, and the facilities that once populated this corner of the 586-square-mile site are slowly dismantled, buried, or sealed, the old gray B Reactor building looks increasingly marooned and stripped of its original context. Most of the post-1943 buildings in the area have been removed. From the highway it could easily be mistaken for an abandoned onion packing plant or a grain storage facility made of the same pale cement as Coulee or nearby McNary Dam.[11]

The nondescript building was of course a linchpin in the Manhattan Project. The B Reactor, in its latest national park incarnation, offers visitors a tightly focused narrative that seldom strays from World War II. As the first large-scale plutonium production facility in the world, it played a critical role in manufacturing the devastating punctuation to that worldwide conflagration. Though Hanford can claim only the Nagasaki bomb, it was one of the two and only nuclear bombs ever used in war on human populations. Together the bombs killed well over 200,000 people. Most died on August 6 and August 9. But many more died of radiation burns and sickness in subsequent days and years. This story is given some of its due at the reactor. And though it is not central, it nonetheless hovers over every other story at the site. Indeed, according to one possible narrative, we might trace the beginning of the nuclear age to the reactor's door.[12]

On the tour we learn of the massive effort to safely construct and operate the reactor during the war while keeping the project a secret; the famous scientists, like Enrico Fermi who worked in cramped offices there; the reactor's alchemical magic tricks; the safety features; the ingenuity and pluck of engineers; the plutonium produced for the Trinity test; and finally, the Nagasaki bomb as an endpoint in the narrative. The B Reactor won the war, we are told and "saved lives." And all of this would seem like enough history for one building. But after visiting a half-dozen times now, I'm always left wondering again why the story has to stop at 1945 and not the next year, the next three decades, or much longer or deeper.[13]

Once when I strayed from the group and the script to walk the polished cement floor of the reactor building, I noticed the yellow tags on decommissioned nozzles and wide pipes that once delivered water from the river. The reactor only ceased making plutonium in 1968. Though significant to the Manhattan Project and World War II, the B Reactor kept chugging right along well after the war. Yet the stretch of history between 1945 and 1968—the megadeaths of thousands of people at Nagasaki, the contested health consequences of many distinct forms of radiation and other by-products left in the wake of "the bomb" and hundreds of subsequent bombs—mostly recede from view in the telling. The long complicated Cold War itself is muffled here. The building instead settles mostly on a nationalist or highly local set of stories that seem suspended between the end of one age and the beginning of another. This choice of time scales is revealing in what it does not reveal.

The emphasis on engineering makes sense. Once inside you can easily imagine you are in an old navy submarine plying the dry scrublands. Many of the tour leaders who narrate this atomic history for us were once nuclear engineers on Navy vessels too. In fact, dozens of mothballed reactors from submarines and ships on which they served sit ready to be buried just a few miles away from the reactor on the other side of the reservation. They run a tight ship.

The technological and scientific sublime has obvious appeal here too. The design of the reactor building interior creates an experience of atomic drama that almost transcends the mundane materials and isolated location. Walking from the hallway into the reactor room, atomic tourists like my group emerge from the green, bat corridor into a vaulted cavern. When you come around the corner, dramatically moving out of a low-ceilinged hallway and into the giant reactor room, the brutal blank face of the twentieth century stares you down.

Intentionally or not, the room suggests a cathedral sanctuary, a sacred space. Your eye moves upward. Even the seating is arranged with an aisle down the center for an imagined procession and sacraments. (Or maybe it's just a classroom waiting for an inflated professor.) The nave of this cathedral of science and technology, the reactor room enacts a different mystery of transubstantiation. In addition to the liturgy of safety, the tour guides and signage underline the building's adjacency with mid-century scientific achievement. The theory of relativity. Fuzzy-headed geniuses. Atomic magic hovers around it all. The tour guides encourage these associations in hopes of adding value and meaning to this place that was once so secret, as if to make up for lost time.[14]

The transcendent is in tension with the practical and the local. This more local story dominates. Cut off from the world and behind the fences, the lonely reactor can seem forgotten and specific to the place. It's easy to forget that once there were thousands of proud laborers and engineers working here around the clock to build this Rube Goldberg contraption. So secret was the project, though, that few who came here to build the reactor actually knew what they were working on until that day in August of 1945 when the newspaper headlines blared it and confirmed their role. In an instant the B Reactor and the hybrid government/company town that supported its mission, became tethered to the profound violence of the twentieth century. The secrecy of the original effort still pervades the space. We all watch ourselves. Watch what we say here as if we are channeling the laborers of the past.

The building is a massive cluster of redundant fail-safe measures, a marvel of pragmatic, site-specific precision engineering. They call the reactor that fills the main room inside "the pile," and it is literally a pile of stacked graphite blocks, although the reactor blocks are not visible. A pattern of gaskets and valves cover one towering side of the reactor, a multistoried cube. You stand for a moment reckoning with the machine that defined the binary metaphors we used to explain our geopolitics and domestic selves for years after, maybe still.

"Pile" sounds like something improvised, like a cairn that hikers might build to mark a trail. In a way that first pile was like a controlled campfire inside a stack of graphite, the work of ingenious Boy Scouts in lab coats at the University of Chicago. The scaled-up version, the B Reactor pile, created tons of plutonium between 1943 and 1968. The reactor supplied the plutonium that filled hundreds of warheads, including for the bomb test at Trinity, the devastation of Nagasaki, and tests at far-flung proving grounds throughout the 1950s. American ingenuity and exceptionalism dominate.

The building's imprint on the local ecology as well as its local material origins gets some attention inside, but not enough. In the process of creating plutonium as noted earlier, the reactor left a legacy of waste and contamination that still resides right here at Hanford, some of which is now being gathered up and stored in the middle of the reservation. Water, minerals, metals, and energy coursed through it and out again, transformed and hot. Thousands of custom-made tubes fed into the thousands of custom-made valves in the face of the reactor. Millions of gallons of glacial melt handy nearby cooled down the heat that split atoms spun out. In a room near the reactor, the stones that laborers lifted from the bed of the Columbia fill giant metal buckets, counterweights at the ready to immediately lever the control rods into the reactor if the power from Coulee Dam goes out. Boron balls in giant hoppers still sit at the ready to cascade into the pile and cease a reaction. These rich material details of the local engineering histories are told at the reactor, but less than you'd think, and ecology is rarely mentioned.

It shouldn't be a surprise either that the B Reactor packages its story so tightly. The DOE, which still oversees the land and manages the historic site, wrote the script. The agency has a vested interest in telling a story as controlled and tidy and final as they hope the interminable waste cleanup might be. The building's very ethos has always rested on the controls necessary to safely avoid a meltdown too. An unchecked chain reaction during those years of operation would have meant total disaster. Such a major failure never happened, a testament to that ethos and the engineers who worked there, and a lot of good luck.

The success of this first plutonium-making machine and the secrecy of the project during the war, and well after, depended on actual and not figurative containment, like the carefully controlled reactions in the graphite core. This story is important to understand. But so is the story of waste and pollution that resulted from this industrial process and that still moves toward the Columbia River in the groundwater nearby. The B Reactor presentation mostly ignores its context of waste cleanup, though I think this present situation, this context indirectly gives shape to the historical narrative.

The reactor is both exceptional and completely recognizable at the same time. It belongs beside other great engineering and production achievements of the twentieth century, such as Ford's River Rouge plant or the Grand Coulee Dam. But to track what happened at the B Reactor it might help instead to see the place as less exceptional. Such an approach might go a long way toward demystifying it and its products. In fact, the activities of the Manhattan Project and then the Atomic Energy Commission were as familiar as other large American industrial concerns that grew ever more powerful and organized during the war and after. Corporations, including DuPont and General Electric lent expertise in their contracts with the government's Manhattan Project. Like other major companies, the nuclear business had its supply chains, R&D divisions, field tests, logistics, and ultimately, consumers too. The reactor was a node in this industrial process, one that included manufacture and testing of jet airplanes, rockets, and explosives, much of it in the US West where deserts and other expanses were treated as sacrifice zones. Abstracting the reactor from these more complicated but also familiar associations diminishes a deeper significance, our ability to see patterns and crucial continuities.[15]

Cronon's paths out of town, or into town, could help to track the enterprise as it unfolded in the late 1940s and the 1950s especially. The industry depended on uranium mined in the Congo and then later the Navajo Nation, Canada, and Namibia. Engineers designed machines and processes to make uranium into plutonium in the midst of war and long after. The Manhattan

Project was only the starting point of a still unfolding story. Plutonium moved out of town, assembled into hundreds of warheads at Rocky Flats, Colorado, between 1952 and 1992. In the 1950s and 1960s at isolated testing sites in Nevada or near atolls in the Pacific Ocean, workers blew stuff up like obnoxious neighbors on the 4th of July. They detonated bombs beneath the blue waters of the Pacific and blasted old Navy ships full of pigs and goats tied to the decks. They built scale model replicas of suburban neighborhoods and concrete bridges and bank vaults and bomb shelters at the Nevada Test Site (NTS). Then they blew them up leaving behind a scattered junkyard of rusted bolts, pockmarked cement abutments, flattened structures, and piles of gravel on the floor of Frenchman Flat. After the ban on atmospheric testing they drilled thousands of feet underground and detonated bombs that collapsed the earth above into massive craters, making Yucca Flats resemble the moon (Buzz Aldrin would even practice driving moon buggies in them in the late 1960s). On these un-ceded Shoshone tribal lands in Nevada, technicians bored into the sides of sacred mountains, put bombs in the holes, and blew them up too. The B Reactor's plutonium was central to each of these stories.[16]

At the NTS the products the B Reactor contributed to the arms race were well advertised, grand performances of power in an overheating Cold War standoff. Some people living in communities closest to these test sites—especially at ground zero in Nevada, Utah, and the Marshall Islands—received more immediate harmful and lethal doses of beta radiation, radiation burns, and then developed thyroid cancers and leukemias. But proximity did not determine the consumer experience in this industrial chain either. Young people as far away from the arid West and the Pacific as well—Wisconsin, Missouri, New York, or the misty hillsides of Wales—were victims of what one scholar terms "slow violence."[17]

Following Langston and Nash's example reminds us that the nuclear industry was no different from other industrial operations that produced postwar chemicals and that changed environments along with the people and nonhuman animals that depended on them. Ultimately, everyone was a potential consumer of radiostrontium and other materials manufactured at Hanford, especially the young.

In this way the ecological and the nuclear are personal too. My mother was one of the potential data points, for instance. She grew up in central Washington near the Grand Coulee Dam that supplied energy to the reactor's pumps and machinery. Kids like my mother ingested sometimes large, sometimes barely traceable amounts of radioactive materials—especially of concern was iodine-131—that drifted across the center of the state. Where radiation might land depended on the weather and hundreds of other variables. Technically she is a Downwinder (we all are to a degree), though she never developed conditions that put her officially in this group. Some of those unstable isotopes that she likely ingested came from Hanford reactors. Hanford's plumes rose mainly from the radioactive products of fission, the process of making plutonium for bombs.[18]

Washingtonians would learn some of the details about all this only later when the government reluctantly opened the Hanford archive in the 1980s. In the meantime, between the 1940s and the 1960s technicians working at Hanford intentionally released various forms of radioactive gas and chemicals to see where they'd go—one part of research and development. They named one of their tests the Green Run, an effort to learn how to trace these new plumes (or to keep track of the Soviet plumes they mirrored) as they wended through the earth's atmosphere. Here the B Reactor could tell another kind of story of failed containment. Along with the story about the release of radioactive materials, the tour might tell a story about the failure to contain secrets that also finally escaped and that left a toxic brew of public distrust of authorities that lingers.

Two Paths of Strontium-90

Intentional releases of radioactive iodine played a central role in the specific history of Hanford Downwinders and their concerns about thyroid cancers in Washington. But it was another isotope—strontium-90—that took center stage among scientists, public officials, and eventually the general public after World War II. It produced new insights, science, cultural meaning, and pain—a set of chain reactions with no clean end in sight. To make the invisible visible at the B Reactor requires following by-products like this, the same radionuclides that will continue to bedevil us for thousands of years to come.

Released from its strict national framework, the reactor provides an opportunity to place this singular building in conversation with a much broader nuclear world that struggled with similar yet specific and local histories of contamination. The reactor need not stand alone. Although American testing has ended, the global story of strontium-90 continues to reemerge too. Both citizens and state-funded scientists have spent each decade since the 1940s following strontium-90's behavior and movements from the B Reactor and the Columbia River to Nagasaki to Chernobyl to Fukushima. These isotopes revealed pathways, connections, and shared suffering.

To my mind strontium-90's cultural and ecological valence makes it one of the most useful materials we might follow out of the reactor's doors. Growing public fears and debate about bombs and atmospheric testing during the 1950s centered most intensely on this material as well, making it a national and even an international story during the Cold War and after. Treated less like an environmental externality then, the pathways of materials such as strontium-90 suggest potential stories and unintended consequences still to explore. This new material was especially important in explaining how ecological systems functioned on local and global scales.

Beginning in the 1940s millions of people around the world inadvertently but predictably consumed strontium-90. One of the most lethal and mobile by-products of plutonium production, the bombs dropped on Japan, and years of atmospheric testing, this new creature called radiostrontium steadily enveloped the planet beginning in 1945. The ubiquity of the material in the stratosphere and settled and absorbed in local ecosystems made it a risk and a revelation. Strontium-90 allowed the bomb makers to study what they had unleashed and the implications for health and for basic science. Biologists and ecologists who are never featured in the story of the B Reactor made strontium-90 their touchstone, a way to measure risk, as they grappled with the consequences of an unfolding atomic era on both local and global scales.

While the B Reactor tour emphasizes the chemists and physicist who made brief appearances at the reactor, hundreds of other scientists working for the Manhattan Project and then the AEC tracked the impact of these new materials and their efficacy as scientific tools employed outside the reactor doors. Practitioners of a new science of radioecology found tracers like strontium-90 especially useful in their study of more contained or local ecosystems. They honed their understanding in the cooling ponds at the B Reactor, or in the Columbia River teeming with salmon, or coral reefs in the Pacific. Strontium-90 ultimately made the idea of a planetary ecosystem legible.[19]

Soon after the Americans used nuclear power to destroy and contaminate Hiroshima and Nagasaki, Manhattan Project officials (and later the AEC) set about studying the results in a new kind of atomic fieldwork. They asked, What dose of radioactive materials was too much? How did these new materials behave? Where would they go? What would they do to us?

That year, federal authorities launched Project Gabriel, the first effort to monitor radioactive debris in the atmosphere. With each test after 1945, and with the reactors running full tilt at Hanford throughout the 1950s, AEC scientists and test planners became ever more aware and worried about the worldwide fallout problem. The tests at NTS or in the Pacific might contaminate people nearest to the test site, but drifting by-products like strontium-90 had a much broader reach.[20]

Once in the stratosphere and then rained down with other materials, strontium-90 could be easily found. This was its benefit and its curse. It settled down far afield. In soil it acted like metal shards to a magnet. It behaved like calcium, its closest kin on the periodic table, attaching to it in soil and staying put. But where soil was less chalky or limey the isotopes kept on moving. Up the food chain—from grass to cow to milk—it searched for connection with animals and people, bonding with their bones and causing cancers of the blood. In the 1950s, an age lousy with baby boomers, strontium-90 especially found plenty of growing, milk-drinking bodies around to absorb this radioactive metal after each release, making it the most notorious and useful radioactive tracer of the age.

In 1953 the AEC then created the secret and deceptively cheery-sounding "Project Sunshine" in order to intensify global efforts begun after 1945 to understand the effects of fallout in the atmosphere, especially the "the hazards associated with the release of radioactive materials in large scale warfare." The first set of questions this new commission posed seemed simple enough: "What are the present levels of radioactive fallout in the soil, in milk, and in people in the United States? In other parts of the world? Are these levels hazardous to health?" AEC investigators made the voracious collection of data central to their work and they debated how to use it and what it told them. In the process they stumbled on a way to imagine a global ecosystem.[21]

Project Sunshine drew primarily on the expertise of investigators in the health sciences and from the government labs that nurtured them, including the Health and Safety Lab and the Naval Research Lab. These were not glamorous or heroic theoretical physicists, but practitioners of basic science drawn beyond the laboratory walls. In the field, like the environmental historians they would later inspire, these investigators explored the intersection where the thousands of variables of materials, places, and bodies met.[22]

As they collected materials and traced paths of radiostrontium, the project scientists assembled a narrative for themselves and with significant alterations for public consumption that suggested an acceptable level of risk, a "body burden" of "sunshine units" that everyone, they thought, could live with. The goal was to allay fears about ongoing testing and to imagine how we could survive a nuclear war and continued above ground testing. It turned out that growing children faced the worst odds.[23]

Because calcium and strontium-90 bond and act similarly, the commission focused on collecting calcium rich materials of all kinds across the globe from which they could later pry strontium-90 and thus, in theory, measure average levels across the planet. They hoped to establish a baseline, "a direct measurement of the world-wide Sr-90 distribution which has resulted from the 40 to 50 nuclear detonations in the last few years." It wasn't so much that strontium-90 was hard to find, or that it would be hard to make visible in the form of numbers, data in the commission's charts that could be manipulated and modeled. It would however take some effort to both hide their work and coordinate such a "world-wide" effort at the same time. They kept most of it from the public until the late 1950s. The residue of successful secrecy from this period continues to flavor the presentation at the B Reactor.[24]

The two super powers dominated the structures of thought and even the landscapes of atomic culture. But atomic information and some secrets were also shared, tenuously and suspiciously, between the US and its allies, especially Britain and Canada. Each country had their variations on atomic culture that didn't always line up with the bipolar equation. In the search for samples and understanding, Project Sunshine's scientists created working relationships with these allies who, like the Soviets, produced plutonium and grappled with the problem of consuming by-products like strontium-90, too. This pursuit and measure of worldwide strontium-90 levels brought scientists together across borders as well. In their pursuit of data worldwide Project Sunshine's leaders cobbled together an agreement between Canada, the UK, and the United States that they called the Tri-Partite Exchange of Information on Nuclear Fallout. The exchange drew scientists deeper into new diplomatic and transnational relationships. In the 1950s the United Kingdom was in the midst of developing its own production program and like the US had begun monitoring the movement of strontium-90 in the stratosphere after each Christmas Islands bomb test and then locally how it moved through UK landscapes, animals, and people. The UK in particular, struggling with catastrophic events, such as the Windscale reactor disaster in 1957, had their own reckoning to do. In the B Reactor's nationalist script this wider world of nuclear exchanges and parallel transnational efforts scarcely exists, as if the United States and Richland lived in a vacuum.[25]

Even when we think of "Downwinders" or victims of radiation poisoning, we've learned to impose certain borders to contain this history. We might imagine sagebrush and sheep, desert or Pacific skies filling with mushroom clouds. We rarely imagine Welsh farmers in rubber boots. But like Downwinders in the Atomic West, Wales was the epicenter of concern in the UK, where its mountains and mists were most likely to collect the drifting strontium-90 from each British test in the Christmas Islands or at the NTS. Welsh farmers and their sheep absorbed higher than normal rates of these isotopes during the 1950s and again in the 1980s with Chernobyl, and they became the focus of intense study. These Welsh farmers and milk producers also raised concerns about what was happening to their animals and human bodies. Many of them soon joined a transnational political movement in Britain and the United States to ban the use of such weapons based on their concerns. (See the work of Joshua McMullan in this volume.)[26]

While the AEC quietly traced strontium-90 and risk worldwide, another set of scientists used the material on a more humble scale. Radioisotopes and strontium-90 in particular would be the important tracers that helped bring ecosystem science into focus at mid-century. You wouldn't know it either from a visit to the B Reactor, but scientists worked throughout the Cold War at Hanford to understand how tracers at the site behaved, their science tethered to the reactor's by-products. Biologists and radioecologists worked on less grandiose but perhaps just as profound basic science in ecosystems that were home to AEC facilities. They performed experiments at test sites, at Hanford, and at other installations of the atomic establishment during the Manhattan Project and the Cold War. In a sense, without the B Reactor and the ongoing experiments, tests, and production of radiostrontium in its wake, we would not have modern postwar ecology as a discipline or influential set of ideas.

Ironically the profound insights that would shape a generation of environmental activists and scientists concerned for the health of ecological systems—and some of the underpinnings of environmental history itself which I am championing here—emerged firmly yoked to one of the most violent and environmentally destructive forces of the twentieth century. The same releases of radioactive chemicals that scientists traced through the stratosphere after each test and deposits in calcium worldwide allowed ecologists (and many other disciplines and investigators) a window into how local and global ecosystems functioned.

One of these scientists, a biologist named Eugene Odum, worked with strontium-90. His work would underpin postwar ecosystem science. The Odum School of Ecology at the University of Georgia is named after him. Under the auspices of the AEC, Odum slogged through radioactive waters in swamps, holding ponds, and lakes near the Savannah River site, at Hanford, at the Nevada Test Site, and in lagoons at Enewetak. In the wake of bomb tests and fission events, Odum studied the way ecosystems metabolized energy by watching how radioactive tracers moved through living coral or ponds in real time with the help of radiation. Aware of the dangers that strontium-90 posed, "on the positive side," he wrote, "radioactive tracers are providing very valuable tools for research. Just as the microscope in all its forms extends our ability to study structure, so tracers in all their forms extend our ability to function." Odum offered a definition of his science in the opening pages of his field-defining *The Fundamentals of Ecology* in 1953: "The word ecology is derived from the Greek oikos, meaning 'house' or 'place to live.'" "Literally," he wrote, "ecology is the study of organisms 'at home'" and "the study of the relation of organisms or groups of organisms to their environment." He developed this powerful notion of home (a ready metaphor in the domestic-obsessed 1950s), or oikos, amid the destruction and pollution threatening human habitation on the earth. Odum's accessible and influential ideas about "ecosystems and biochemical cycles," according to a recent overview of the history of this field, would become "pervasive within ecology."[27]

You wouldn't know it either from a visit to the B Reactor, but scientists like Odum studied salmon and ducks in cooling ponds in the shadow of the reactors or downstream of facilities. Odum is the most famous and influential of these scientists who worked for a time at Hanford, but there were many others. A University of Washington fisheries scientist named Laurence R. Donaldson, for instance, worked with the Manhattan Project beginning in 1943 and focused on aquatic ecosystems after the war at Hanford. He also worked in the Pacific proving grounds as part of the government's radiation studies in the 1950s. Historian Matthew Klingle explains that in Laurence's work with radiation in the Pacific, the Columbia River, and in mountain lakes of the Pacific Northwest "the atom became a midwife to postwar ecology and fishery management."[28] This deadly by-product of the nuclear arms race then offered unprecedented insights and drew scientists and the AEC into new relationships. Radiation may have even pushed scientists and ecologists such as Odum (and then the rest of us) to apprehend the world in new ways, to go to new places for field research beyond the confines of the laboratory, to see real world connections they might not have seen before.

There is plenty of strontium-90 still on site near the B Reactor, left over from the production of plutonium. Now strontium-90, as well as a brew of other chemicals and a variety of radioactive materials with very different and specific half-lives, is the main concern of engineers. They must manage and replace 177 underground tanks that have leaked for decades. The control and containment of this pollution and its potential seepage into the Columbia River remains one of the priorities at the site. If we include radioecologists and biologists and health safety workers in the B Reactor story, the reactor takes its rightful place in the emergence of postwar ecology itself, but the reactor is also the place where we see the threat that these chemicals, and chemicals like them, still pose. The B Reactor created the very methods by which we measure and understand many of our contemporary environmental concerns.

I'm not arguing that the story of strontium-90 or Project Sunshine or the entire extent of radiation that flowed from the B Reactor must be accounted for within the walls of this one old building. That would be unreasonable. But this brief and selective history of one of the most important by-products does suggest how materials made in the reactor had significance far beyond the doors and pinched time frame of its exhibits. The B Reactor could offer an opportunity to interpret some of these crucial stories for the general public. I can think of few places more appropriate for telling such stories in this place that has been unfortunately so preoccupied with a narrative of isolation, containment, and clean up.

Locked in the Reactor

The B Reactor and the Hanford site sits on the outskirts of the Tri-Cities, a region encompassing the cities of Richland, Pasco, and Kennewick. Today it's the fastest-growing region in Washington. House farms, golf courses, and retirees move closer to the edges of the open space, finding peace that the Hanford reservation now provides. The hundreds of superfund acres are considered an amenity, like communities near the "open space" of Rocky Flats and other similar sites in the United States. The government originally chose the area because it was so far off the beaten track. Of course it never really was. Hanford continues to be an economic engine for the community where at least as many people work at Hanford cleanup now as did making the materials of war and peace, another amazing story about the legacy of World War II and the Cold War. It's also a town that, like the B Reactor and its residue of Cold War secrecy, nurtures an interesting combination of fierce provincialism mixed with a prideful sense of scientific and world historic self-importance. It's a local story, but always also everyone's story.

At the "The Hanford Reach" near the B Reactor, the Columbia takes a big swing around the old town of Hanford. It was evacuated, sealed, and made off limits in the 1940s, now part of the nuclear reservation upstream from town. The federal government forced the inhabitants of these towns on the river to sell their orchards and homes. You can still see the rows of stumps left behind after government workers removed the stone fruit trees, ensuring inhabitants wouldn't be tempted to return for the harvest that summer. One of my colleague's relatives were buried in the graveyard at White Bluffs. They were exhumed and relocated during the war. After that the security-conscious reservation prevented any development of the river along this stretch for the last seventy-five years. The pre-atomic history at the site, including the deeper history of Indigenous people, as well as the more recent history of settlers in the towns of White Bluffs and Hanford, is still muted at the B Reactor itself. It doesn't fit the World War II time frame either. But such stories are emerging from the oral histories and scholarship of Washington State University professors Robert Franklin and Bob Bauman who challenge what they term the "binary narrative" that has framed the story of this place between the poles of war or environmental disaster alone. Their recent work will add to our understanding of "how this place cycled" by including the people who have lived here since time immemorial, and more recently as irrigation-dependent farmers.[29]

The last "free-flowing" part of the river along "the reach" is one of the few perks of locking up land, even toxic land, in this way. Here containment created its opposite again. Elk and other critters have thrived under the gated national security regime out among the cocooned reactors and one of the largest undisturbed scrub-steppe ecosystems in the country. As historian David Bolingbroke has shown, both wild and domestic animals were also intrinsic to the events at Hanford. From elk and salmon to beagles and people, creatures moved across the security barriers at will, too often carrying radiation with them. A frightening brew of wastes is seeping from aging storage tanks underground, making their way into the groundwater and creating a remediation race against time as it creeps closer to the Columbia River (another story that is absent from the B Reactor tour, unless you make a point to ask about it).[30]

The reality of the polluted site's legacy struck close to home for me a few years ago, before the pandemic contained us all. The last time I visited the B Reactor I was joined by my colleague, an early American historian with no professional interest in the Atomic West. Larry was born in Sussex on the south coast of England where his grandfather had helped build the gas-cooled Dungeness nuclear reactor there in the early 1960s. Nuclear energy loomed large in his life as it did in mine.

As we walked the government green passageways together, poking fun at the historical presentation, we started to become aware of a nervous discussion among our tour leaders near the entryway. "What's happening?" we asked. There had been a tunnel collapse in another part of the reservation, they explained. They said little more. We would learn later that the collapsed rail tunnel had served as a hasty burial site for various grades of radioactive wastes from the Cold War era, one of the many haphazardly dumped leftovers after years of production. Much of the earlier cleanup at the site involved tracking down and containing all manner of radioactive garbage, contaminated bottles, tools, masks, and buckets. Back in the day they had simply sealed up the tunnel with grout, leaving it to decay and wait its turn for future cleanup. That grout had broken down over the years. When the tunnel collapsed during our visit, dust escaped into the semi-arid breezes. Nothing to be worried about. Nonetheless, I felt a little like my luck had run out.[31]

As we watched the docents' faces grow more serious, two Hanford safety officers suddenly appeared with an employee who'd been working near the tunnel. They used one of the reactor display rooms—the one that emphasized the history of safety—to check them for contamination. After years of watching nuclear engineers condescendingly demonstrate Geiger counters for me and my students, showing how common background and low-level radiation is in everything from fiesta-ware plates to smoke detectors, the sight of these health officers in uniforms scrambling to actually use one on a person was, to say the least, disturbing.

We were told we would need to shelter in place for the time being. Larry wasn't amused. Neither was I, but mainly because I also knew that the bathroom inside this Cold War relic—the latest addition to the National Historic Park inventory—was only for show. It gets hot in the eastern Washington summers. We'd been drinking a lot of water. The portable toilets sat just outside the glass doors that were now locked as safety officers assessed the risk. The bathroom predicament heightened the tension. We too wondered for a moment if we might be sealed in forever, cocooned like the reactors nearby, waiting for our own bodies' radioactive decay.

Sometimes history can seem distant from us, full of ironies and tragedies that clever historians point at from a comfortable distance. But this time the history of the B Reactor and radiation suddenly felt more visceral than ever, present in the dirt, maybe in our bodies. It was not likely that any radiostrontium would have been in that release. But there is plenty of radioactive waste at Hanford with longer half-lives ticking out near the river. Considering these time scales and enduring presence, history is still very alive, unfolding, or collapsing in on itself like the hastily sealed old train tunnel.

It would be nice to write the definitive tidy ending for a place like the B Reactor. Here, where containment is actually crucial to our survival, the impulse to do the same with history is strong. We shouldn't confuse the two. Radiation and history are both dangerous. But there is a difference between that real danger and acknowledging the porous nature of this place and the stories it might tell. But only by opening up the narrative doors of the reactor and tracing the pathways in and out can we understand the reactor's potential to reach deeper and broader, which will ultimately be more useful for facing the stark challenges ahead.

I still see the B Reactor as porous as ever, but the reality of being locked into this place that I'd entered and exited over the years, like that bat I first encountered, spooked me. After an hour they released us back to the gated courtyard. I visited the port-a-potty. Then we climbed aboard the bus and rolled downwind toward home.

Notes

1 *Our Friend the Atom* (Walt Disney Productions, 1957); "Manhattan Project National Historical Park," https://www.nps.gov/mapr/hanford.htm.

2 During the time of our first visit, wildlife managers on the site were studying a colony of bats living in one of the old facility buildings, including a waste site scheduled for demolition. See K. A. Gano, J. G. Lucas, and C. T. Lindsey, "Identification and Protection of a Bat Colony in the 183–F Clearwell: Mitigation of Bat Habitat on the Hanford Site," https://www.osti.gov/servlets/purl/945221 (accessed August 15, 2022),

3 Alan Nadel, *Containment Culture: American Narratives, Postmodernism, and the Atomic Age* (Durham, NC: Duke University Press, 1995), 1–9.

4 William Cronon, "Kennecott Journey: The Paths Out of Town," in *Under an Open Sky: Rethinking America's Western Past,* ed. William Cronon, George A. Miles, and Jay Gitlin (New York: W. W. Norton, 1992), 35, 28–51. See also Andrew Needham, *Power Lines: Phoenix and the Making of the Modern Southwest* (Princeton, NJ: Princeton University Press, 2014).

5 See Linda Loraine Nash, *Inescapable Ecologies: A History of Environment, Disease, and Knowledge* (Berkeley: University of California Press, 2006), 1–15, and Nancy Langston, *Toxic Bodies: Hormone Disruptors and the Legacy of DES* (New Haven, CT: Yale University Press, 2010), 85–111. For other examples of historians who explore the relationship of landscape, disease, and toxics see Conevery Bolton Valenčius, *The Health of the Country: How American Settlers Understood Themselves and Their Land* (New York: Basic Books, 2002); Greg Mittman, *The State of Nature: Ecology, Community, and American Social Thought, 1900–1950* (Chicago: University of Chicago Press, 1992).

6 Timothy J. LeCain, *The Matter of History: How Things Create the Past* (Cambridge: Cambridge University Press, 2017), 20.

7 "Reinterpreting High–Level Nuclear Waste," Washington State Department of Ecology, https://ecology.wa.gov/Waste-Toxics/Nuclear-waste/Hanford-cleanup/High-level-nuclear-waste-definition (accessed August 15, 2022).

8 The Washington Department of Ecology's website also includes a detailed description of each of the "the primary contaminants of concern" present in the groundwater at Hanford. Much of it is in the 100 area where the B Reactor and several other Cold War reactors and associated chemical separation plants were located, https://ecology.wa.gov/Waste-Toxics/Nuclear-waste/Hanford-cleanup/Protecting-air-water/Groundwater-monitoring.

9 The Environmental Protection Agency publishes a helpful website to help you locate superfund sites near you, https://www.epa.gov/superfund/search-superfund-sites-where-you-live (accessed August 25, 2022).

10 For examples of how to think of buildings as environmental and public history, see Richard White, "The Nationalization of Nature," *Journal of American History* 86 (December 1999): 976–986, and Dolores Hayden, *The Power of Place: Urban Landscapes as Public History* (Cambridge, MA: MIT Press, 1995), 15–22.

11 For the most thorough history of Hanford see John Findlay and Bruce Hevly, *Atomic Frontier Days: Hanford and the American West* (Seattle: University of Washington Press, 2011); "About Hanford Clean Up," https://www.hanford.gov/page.cfm/AboutHanfordCleanup (accessed August 25, 2022).

12 The number of dead is an estimate, and there is disagreement among historians about precise numbers. See Alex Wellerstein's discussion of the range of estimates in "Counting the Dead at Hiroshima and Nagasaki," *Bulletin of the Atomic Scientists,* August 4, 2020, https://thebulletin.org/2020/08/counting-the-dead-at-hiroshima-and-nagasaki/ (accessed August 25, 2022).

13 My discussion here is based on several visits to the reactor over a decade and the script repeated there. I have also benefited greatly by visiting the site numerous times with Robert Franklin of the Hanford History Project. See also S. L. Sanger and Craig Wollner, *Working on the Bomb: An Oral History of WWII Hanford* (Portland, OR: Portland State University, Continuing Education Press, 1995) and Michele Stenehjem Gerber, *On the Home Front: The Cold War Legacy of the Hanford Nuclear Site* (Lincoln: University of Nebraska Press,

1992). The Reach Museum in Richland, Washington, also includes much more information about the history of the region, including Hanford and the production of plutonium, https://www.visitthereach.org/plan-your-visit/ (accessed August 25, 2022).

14 For more on the idea of the technological sublime and a history of how the bomb fit into a longer story of American technological innovations, see David E. Nye, *American Technological Sublime* (Cambridge, MA: MIT Press, 1996).

15 For an overview of the history of technology and the Manhattan Project in this national context, see Thomas P. Hughes, *American Genesis: A Century of Invention and Technological Enthusiasm, 1870–1970* (New York: Viking, 1989), 353–442. For the Manhattan Project, and Cold War atomic culture see Richard Rhodes, *The Making of the Atomic Bomb* (New York: Simon and Schuster, 1986), and Paul Boyer, *By the Bomb's Early Light: American Thought and Culture at the Dawn of the Atomic Age* (Chapel Hill: University of North Carolina Press, 1985), xvii–xxii, 275–287.

16 During the Western History Association conference in Las Vegas in 2019, I visited the Nevada Test Site, and this section is based on notes from the daylong tour with a group of historians. Some of my observations are taken from that tour. For the history of the Nevada Test Site see Andrew Kirk, *Doom Towns: The People and Landscapes of Atomic Testing: A Graphic History* (New York: Oxford University Press, 2017), 291–300; Andrew Kirk, "Rereading the Nature Atomic Doom Towns," *Environmental History* 17, no. 3 (July 2012): 635–647. For a discussion of the Cold War West as wasteland and atomic space see Patricia Nelson Limerick, *Desert Passages: Encounters with the American Deserts* (Boulder: University of Colorado Press, 1989); Valerie L. Kuletz, *The Tainted Desert* (New York: Routledge, 1998), xiii; Kevin J. Fernlund and University of New Mexico, Center for the American West, *The Cold War American West, 1945–1989*, Historians of the Frontier and American West (Albuquerque: University of New Mexico Press, 1998), 1–7, 9–210; Peter B. Hales, *Atomic Spaces: Living on the Manhattan Project* (Urbana: University of Illinois Press, 1997).

17 Robert Nixon describes "slow violence" as "a violence that occurs gradually and out of sight, a violence of delayed destruction that is dispersed across time and space, an attritional violence that is typically not viewed as violence at all." Robert Nixon, *Slow Violence and the Environmentalism of the Poor* (Cambridge, MA: Harvard University Press, 2011), 2.

18 For more on the history of some western Downwinders see Sarah Alisabeth Fox, *Downwind: A People's History of the Nuclear West* (Lincoln: University of Nebraska Press, 2014); Philip L. Fradkin, *Fallout: An American Nuclear Tragedy* (Tucson: University of Arizona Press, 1989).

19 Emory Jerry Jessee makes a compelling argument about the new science of radioecology and the role of state science in preceding "popular ecology" in "Radiation Ecologies: Bombs, Bodies, and Environment During the Atmospheric Nuclear Weapons Testing Period, 1942–1965," (PhD diss., University of Montana, 2013), http://search.proquest.com/docview/1319509193/.

20 Strontium was only one of many forms of radiation released in manufacture and tests. Iodine-131 was the most immediate concern at Hanford. See Findlay and Hevly, *Atomic Frontier Days*, 57–60; Willard F. Libby, "Distribution and Effects of Fall-out," *Bulletin of the Atomic Scientists* 14, no. 1 (1958): 27–30. See Angela N. H. Creager, *Life Atomic: A History of Radioisotopes in Science and Medicine* (Chicago: University of Chicago Press, 2013).

21 Rand Corporation, "Worldwide Effects of Atomic Weapons: Project Sunshine" (Santa Monica, CA: Rand Corporation, 1953), https://www.rand.org/content/dam/rand/pubs/reports/2008/R251.pdf (accessed August 16, 2022); Harriet Alders to John Burgher, July 29, 1953, Sunshine Gabriel General Files, Records Relating to Fallout Monitoring and Studies, 1953–1964, RG 326, National Archives and Records Administration (hereafter NARA).

22 David Cleve Bolingbroke follows the work of several of these scientists who worked with animals in particular in his "Nuclear Animals and an Atomic Restoration: An Environmental History of the Hanford Nuclear Site" (PhD diss., Washington State University, 2020). See Sunshine Gabriel General Files, Records Relating to Fallout Monitoring and Studies, 1953–1964, Box 1, RG 326, NARA.

23 For a discussion of Project Gabriel and the subsequent work of Project Sunshine, see Barton C. Hacker, *Elements of a Controversy: The Atomic Energy Commission and Radiation Safety in Nuclear Weapons Testing, 1947–1974* (Berkeley: University of California Press, 1994), 180–184. For more on efforts to negotiate the meanings of risks in the mid-1950s, see Jacob Darwin Hamblin, "'A Dispassionate and Objective Effort': Negotiating the First Study on the Biological Effects of Atomic Radiation," *Journal of the History of Biology* 40, no. 1 (2007): 147–177.

24 Letter to Dr. James Scott, December 9, 1953, Sunshine Gabriel General Files, Records Relating to Fallout Monitoring and Studies, 1953–1964, RG 326, NARA. For secrecy of Project Sunshine see Hacker, *Elements of a Controversy*, 183; for secrecy at Hanford see Findlay and Hevly, *Atomic Frontier Days*, 252–253.

25 Simone Turchetti, *Greening the Alliance: The Diplomacy of NATO's Science and Environmental Initiatives* (Chicago: University of Chicago, 2019). For a comparison of Soviet and American plutonium production, see Kate Brown, *Plutopia: Nuclear Families, Atomic Cities, and the Great Soviet and American Plutonium Disasters* (Oxford: Oxford University Press, 2013).

26 For concerns of activists see Elizabeth S. Watkins, "Radioactive Fallout and Emerging Environmentalism: Cold War Fears and Public Health Concerns, 1954–1963," in *Science, History, and Social Activism: A Tribute to Everett Mendelsohn*, ed. G. Allen and R. M. MacLeod (Dordrecht: Springer, 2001), 291–306; for a discussion of "politico scientists" and the significance of strontium-90 in the test ban debate, see Michael M. Egan, *Barry Commoner and the Science of Survival: The Remaking of American Environmentalism* (Cambridge, MA: MIT Press, 2007), 10, 47–50, 52; Kendra Smith-Howard, *Pure and Modern Milk* (Oxford: Oxford University Press, 2013). Toshihiro Higuchi, *Political Fallout: Nuclear Weapons Testing and the Making of a Global Environmental Crisis* (Stanford, CA: Stanford University Press, 2020), 1–15; 144–161. For more on the development of environmental politics, and citizen science as a challenge to the AEC's lock on information, see Egan, *Barry Commoner and the Science of Survival*, 64–66; Hacker, *Elements of a Controversy*, 226–230; Scott Kirsch emphasizes what he calls "spatially contested knowledge" regarding the policy of the AEC to ignore local places and local people in their approach to radiation science and discusses the efforts of scientists, including Harold Knapp and Robert Pendleton, who pushed back against this status quo; see Scott Kirsch, "Harold Knapp and the Geography of Normal Controversy: Radioiodine in the Historical Environment," *Osiris* 19 2004: 167–181. By the late 1950s books like Jack Schubert and Ralph E. Lapp, *Radiation: What It Is and How It Affects You* (New York: Viking Press, 1957), and Paul Jacobs, "Clouds from Nevada," *The Reporter*, May 2, 1957, helped to set off alarm bells, followed by intensifying debate over the effects of strontium-90 and testing during the election that same year. See Brian Wynne, "May the Sheep Safely Graze?: A Reflexive View of the Expert–Lay Knowledge Divide," in *Risk, Environment and Modernity Towards a New Ecology*, ed. Scott M. Lash, Bronislaw Szerszynski, and Brian Wynne (London: SAGE Publications; Ann Arbor, MI: ProQuest, 2012), 44–83.

27 Libby Robin, Paul Warde, and Sverker Sorlin, eds., *The Future of Nature: Documents of Global Change* (New Haven, CT: Yale University Press, 2013), 208. See Mark Fiege's discussion of the scientists who worked in Los Alamos, New Mexico, in *The Republic of Nature: An Environmental History of the United States* (Seattle: University of Washington Press, 2012), 281–317; Donald Worster, *Nature's Economy: A History of Ecological Ideas* (Cambridge: Cambridge University Press, 1994); Eugene P. Odum, *Fundamentals of Ecology* (Philadelphia: Saunders, 1959), 451, 3.

28 Matthew W. Klingle, "Plying Atomic Waters: Lauren Donaldson and the 'Fern Lake Concept' of Fisheries Management," *Journal of the History of Biology* 31, no. 1 (1998): 1–32.

29 Robert Bauman and Robert R. Franklin, eds., *Nowhere to Remember: Hanford, White Bluffs, and Richland to 1943* (Pullman: Washington State University Press, 2018), 7, 2. See also Robert Bauman and Robert R. Franklin, eds., *Echoes of Exclusion and Resistance: Voices from the Hanford Region* (Pullman: Washington State University Press, 2020). See also The Hanford History Project, https://tricities.wsu.edu/hanfordhistory/; David Bolingbroke's forthcoming book on animals and the history of Hanford and my discussions with him have been invaluable to thinking about the site; see also Jessee, "Radiation Ecologies."

30 The script for the B Reactor is changing and updating all the time. This aspect of the site's history is made available from tour guides upon request .

31 See David Kramer, "No Injuries or Contamination Reported in Hanford Tunnel Collapse," *Physics Today,* May 10, 2017, https://physicstoday.scitation.org/do/10.1063/pt.5.1122/full/ (accessed August 15, 2022).

Playing Games on the Graves of the Dead

Commemoration, Forgetting, and Ways of Knowing in Richland, Washington

SARAH FOX

In March of 2018, retired schoolteacher Mitsugi Moriguchi stood on a balcony overlooking the Richland High School Gymnasium in eastern Washington, home of the "Bombers" athletic program. Below him, several students were playing basketball on a court emblazoned with the school's logo: a giant mushroom cloud rising from the letter R. Above the basketball court, a banner affirmed that the local Nissan dealership was "proud of the cloud." A reporter asked Moriguchi, a survivor of the atomic bombing of Nagasaki (survivors are known as *hibakusha*), for his reaction. Through the translator, Dr. Norma Field, Mr. Moriguchi said quietly that people he cared about had died under a mushroom cloud, and to his mind, playing games on the image was like stepping on the graves of the dead. Outside the auditorium, Mr. Moriguchi met a Richland High student who explained that she is proud of the cloud symbol because it reminded her of how her town came together to help win a war. She added that while she knew people suffered after the use of atomic bombs against Japan, she thought the mushroom cloud symbol was important because it helped people to remember the past.[1]

These disparate ways of "remembering" the mushroom cloud do different kinds of work. The Richland High Bombers mascot frames the invention and deployment of atomic bombs as a triumph of national-scientific-technological-military mettle, a tidy story in which merit and democracy earned the power to control the ultimate weapon. Invisible in this narrative are the people whose lives were devastated beneath that cloud, along with any and all injuries caused to Hanford workers and community members whose bodies have been endangered in the course of plutonium production. The stories of hibakusha like Mitsugi Moriguchi surface much messier historical realities, revealing nuclear weaponry as a form of violence that refuses containment in places, bodies, eras, or "sides" in discrete conflicts, unleashing damage and suffering

Mushroom Cloud Logo on the floor of Arthur Dawald Gymnasium, Richland High, Richland Washington, March 2018. Photograph by author.

across space and into the future at every stage in its development, rendering the lives of all current and future living things precarious.

Listening to the Richland High student speak, I was struck by the similarity between her argument and the one some were making at the time in favor of the preservation of Confederate monuments in the southern United States. While defenders of the statues insisted that history would be "forgotten" if the monuments were removed, historians and commentators pointed out that the monuments had been installed to shape contemporary attitudes as much as commemorate the past.[2] Noting that most Confederate statues in the South were erected during the Jim Crow era, decades after the Civil War, historian Sarah Gardner wrote that Confederate statues "tell us a great deal about how a cultural industry worked in tandem with a political movement to legitimate white supremacy. . . . The purpose of these statues was not to honor the Confederate dead but to assert and celebrate white supremacy in the present."[3]

Similar controversies have adhered repeatedly to commemorations of nuclear history in the United States, which tend, in the words of Conny Bogaard, to "encourage awe for the bomb scientists while marginalizing controversial issues such as the human toll, as well as the political and environmental costs of the Manhattan Project."[4] In 1995 the US National Air and Space Museum canceled a planned exhibit which would have linked the *Enola Gay* bomber to the devastation of Hiroshima, in response to political pressure, which "charged that the exhibition script dishonored the Americans who fought the war."[5] In 2005, Downwinders of nuclear tests in Nevada protested the Las Vegas Atomic Testing Museum's omission of their stories of radiation health impacts, to no avail.[6] In 2018, the Los Alamos Historical Museum decided against hosting "a travelling exhibit organized by the Hiroshima Peace Memorial Museum and Nagasaki Atomic Bomb Museum" because museum directors "felt uncomfortable about the exhibit's call to abolish nuclear bombs."[7] The Richland Bombers' mushroom cloud has been the subject of debate since the 1980s, but the town and the school remain dedicated to the symbol. While each of these controversies is ostensibly a disagreement about the interpretation of the past, they are also fundamentally about contemporary politics. The United States' ongoing commitment to maintaining nuclear weapons infrastructure (at a projected cost of over sixty billion dollars a year, according to the Congressional Budget Office) is threatened by thorough reckonings with nuclear history, which reveal incalculable costs both at home and abroad.[8]

Historian Michel-Rolph Trouillot argued that "silences are inherent in history because any single event enters history with some of its constituent parts missing. Something is always left out while something else is recorded."[9] Trouillot explained this leaving out and recording has always been shaped by "the many ways in which the production of historical narratives involves the uneven contribution of competing groups and individuals who have unequal access to the means for such production."[10] Like the production of history, the production of scientific knowledge about toxic exposure is also a project shaped by disparities in power, disparities which condition what forms of exposure are studied and legitimated, which are dismissed or ignored, and how the impacts of a particular industry are understood.[11]

Richland High School is a short drive away from one of the most contaminated sites in North America, the decommissioned Hanford nuclear facility. This proximity is what brought Mitsugi Moriguchi to town. Although few of them knew it at the time, Richland residents working at the Hanford facility during World War II were producing plutonium, the fuel for a new superweapon that would change the course of Moriguchi's life. The bombing of Nagasaki and the multi-decade production of plutonium at the Hanford site resulted in generations of radiation-related health problems for local people in both places, a shared experience obscured by the Richland High Bombers tradition and popular understandings of the history of the Hanford site. Reliant on nationalist justifications for nuclear weapons production and use and a robust local culture of faith in (and economic reliance on) the ability of scientists and technicians to administer the Hanford site safely, the Bombers tradition has for over seventy-five years enshrined the fallacious idea that nuclear technology can be controlled, targeted, or contained, while silencing those stories that undermine this idea. This essay will historicize the Richland High School "Bombers" mushroom cloud symbol, exploring its persistence alongside the stories of two radiation-impacted people, one from Nagasaki and one from Richland, in an effort to cast light on the formation of different ways of knowing about nuclear history and the consequences of radiation exposure.

When school let out for the summer in 1945, no one in Richland had ever heard of an atomic bomb, save perhaps a few top-level administrators at the Hanford site. Richland High was known as Columbia High back then, named for the massive river that wound past the townsite. Constructed the previous year by the Army Corps of Engineers, the high school had chosen beavers for their school mascot. It was a fitting choice; beavers manipulate rivers for their needs, and upstream from Richland, the Army Corps of Engineers was busily doing just that. In 1943, under the War Powers Act, the federal government had seized nearly 600 square miles of semi-arid shrub-steppe land along the river. The Columbia River's dams would provide the massive amounts of electricity needed to power plutonium production, and its waters would provide the coolant.[12]

The 1943 land seizure expulsed local farmers and denied access to hunting and gathering areas and sacred sites utilized by the Wanapum Indian Nation, the Yakama Nation, the Confederated Tribes of the Umatilla Reservation, and the Nez Perce Tribe.[13] In this way, Hanford is a typical nuclear site. Overwhelmingly, nuclear weapons production worldwide has relied on the land, resources, and labor of colonized regions, a phenomenon rarely acknowledged in the commemorative, nationalistic public histories that tend to be generated around these sites, particularly in the United States. The 1943 expulsion of local Indigenous people from the Hanford Reach area fell into a lengthy pattern of settler colonial violence and displacement those communities had survived for generations.[14]

Almost immediately, the Army Corps of Engineers and the DuPont Corporation began to fill the freshly emptied lands of the Hanford Reach with the structures and secretive agendas of military-industrial production. Site managers were under tremendous pressure to keep the site's purpose and workings secret, making the management of the tens of thousands of laborers and technicians streaming in to staff the site a paramount concern. Officer-in-charge Colonel Franklin Matthias's demand that plant operators living in the new Richland townsite "be kept under control for security reasons" posed a challenge for DuPont executives, who believed white-collar workers would be reluctant to live on anything resembling a military base. Eschewing tall fences and security checkpoints, DuPont settled with a more subversive form of control: what historian Kate Brown describes as "a new regime that equated security with white middle-class families in a new upscale, exclusive bedroom community."[15] Hanford's nuclear secrets would be protected by the corporate company town model.[16] Years later, Richland resident Muriel Sears remembered the way townsfolk related to Hanford, which they called "the Site" or "the Area." "Well you didn't talk about what's out north, and you knew that. You just did not talk about it. And boy they were right on your tail. . . . If your kid got in some pretty deep trouble they moved you out. . . . They packed up your stuff and moved you out because you would be concentrating more on the trouble that the kid had than you would on your job."[17]

Sears was born Muriel Olson on January 18, 1930, in the small town of Opheim in northeastern Montana. Her father lost his job as a banker after the stock market crash of 1929, and the family moved to the temporary boomtown of New Deal, Montana, where the Fort Peck Dam was being constructed. From there, her family moved to Spokane, Washington. Muriel Olson attended teachers' college in Cheney, and moved to Richland as a student teacher in the fall of 1950. During her student teaching, she lived with the family of Arthur Dawald, the Richland high basketball coach whose name would later be given to the Richland High Gymnasium. Eventually, Olson met and married one of Dawald's former players, Richard Sears, and they moved to Seattle, where he attended architecture school. Eight years later, the Searses returned to Richland with their son David and daughter Gail; their son Greg would be born in Richland. "I loved it here," Muriel related in 2017. "It was a wonderful place to raise kids."

Richland, intended primarily for Hanford site management and technicians, was constructed in a matter of weeks in 1943, a neat grid of streets named for military generals and studded with prefab homes. Intolerance for disloyalty or questioning was built into the very foundations of the new town, as was an initial policy of racial segregation, established by the DuPont company. Richland became known as a "sundown town": Indigenous, Latinx, and African American people were expected to leave town by the end of the workday, and were restricted to living in the nearby town of Pasco.[18] While the surrounding region may have appeared dusty and desolate to new arrivals, residence in Richland promised membership in a modern, patriotic, white middle-class community. Brown describes how those locals with ties to the old farming economy who remained in the area "at first haltingly and then passionately hitched their fortunes and prosperity to an expanding military-industrial complex backed by private corporations and sustained by federal subsidies," the alliance that resulted in the 1944 construction of Columbia High School.[19]

On August 6 and August 9 of the following summer, the United States detonated nuclear weapons over the Japanese cities of Hiroshima and Nagasaki. The weapon that devastated Nagasaki, code named "Fat Man," was a plutonium implosion bomb; the plutonium for that bomb was produced by Richland workers at the Hanford facility. Mitsugi Moriguchi was nine years old at the time. Just prior to the bombing, his mother had evacuated him, his sister, and younger brother roughly 40 kilometers from the city. On his 2018 visit to Richland, Moriguchi related his memories of August 9, with Dr. Norma Field translating. With his siblings and mother, he heard "a huge explosive sound, smoke started rising from all over the city, and then the mushroom cloud. We didn't know what it was. . . . Someone near us, strangely enough, was able to hear the radio, which just repeated the announcement 'citizens of Nagasaki, get out, get out,' and then it just went silent." Moriguchi's mother "realized that under all that smoke was the city of Nagasaki, and that her husband and two older children were there. She immediately decided to return." Giving the money she had with her to her sixth grade daughter, the children's mother instructed them to remain in place, telling them: "You keep this and if none of us survives to come back you must give this to an acquaintance and do your best to live on." Seven decades later, Moriguchi's voice welled with emotion as he recounted:

> We kept trying to stop her but she left. Every day the three of us went to the train station waiting for her to come, but no trains came. And finally a station attendant told us, "Nagasaki has been totally destroyed." On the third day since no one had arrived, we decided maybe we should give up . . . we lay down on the grass [to sleep]. Late that night someone was grabbing me by the shoulder, I thought it was a ghost, a person with all her clothing in shreds, hair falling over her, it was a ghost, and then there were other forms near her. It was my mother and brother and sister. My brother and sister were injured, but she had managed to find them and bring them back. My brother sustained injuries all over his body, but there has never been a happier moment in my life. We clung to each other and sobbed. [20]

A few days later, on the other side of the world, Richland residents were informed their labor had contributed to the end of the war. Under a massive "PEACE!" headline, the August 14 issue of Richland's newspaper, *The Villager*, announced: "OUR BOMB CLINCHED IT!" followed by news that the Hanford "Plant Will Not Close!" A fourth headline announced the Japanese had surrendered, employing a slur normalized by years of virulent wartime media coverage subsequent to the bombing of Pearl Harbor.[21] Richland, a DuPont company town originally constructed for purposes of wartime industry, would now persist as a Cold War scientific-industrial community, producers of weapons grade plutonium funded by lucrative government contracts.

Historian Peter Hales has argued that the distinctive atomic mushroom cloud arrived as "a visual icon so unprecedented that, for the moment at least, it lay outside the webs of signification that comprised a watching culture." Grasping to make sense of that icon for the American public, journalists and cultural commentators often defaulted to the language of sublimity, a state that Hales describes as "that combination of terror and wonder that accompanied confrontation with the Infinite."[22] Almost immediately upon its introduction, the atomic mushroom cloud in the United States had become associated with the mastery of infinite power. As classes began at Columbia High School that fall, conversations were already underway about changing the mascot from "the Beavers" to something more representative of Richland's new nuclear identity. A variety of possibilities, including the "Richland Atoms," the "Atomizers" and the "Richland Bombers" were proposed; the Bombers stuck.[23]

The following year Columbia High School's yearbook affirmed, "For memories sake, and because of its greatness, we have carried the 'Atomic Bomb' theme through the annual in an effort to symbolize the world history, which has been in progress here in Richland, in which we and our parents have a part."[24] The yearbook statement reveals how young people in Richland were learning to conceptualize the atomic bomb, and simultaneously, the national and global significance of their town. It also allows us an opportunity to imagine the mascot through the gaze of parents and community members rooting for the Bombers. Anyone who has spent time in a small town in the United States has likely witnessed the prominent role of high school sports in community identity. Rooting for the Bombers allowed Richland residents to cheer simultaneously for their young people, for their town's major industry, for their own roles in supporting nuclear production, and for the military clout of the United States on the global stage. Sociologist Todd Callais has written that mascots "are supposed to allow violence and aggression to be played out symbolically," a way for participants to project their ideas about power and dominance.[25] It is difficult to imagine a symbol with greater connotations of power and dominance than the mushroom cloud, or the bomber which drops a nuclear weapon. Invoking these symbols necessarily obscures the view on the ground: the view of those susceptible to the harms of radiation exposure, at both the site of weapons production and the site of weapons use.

In the months after the bombing of Japan, military and scientific experts maintained tight control over information about fallout and radiation sickness, even as American medical professionals descended on the devastated Japanese cities to document the health effects of the new weaponry for the Atomic Bomb Casualty Commission (ABCC). Overseen by the National Academy of Sciences and funded by the newly organized Atomic Energy Commission (hereafter AEC)'s Biology and Medicine division, the ABCC did not treat radiation-related ailments among the Japanese; it simply monitored them. After Mitsugi Moriguchi's older brother recovered from his injuries and was married, his wife bore three infants with serious birth defects. Mitsugi Moriguchi described his brother and sister-in-law's hardships in 2019:

> The first baby had no shape. And the second baby had no arms or legs and died. The staff of the ABCC came to our house to check that the babies, you know, had effects of the atomic bomb or not. . . . I was very angry at the ABCC, and said, "Get out of here!" The ABCC staff were coming to get only material or information of disease. They did not give any medical treatment, only to get research information.[26]

Other members of the family also experienced health problems in the ensuing years, including Moriguchi himself. His older sister "suffered from cancer all over her body" and died in her thirties.[27]

Historian Susan Lindee has categorized the ABCC's structure as "a form of science comparable to colonial indirect rule . . . a science conducted by outsiders, that depends on local knowledge, particularly when that knowledge is invisible to the colonizers themselves."[28] A small number of American doctors, scientists, and scholars oversaw several hundred Japanese medical professionals in carrying out the ABCC's research, which Lindee characterizes as "the largest epidemiological study of its kind up to that time."[29] Data was funneled back to the Atomic Energy Commission's Biology and Medicine Division, which was simultaneously gathering data from radiation health research conducted around the globe: on Indigenous Marshall Islanders exposed to Pacific nuclear tests; on uranium miners in the southwestern United States; on workers at Hanford; on residents of the rural areas impacted by nuclear production sites.[30] The AEC's biomedical research would prove foundational to modern understandings of radiation health effects, yet it was frequently conducted without the informed consent of research subjects, without the provision of warnings about ongoing radiation exposure risks, and without medical care for radiation health impacts.[31]

Reports did begin to emerge about the horrific impacts of atomic weapons in the wake of the bombings of Hiroshima and Nagasaki, but assertive US government propaganda and lingering racist fervor about the Japanese helped soothe anxieties that might have arisen in Richland.[32] In February of 1947 former Secretary of War Henry Stimson published an explanation titled "The Decision to Use the Atomic Bomb" in *Harper's Magazine*. He informed the public that if the atomic bomb had not been developed and employed, American troops would have been sent to invade Japan, affirming: "The total US military and naval force involved in this grand design was of the order of 5,000,000 men." Stimson claimed the invasion had been predicted to cause "over one million casualties, to American forces alone."[33] At the end of four years of war, Stimson's casualty numbers were all many U. S. citizens needed to set their minds at ease about the use of atomic bombs.

In actuality, the chiefs of staff had estimated losses in a land invasion at between 25,000 and 50,000 men, and the land invasion had never been certain.[34] By July 1945, intercepted messages had revealed to U. S. officials that the Japanese emperor was committed to ending the war, and numerous Japanese officials were prepared to discuss terms of surrender.[35] Other factors motivated the decision to use atomic bombs against Japan: President Truman's advisor on atomic issues, James F. Byrne, believed that "possessing and demonstrating the bomb would make Russia more manageable in Europe."[36] Of course, few members of the public had access to this information, cementing the widespread perception that use of atomic weaponry had saved the lives of more than one million American soldiers.

In Richland, this justification was readily accepted, and atomic technology grew more inextricably linked to patriotism and a thriving local economy. Management of Hanford site activities periodically shifted to different contractors after the end of WWII, but oversight authority and copious federal funds flowed from the federal Atomic Energy Commission for most of the Cold War era. Local businesses began incorporating atomic symbols and references, and Columbia High students opted to copyright the mushroom cloud as part of their school's coat of arms in 1965. Amateur historian and Richland High alumnus Keith Maupin notes the cloud began to appear on athletic uniforms beginning in the 1970s.[37]

In 1960, Muriel Sears, by then a single mother of three, went to work as a clerk for AEC contractor General Electric Medical (later renamed the Hanford Environmental Health Foundation), the contractor at the Hanford site responsible for worker physical and psychological health evaluations. In the course of her work administering psychological and vocational testing, Sears routinely traveled with her supervisor all over the Hanford site. She found the work interesting, though she was troubled by pervasive sexism in the analysis of the tests.[38] One day she was approached by Dr. G. H. Crook, a medical researcher whose office was across the hall. "He said how would you like to earn two hundred and fifty dollars?" Dr. Crook explained he was hoping to enroll her in an experiment studying the radioactive isotope promethium. He assured Sears the experiment was safe, telling her "all you have to do is drink a glass of orange juice with this in it and we want to see how long it takes to go through the liver." Sears remembered thinking: "That sounded pretty good. Two hundred and fifty bucks, I can buy a dishwasher." She participated in monitoring after drinking the orange juice, leaving biological samples in a black box on her home's doorstep for collection and traveling out to the site to lie in a full body radiation counter. For many years after the experiment, Sears felt perfectly healthy. In the 1970s she became a manager of Equal Employment Opportunity and Affirmative Action for Energy Northwest, a job she loved.[39]

In the 1980s Columbia High School was renamed Richland High. By this time, the growth of the global nuclear disarmament movement and meltdowns at nuclear power plants in Chernobyl and Three Mile Island had made it increasingly difficult to suppress public knowledge of radiation effects. The "proud of the cloud" narrative at Richland High began to attract controversy and press coverage, and in 1988, two Hiroshima hibakusha visited the high school to see the mascot for themselves. According to one account, during their meeting with Principal Gus Nash, hibakusha Sakae Itoh "became overcome with emotion and began talking to Mr. Nash like a mother scolding a child. Nash, a World War II veteran, said, 'We didn't start that war,' and walked out. The *Tri-City Herald*'s photograph of the confrontation appeared in newspapers in Japan."[40]

This exchange, and the media attention it generated, prompted heated conversations about the appropriateness of the mascot, conversations that continue to the present day. Since the controversy began, many local residents have vocally defended their mascot, using language about tradition and local pride and pushing back on what they see as a move to enforce political correctness. By the early 1990s staff at the high school began to suggest "the Bombers" had never been intended to refer to the bombing of Japan. They claimed it a was a reference to a B-17 Bomber, known as "The Day's Pay," purchased in support of the war effort in 1944 by Hanford workers who donated a day's pay each. In 1992 a massive mural of the Day's Pay was added to the side of the gymnasium, and from that point forward, the Day's Pay story became the official Richland High explanation for their mascot.[41] Alumni of the high school are still fiercely divided over what the original mascot selection was intended to represent. Some, like alumnus Keith Maupin (class of '47), a former Hanford lab technician, fiercely maintain the "Bombers" was always meant to signify the atomic bombing of Japan.[42] Others, like Ray Stein (class of '64) continue to insist "the genesis of the name was (and still is) that B-17, Day's Pay Bomber."[43]

Since the 1980s, a small minority of staff, students, and alumni at Richland High School have voiced concerns over the ethics and appropriateness of the mascot; some have reported facing retribution in the form of threats or social media harassment.[44] The controversy has reanimated in recent years, likely a result of the United States National Park Service's 2015 decision to make the Hanford B-Reactor part of a National Historic Park commemorating the Manhattan Project, deemed safe for guided tours. The remainder of the site remains extremely dangerous, home to what the Washington State Department of Ecology calls "some of the nation's most complicated nuclear and mixed dangerous waste." In February of 2020, the *Tri-City Herald* reported that the high school was debuting its first character mascot, "Archie," a bomber pilot intended to honor the "Day's Pay" bomber, created by Richland High student Chris Morano.[45]

The persistence of the Richland Bombers mushroom cloud relies on particular ethical, spatial, and temporal understandings of nuclear weapons technology. These understandings situate atomic bombs as a devastating but necessary technology that came into American hands because of American ingenuity and moral authority, technology only ever deployed on two occasions, against vicious, inhuman enemies far away. In this narrative, nuclear weapons technology is not only justified, it is contained: its effects remain limited to a specific moment in history (1945), and a specific geography (Hiroshima and Nagasaki). This sort of thinking obscures the domestic realities of pervasive and ongoing radiological contamination generated by sites like Hanford.

Beginning in 1944, back when Richland still rooted for the Beavers, and continuing through 1971, radioactive isotopes were released en masse into the Columbia River, to flow through Washington and Oregon farmland (producing agricultural goods for export all over the world) and into the Pacific Ocean. Airborne releases of radiation were also a regular part of the plant's operation. Massive quantities of nuclear waste, from barrels to the radioactive corpses of animals used in experiments were buried at the site in unlined trenches, riddles for future generations to solve once the Cold War was won. Today, nearly a million gallons of radioactive waste are known to have leaked from Hanford's underground tanks, and roughly fifty-six million gallons remain. Tunnels sealed around hastily buried radioactive waste have collapsed, and as recently as 2018, workers involved in the Hanford cleanup were found to have ingested plutonium particles, and to have brought radioactive particles home to Richland in the air filters of their personal vehicles. Because some radioactive isotopes remain dangerous for extended periods of time, from mere hours to tens of thousands of years, this historic contamination and the genetic changes it causes remain contemporary processes, even as nuclear sites like Hanford are being "cleaned up."[46]

When locals living downwind of the site began to raise concerns about safety practices at the site or health problems in the downwind area in the 1980s, their voices were drowned out by reassurances from site officials and Department of Energy authorities, who had assumed oversight of nuclear projects after the AEC's dissolution in the 1970s. This dismissal of local health concerns has historically been a common phenomenon in communities near nuclear facilities, where conversation about the risks of nuclear industries is seen as bad for business.[47] The narratives of people who have been impacted by radiation exposure also tend to go unmentioned in public and academic histories of nuclearism, which accord greater weight to technologies, diplomacies, and "Great Men" than the experiences of ordinary people, which are less well documented and more difficult to substantiate.

Some ordinary Richland residents were actually documented quite heavily during the decades of Hanford's operation. Their bodies were surveilled in dozens of studies, from occupational epidemiological studies to studies of the incidence of specific cancers, like those of the breast and lung, among Hanford workers. These data pointed to "increased mortality within the Hanford work force from several cancer types." Researchers also tracked "congenital malformations among children . . . of Hanford workers and . . . children born in [neighboring] Benton and Franklin counties."[48] Prisoners from the nearby state penitentiary had their testicles irradiated for one study; Muriel Sears remembered seeing some of those prisoners brought into her workplace for testing.[49] When she retired from Hanford-related work in 1995, after a career which spanned work for contracting companies including Battelle, Pacific Northwest National Laboratory, Washington Public Power Supply System, and Westinghouse, Muriel asked the doctor at her exit physical if he could tell her anything about the promethium study she participated in three decades prior. She'd never received a follow-up, and the doctor at her exit exam was unable to provide her with any information. When she later became ill with the blood disorder Myelodysplastic Syndrome, or MDS, she did research online and learned that promethium could accumulate in the bones, a fact confirmed by a 1970 report her son David tracked down on the study she'd been involved in, published in the journal *Health Physics*.[50]

In that report, G. H. Crook and coauthors H. E. Palmer and I. C. Nelson described work "performed under United States Atomic Energy Commission Contract AT(45-1)-1830." Using promethium produced at Oak Ridge Laboratory and further refined by Battelle-Northwest company, the researchers injected six subjects and gave it to two others to ingest. Their monitoring tracked the rate at which the test subjects excreted promethium in their bodily wastes, and concluded: "The 143Pm which does not go immediately to the liver appears to remain in the blood for a short time and gradually deposits in the bone." While the report did claim that "essentially none of the material" consumed by participants who drank the orange juice "was absorbed by the body and all of it was excreted," Muriel wondered how it might have impacted her till the end of her life.[51] In an interview several months before she died, when her MDS had progressed to acute leukemia, she reflected that her illness could just as easily have been from radiation exposures received at other times during her employment, given that her job required her to spend time at various locations at the Hanford site. "I'd been at the B-Reactor and stood right in front of it, you know," she reflected.

Researchers conducting studies connected to the Hanford site provided their raw data and initial conclusions to the Atomic Energy Commission (AEC) in regular reports, but in the interest of national security, "did not submit their findings for review and publication in professional or medical journals." Findings that pointed to the intrusion of radiation into regional agricultural systems and human bodies were not shared with agricultural producers or communities at risk of health impacts.[52]

The strength of Richland's adherence to their patriotic "proud of the cloud" tradition may seem peculiar to outsiders if residents were simultaneously participating in mysterious, federally sanctioned medical studies, but for residents of Richland, participation in secretive biomedical research was normal during the Cold War. For some residents, this monitoring may have even cultivated a sense of safety; many assumed that local scientific and medical authorities would alert them if there was anything to be concerned about. Leaving biological samples in a box on the doorstep was ordinary practice in Richland, as Muriel Sears recalled: "Nobody paid any attention because everybody had that at one time or another." Former Washington poet laureate Kathleen Flenniken's father worked at Hanford, and as an adult she worked there as well. In her poem "Bedroom Community," she writes of mornings in her childhood:

> . . . our fathers rose, dressed, and boarded
> blue buses that pulled away, and men
> in milk trucks came collecting bottled urine
> from our doorsteps.[53]

DuPont's nuclear company town model proved extraordinarily effective in Richland; as one federal contractor has ceded to another over the years as the main employer in town, scientific authority has remained a paramount force in the community, entwined with a patriotic local identity deeply invested in the rightness of dropping the bomb on Japan: an investment exemplified by the persistence of the Richland High Bombers mushroom cloud.

Mitsugi Moriguchi came to Richland in part because he wanted to meet American survivors of radiation exposure, and to understand how Richland residents viewed the production of nuclear weapons today. His visit occurred as part of the Hanford-Nagasaki Bridge project, a collaboration between the City of Nagasaki and the Washington-based nonprofit group Consequences of Radiation Exposure (CORE).[54] Most of the local people Moriguchi encountered expressed little anxiety over their town's legacy, or the Richland High mushroom cloud for that matter. Listening to Moriguchi speak at the Richland Library on March 8, I found myself wishing he could meet Muriel Sears; when I stopped by her home on the 9th to say hello, her daughter Gail shared with me that Muriel had died the previous day.[55]

Atomic imagery in Richland—from the mushroom cloud athletic logo to the "Plutonium Porter" on tap at the brewery—is unlikely to be retired anytime soon, an ironic loyalty in a town where many have suffered tremendously, often from some of the same health ailments Nagasaki hibakusha experience. The scale of remediation projects at the site means it remains a major employer in the area: In 2021, approximately 11,000 people worked there.[56] These jobs represent "about a quarter of the total wages earned in the region."[57] In a June 2021 report from the Washington State Department of Commerce, 57 percent of workers surveyed "reported being in an exposure incident, which could include the release of radioactive material into the air. . . . 32% reported they had long-term exposure to hazardous materials at the nuclear reservation."[58] Workers' narrative responses about their experiences seeking health care and navigating workers' compensation were marked by "despair and distrust," according to report authors, and many noted they were afraid to report their exposures and health problems for fear of losing their jobs.[59] Silence about radiation health effects is still an important part of life in Richland.

Dr. Norma Field, a professor emerita from the University of Chicago, translated for Mitsugi Moriguchi on his visit to Richland in March of 2018. Earlier that year, Field published an open letter about an art installation at the University of Chicago which made use of mushroom cloud imagery to mark the anniversary of the first self-sustained nuclear reaction there. "The very nature of radioactivity means that the nuclear industry (military and civil) entails potential harm for living things at every stage of its operations, from uranium mining to weapons and fuel (nuclear power) production to testing to waste storage," Field wrote. "The production of plutonium in the name of national security in Hanford, Washington, for instance, exposed, at times deliberately, US citizens, workers, farmers, and residents, including African Americans and Native Americans, long after WWII was over."[60] Remediation at Hanford is anticipated to drag on well past the current 2047 deadline, and any plutonium that cleanup leaves buried on the Hanford Reach will take twenty-four thousand years to decay to half of its current radiation level, just as it will at Rocky Flats, Colorado, Savannah River, South Carolina, and everywhere else humans have produced and tinkered with plutonium.[61] Genetic changes experienced by impacted populations will ripple through generations into the future, and the presence of nuclear technology will always mean catastrophe is only an accident away. These are only some of the stakes of nuclear production.

What is the risk of rendering the mushroom cloud a form of historical kitsch? Dr. Field writes:

> The mushroom cloud, with its huge reach, was a manifestation of atmospheric tests conducted around the world by the nuclear powers—the US, UK, Soviet Union, France, and China—mostly in areas remote from the centers of power, preferably colonial or quasi-colonial sites inhabited by people who didn't matter and who in any case had no say: Nevada, the Marshall Islands, Polynesia, Christmas Island, Algeria, Xinjiang, Kazakhstan, to give just a few examples. And under each of those clouds—and this is what dismays me about the circulation of the mushroom cloud sprung loose as an iconic image—were human beings and other living things whose health and habitat and livelihood were ravaged, often with intergenerational implications relevant to this day.[62]

When I asked her about the Richland Bombers mushroom cloud, Muriel Sears told me, "I wish they'd get rid of it. For me that's death and destruction." The day after his visit to the Richland High gymnasium, Mitsugi Moriguchi was asked again what he thought of the school's logo. With Dr. Field translating, he reiterated "it's like stepping all over graves, I can't forgive that." He recalled the student he met outside of the gymnasium. "She said that it's a good thing that we have that image of the mushroom cloud so that we don't forget it, but that's not sufficient knowledge. I want them to learn, from materials, from historical writings . . . nuclear weapons have not benefitted human beings."[63]

Stripped of its context and plastered onto high school football helmets, the Richland Bombers mushroom cloud represents nuclearism as an event contained in time and space, rather than its actuality: a disorderly, ongoing, global dispersion of indefinite toxicity across geographies, genomes, and generations. This actuality becomes apparent when we consider the symbol of the cloud through the gaze of people like Muriel Sears and Mitsugi Moriguchi, for whom radiation exposure is a lived experience. Their stories interrupt the silence around the radiation health effects of nuclear weapons production and use, offering the possibility of a more inclusive history which can help us make informed decisions about nuclear weapons regimes in the present day.

Notes

1 Unless otherwise indicated, descriptions of Mitsugi Moriguchi's visit to Richland are taken from the author's field notes and video recordings (in the author's possession). I am indebted to Dr. Norma Field, Dr. Jacob Hamblin, Dr. Linda Marie Richards, Gail Sears, Nila Utami, Nicole Yakashiro, Mercedes Peters, and the attendees of the 2019 Oregon State University Conference "Ways of Knowing and Radiation Exposure International Research Workshop," for their thoughtful comments on this piece.

2 For examples of arguments that removing Confederate monuments would cause the forgetting of history, see Douglas Jackson, "Confederate Statues: 'When We Remove and Forget Our History, We Are Doomed to Repeat It," *Des Moines Register*, August 17, 2017, https://www.desmoinesregister.com/story/opinion/readers/2017/08/17/removing-confederate-statues-forget-history-doomed-to-repeat/574396001/ (accessed August 14, 2021); Rob Natelson, "Why Removing Historical Monuments Is a Bad Idea," *The Hill*, September 20, 2017, https://thehill.com/opinion/civil-rights/351227-why-removing-national-monuments-is-a-bad-idea (accessed August 15, 2021); Melba Newsome, "Is Removing Confederate Monuments Erasing History?" *NBCnews.com*, April 25, 2017, https://www.nbcnews.com/news/nbcblk/are-removing-confederate-monuments-erasing-history-n750526 (accessed August 14, 2021). For a collection of articles analyzing this controversy, see Lena Felton and Taylor Hosking, "The Legacy of Confederate Symbols," *The Atlantic*, August 17, 2017, https://www.theatlantic.com/politics/archive/2017/08/charlottesville-confederate-monuments/537177/ (accessed July 15, 2021).

3 Sarah E. Gardner, "What We Talk About When We Talk About Confederate Monuments," *Origins*, November 2017, http://origins.osu.edu/article/what-we-talk-about-when-we-talk-about-confederate-monuments (accessed January 15, 2019). For an excellent discussion of controversies around heritage and commemoration, see David Lowenthal, *Possessed by the Past: The Heritage Crusade and the Spoils of History* (New York: Free Press, 1996).

4 Conny Bogaard, "Problems of Inclusiveness: The Knowledge Frames of Nuclear Museums in the United States," *International Journal of the Inclusive Museum* 8, no. 1 (2014): 30. See also Jessie Boylan, "Grievability and Nuclear Memory," *American Quarterly* 71, no. 2 (June 2019): 379–388.

5 Richard H. Kohn, "History and the Culture Wars: The Case of the Smithsonian Institution's *Enola Gay* Exhibition," *Journal of American History* 82, no. 3 (December 1995): 1036.

6 Richard Lake, "Atomic Testing Museum: Downwinders Feel Left Out," *Las Vegas Review-Journal*, February 26, 2005, 2B.

7 Russell Contreras and Mari Yamaguchi, "Museum Stalls Hiroshima Exhibit over Nuclear Weapons Ban Push," *Skagit Valley Herald*, April 4, 2018, A9. Controversy has also adhered to the new Manhattan Project National Historic Park, which in 2023 is still under pressure from *hibakusha* and Downwinder groups to include discussion of health and environmental impacts of nuclear technology. For a discussion of the absences in the MPNHP's B Reactor exhibit at Hanford, see Linda Richards, "Exhibit Review: The B Reactor National Historic Landmark," *Public Historian* 38, no. 4 (November 2016): 305–317.

8 Congressional Budget Office, "Projected Costs of US Nuclear Forces, 2021 to 2030," May 24, 2021, https://www.cbo.gov/publication/57130 (accessed August 12, 2021).

9 Michel-Rolph Trouillot, *Silencing the Past: Power and the Production of History* (Boston: Beacon Press, 1995, 2015), 49.

10 Trouillot, xxiii.
11 I discuss the production of knowledge and ignorance around radiation exposure related to nuclear weapons testing and uranium production in the western United States in *Downwind: A People's History of the Nuclear West* (Lincoln: University of Nebraska Press, 2014). For related discussions, see Olga Kuchinskaya, *The Politics of Invisibility: Public Knowledge about Radiation Health Effects after Chernobyl* (Cambridge, MA: MIT Press, 2014); Frank Fischer, *Citizens, Experts, and the Environment: The Politics of Local Knowledge* (Durham, NC: Duke University Press, 2000); Michelle Murphy, *Sick Building Syndrome and the Problem of Uncertainty: Environmental Politics, Technoscience, and Women Workers* (Durham, NC: Duke University Press, 2006).
12 Ray Stein, "Beavers-Atoms-Bombers Research Project," *Alumni Sandstorm* (undated), http://alumnisandstorm.com/Mascot/BvB/Report-Stein.htm (accessed November 30, 2018). "Hanford History," https://www.hanford.gov/page.cfm/hanfordhistory (accessed December 6, 2019). Jim Kershner, "Richland: Thumbnail History" (January 8, 2008), http://historylink.org/File/8450 (accessed December 1, 2019). Historian Richard White discusses the enlistment of the Columbia River for nuclear aims in *The Organic Machine: The Remaking of the Columbia River* (New York: Hill and Wang, 1995), 81–88.
13 "Department of Energy's Tribal Program," https://www.hanford.gov/page.cfm/inp (accessed December 12, 2018.
14 For discussion of the colonial legacy of global nuclear production see Robert Jacobs, "Nuclear Conquistadors: Military Colonialism and Nuclear Test Site Selection during the Cold War" *Journal of Peacebuilding* 1 (November 2013): 157–177; Gabrielle Hecht, *Being Nuclear: Africans and the Global Uranium Trade* (Cambridge, MA: MIT Press, 2012); Gabrielle Hecht, ed., *Entangled Geographies: Empire and Technopolitics in the Global Cold War* (Cambridge, MA: MIT Press, 2011).
15 Kate Brown, *Plutopia: Nuclear Families, Atomic Cities, and the Great Soviet and American Plutonium Disasters* (Oxford: Oxford University Press, 2013), 109–110.
16 For discussion of the company town model, see Neil White, *Company Towns: Corporate Order and Community* (Toronto: University of Toronto Press, 2012).
17 Muriel Sears, interview with author, December 9, 2017, 1346 Haupt Avenue, Richland, Washington. All subsequent quotes from Muriel Sears are from this source.
18 Robert Bauman, "Jim Crow in the Tri-Cities, 1943–1950," *Pacific Northwest Quarterly* 96 (Summer 2005): 124–131.
19 Brown, 54.
20 Author's video recording of Mitsugi Moriguchi's comments, translated by Dr. Norma Field, March 8, 2018, Richland Library, Richland, Washington.
21 *Richland Villager*, August 14, 1945, 1.
22 Peter B. Hales, "The Atomic Sublime," *American Studies* 32, no. 1 (Spring 1991): 5, 12. For an extended discussion on the fraught relationship of the American press to nuclear reportage see Sarah Fox, *Downwind: A People's History of the Nuclear West* (Lincoln: University of Nebraska Press, 2014), especially 50–51.
23 Stein, "Beavers-Atoms-Bombers Research Project."
24 1946 *Columbian* yearbook quoted by Keith Maupin in "The Bomber, the Bomb, and the Bombers: Myth, History, and Traditions," *Alumni Sandstorm* (February 7. 2001), http://alumnisandstorm.com/Mascot/MaupinPaper.htm (accessed November 15, 2018).
25 Todd Callais, "Controversial Mascots: Authority and Racial Hegemony in the Maintenance of Deviant Symbols," *Sociological Focus* 43, no. 1 (2010): 61.
26 Mitsugi Moriguchi, interview with Cindy Kelly, February 20, 2019, Nagasaki, Japan. Transcript available online at https://www.atomicheritage.org/tour-stop/mitsugi-moriguchis-interview#.X1KVxi2z3ok (accessed August 31, 2020).
27 Moriguchi, interview with Cindy Kelly.

28 Susan Lindee, *Suffering Made Real: American Science and the Survivors at Hiroshima* (Chicago: University of Chicago Press, 1994), 20.

29 Ibid., 78.

30 Barbara Rose Johnston's "'More Like Us Than Mice': Radiation Experiments with Indigenous Peoples" examines the AEC's "targeted and opportunistic use of indigenous peoples in the Marshall Islands, Artic, Andes, and Amazon" (26). Article in Barbara Rose Johnston, ed., *Half-Lives and Half-Truths: Confronting the Radioactive Legacies of the Cold War* (Santa Fe, NM: School for Advanced Research Press, 2007), 25–54. I discuss research conducted on uranium miners in *Downwind*, especially pages 35–40. See also Advisory Committee on Human Radiation Experiments, "Final Report of the Advisory Committee on Human Radiation Experiments," US Government Printing Office (1995). I discuss studies conducted at Hanford later in this essay.

31 Data gathered by the ABCC became part of the Life Span Study, or LSS. Robert Jacobs notes, "The LSS is frequently referred to today as the 'gold standard' database correlating radiation exposures with health outcomes . . . yet there are problems with its design and use." Jacobs explains that the LSS struggled to come up with accurate dose reconstructions, and it did not take internal radiation exposures (from food, drink, or inhalation) into account, making it a problematic standard, given that internal exposures remain one of the primary modes via which people are exposed to radiation. Despite these issues, LSS data is regularly invoked to dismiss concerns of people in radiogenic communities. Robert Jacobs, *Nuclear Bodies: The Global Hibakusha* (New Haven, CT: Yale University Press, 2022), 10. See also Fox, *Downwind*, especially chapters 1, 2, 4, and 6 for discussions of AEC research agendas related to nuclear testing and uranium extraction in the American West and the tendency of those agendas to overlook local knowledge.

32 Australian journalist Wilfred Burchett was the only journalist to successfully publish an uncensored story from the immediate aftermath of Hiroshima; it appeared in the *London Daily Press* in early September and was subsequently reprinted globally. American authorities were prepared; they had preemptively engaged journalists to shape the American response to the news. Sven Lindqvist notes that the same day Burchett's report was published in the US, articles were released depicting Japanese atrocities against prisoners of war, and the American view of the bombing of Nagasaki, authored by William Laurence, a *New York Times* science journalist secretly on the payroll of the Manhattan Project. Laurence wrote, "Being close to it and watching it as it was being fashioned into a living thing so exquisitely shaped that any sculptor would be proud to have created it, one somehow crossed the borderline between reality and non-reality and felt oneself in the presence of the supernatural." See Sven Lindqvist, *A History of Bombing*, trans. Linda Haverty Rugg (New York: New Press, 2001), 114–115. Most Americans' first encounter with extended descriptions of the devastation of the atomic bombs in Japan—or with Japanese individuals who weren't depicted as violent, inhuman monsters—was John Hersey's "Hiroshima," serialized in the *New Yorker* in 1946. For a thorough treatment of this subject, see Leslie Blume, *Fallout: The Hiroshima Cover-Up and the Reporter Who Revealed It to the World* (New York: Simon and Schuster, 2021).

33 Henry Stimson, "The Decision to Use the Atomic Bomb," *Harper's Magazine*, February 1947, https://www.atomicheritage.org/key-documents/stimson-bomb (accessed December 15, 2018).

34 Sven Lindqvist discusses the chiefs of staff report in *A History of Bombing*, on page 118.

35 See Gar Alperovitz, *The Decision to Use the Atomic Bomb* (New York: Vintage Books, 1995), chap. 2.

36 Byrne expressed this belief to atomic scientist Leo Szilard in May of 1945; Szilard's recollection of Byrne's opinion is quoted in Alperovitz, 6. U.S. military leaders knew they could not maintain a monopoly on nuclear weapons technology for long, and some began stoking public fears about the possibility of a future nuclear war with the Soviet Union within

months of the end of WWII, rallying support for the development of an American nuclear weapons arsenal. See, for example, General Hap Arnold, "36-Hour War," *Life Magazine*, November 19, 1945, 27–35.

37 Maupin, "The Bomber, the Bomb, and the Bombers: Myth, History, and Traditions."

38 "In those days it was very discriminatory," Sears remembered in 2017. She recalled administering the same vocational test to men and women, and sending the results to Minneapolis to be evaluated. "When the results came back, if you were a woman you were suited to be a nurse or a teacher. . . . If you were a man, you're an engineer, head of the company." Sears and her supervisor tested their theory of institutional sexism: "We took one of the women and we sent her test results in as a woman . . . nurse, teacher you know. We sent the same thing in as a man. Boy I'll tell you. Top money maker."

39 Based in the Tri-Cities, Energy Northwest [EN] is a joint operating agency established in 1957 by the Washington state legislature. Today EN still supervises nuclear power production at the Columbia Basin Generating Station on the Hanford site, as well as training for nuclear industry professionals. In spring 2023 EN announced they are "actively working with X-energy to pursue funding for an advanced reactor project" on the Hanford site, in the location where EN has recently finished demolishing Washington Nuclear Project 1, a nuclear station abandoned shortly before completion of construction after the Washington Public Power Supply System bond default. See Savannah Tranchell, "EN closes chapter on restoration, opens a new one" Energy Northwest.com, June 1 2023, https://www.energy-northwest.com/whoweare/news-and-info/Pages/EN-closes-chapter-on-restoration,-opens-a-new-one.aspx (accessed July 18, 2023).

40 See Associated Press, "School's Mushroom Cloud Symbol Stirs Up Controversy," *AP News*, December 11, 1987, https://apnews.com/article/d5ca2070ca488aedaa9553861519f614 (accessed August 18, 2021); Timothy Egan, "Richland Journal; Little Sentiment Here to Ban the Bomb," *New York Times*, January 14, 1988. Description of the 1988 *hibakusha* visit to the high school is from Jim Stoeffels, "World Citizens for Peace and the Bomb," World Citizens for Peace, April 2005, http://wcpeace.org/History/WCPeace/wcpeace&bomb.htm (accessed December 10, 2018).

41 Maupin, "The Bomber, the Bomb, and the Bombers: Myth, History, and Traditions."

42 Ibid. See also http://alumnisandstorm.com/Obits/pics09/RIP47MaupinKeith09.htm (accessed December 12, 2018).

43 See Ray Stein, "Beavers-Atoms-Bombers Research Project," *Alumni Sandstorm* (undated), http://alumnisandstorm.com/Mascot/BvB/Report-Stein.htm (accessed November 30, 2018); Charles "Bud" Row, "Re: As I Remember," message posted on *Alumni Sandstorm* (November 15, 1999), http://alumnisandstorm.com/Mascot/BvB/1999-11-15-AS.htm (accessed December 12, 2018); Charles "Bud" Row, "RE: The Bombers," message posted on *Alumni Sandstorm* (November 22. 1999), http://alumnisandstorm.com/Mascot/BvB/1999-11-22-AS-br.htm (accessed December 12, 2018).

44 See Leah Sottile, "In a Small Town in Washington State, Pride and Shame over Atomic Legacy," *Al Jazeera America*, July 21, 2015, http://america.aljazeera.com/multimedia/2015/7/richland-washingtons-atomic-legacy.html (accessed December 9, 2019); Samantha Frost, "The Suggestion of Changing the Bomber Mascot Made People Lose Their Minds," *Odyssey Online*, May 8, 2017, https://www.theodysseyonline.com/suggestion-changing-bomber-mascot-people-lose-minds; Cameron Probert, "Richland High School's Bomber Mascot in the News Again," *Tri-City Herald*, May 27, 2017, https://www.tri-cityherald.com/news/local/education/article153099819.html (accessed November 16, 2018).

45 National Park Service, "Manhattan Project National Historic Park" (no date), https://www.nps.gov/mapr/index.htm (accessed December 12, 2018); Washington State Department of Ecology, "Hanford Cleanup Oversight" (2019), https://ecology.wa.gov/Waste-Toxics/Nuclear-waste/Hanford-cleanup (accessed January 14, 2019). In 2019, Nonoka Koga, a Japanese exchange student attending Richland High made a short video statement on the

high school's daily "AtomicTV" program, revealing how close her own family had lived to Nagasaki and asking "Should we have pride in killing innocent people? . . . I am not trying to change your mascot, but just help you consider a perspective that is more personal." Although some locals supported Nonoka Koga, other students and locals protested her interpretation of the cloud. Annette Cary, "Richland High's Mushroom Cloud Logo Surprised a Japanese Student. She Finally Spoke Up," *Tri-City Herald*, June 5, 2019, https://www.tri-cityherald.com/article231187523.html (accessed June 7, 2019); Cameron Probert, "Richland High Has a Mascot for the First Time, and It's Soaring into History," *Tri-City Herald*, February 24, 2000, https://www.tri-cityherald.com/article240527446.html (accessed March 1, 2020).

46 See Arjun Makhijani, Howard Hu, and Katherine Yih, eds., *Nuclear Wastelands: A Global Guide to Nuclear Weapons Production and Its Health and Environmental Effects* (Cambridge, MA: MIT Press, 1995); Steve Gilbert, "Columbia River Contamination," *Particles on the Wall*, http://particlesonthewall.org/display/potw/Columbia%2bRiver%2bContamination.html (accessed December 1, 2018;) Columbia Riverkeeper, "Hanford and the River," report funded by Washington State Department of Energy (2010), p. 6, https://www.columbiariverkeeper.org/sites/default/files/2011/10/hanford_and_the_river_final2.pdf (accessed December 14, 2018); Oregon Department of Energy, "Hanford Tank Waste," (no date) https://www.oregon.gov/energy/safety-resiliency/Pages/Hanford-Tank-Waste.aspx (accessed 30 July 2023); Jacobs, *Nuclear Bodies*, 240; Annette Cary, "Second Hanford Radioactive Tunnel Collapse Expected. And It Could Be More Severe," *Tri-City Herald*, August 28, 2018, https://www.tri-cityherald.com/news/local/hanford/article217470425.html (accessed August 28, 2018); and Annette Cary, "Radioactive Contamination Found in Workers' Car Filters, Says Hanford Watchdog," *Tri-City Herald*, September 4, 2018, https://www.tri-cityherald.com/news/local/hanford/article217827665.html (accessed September 5, 2018).

47 For discussion of the "bad for business" phenomenon see Fox, *Downwind*, chap. 5. Historian Robert Jacobs has written about the commonalities of experience he has observed in his work with radiation-impacted people from around the world. Regardless of the nationality of the impacted individual or the type of nuclear technology that caused the exposure, Jacobs finds that "people who have been exposed to radiation, or even those who suspect that they have been exposed to radiation, including those who never experience radiation-related illnesses, may find that their lives are forever changed—that they have assumed a kind of second-class citizenship. They may find that their relationships to their families, to their communities, to their hometowns, to their traditional diets and even traditional knowledge systems have been broken. They often spend the remainder of their lives wishing that they could go back, that things would become normal. They slowly realize that they have become expendable and that their government and even their society is no longer invested in their wellbeing." Robert Jacobs, "The Radiation That Makes People Invisible: A Global Hibakusha Perspective," *Asia-Pacific Journal* 12 (July 2014), https://apjjf.org/2014/12/31/Robert-Jacobs/4157/article.html (accessed December 19, 2019).

48 United States Department of Energy, "Hanford Site: A Guide to Record Series Supporting Epidemiologic Studies Conducted for the Department of Energy," https://ehss.energy.gov/ohre/new/findingaids/epidemiologic/hanford/intro.html (accessed December 12, 2018).

49 Karen Dorn Steele reported on Dr. C. Alvin Paulsen's 1960s experiments on prisoners' testicles in "Radiation Experiments Raise Ethical Questions," *High Country News*, April 4, 1994, https://www.hcn.org/issues/8/250 (accessed December 6, 2018).

50 H. E. Palmer, I. C. Nelson, and G. H. Crook, "The Uptake, Distribution, and Excretion of Promethium in Humans and the Effect of DTPA on These Parameters," *Health Physics* 18 (1970)53-61. In 1994, Muriel's daughter Gail Sears learned that *Tri-City Herald* reporter John Stang was hoping to speak to local people who had been involved in radiation experiments at Hanford. She connected the reporter to her mother; Stang's account about

Muriel's participation in the promethium experiment appeared in his piece "Human Tests at Hanford," *Tri-City Herald*, February 6, 1994, A1, A2. This article appeared in the wake of Energy Secretary Hazel O'Leary's highly publicized declassification of DOE records related to US human radiation experiments in 1994.

51 Palmer, Nelson, and Crook, 56, 58.

52 United States Department of Energy, "Hanford Site: A Guide." The AEC maintained this degree of oversight for medical, environmental, and scientific research into radiation effects at research institutions and sites around the United States and beyond. See discussion of AEC protocols for radiation research conducted on livestock by cooperating agricultural colleges in Fox, *Downwind*, chapter 3, as well as the chapter 4 discussion of AEC pushback against scientists such as Robert Pendleton, who violated AEC chain of command related to public health hazards. David Price's "Earle Reynolds: Scientists, Citizen, and Cold War Dissident" explores the topic of dissenting scientists in *Half-Lives and Half-Truths: Confronting the Radioactive Legacies of the Cold War*, ed. Johnston, 55–75; see also Devra Davis, *When Smoke Ran Like Water: Tales of Environmental Deception and the Battle Against Pollution* (New York: Basic Books, 2002), esp. chap. 10.

53 Kathleen Flenniken, "Bedroom Community," in *Plume* (Seattle: University of Washington Press, 2012), 9.

54 Yuki Miyamoto, "Nagasaki-Hanford Bridge Project: Joining the Victims of Production and Use of Nuclear Weapons via CORE" (2018), https://lucian.uchicago.edu/blogs/atomicage/nagasaki-hanford-bridge-project/nagasaki-hanford-bridge-project-joining-the-victims-of-production-and-use-of-nuclear-weapons-via-core/ (accessed February 2, 2018). At the time of Mitsugi Moriguchi's visit I was newly involved with CORE; I have since become a board member. To learn more about CORE, visit https://nuclearharm.org.

55 See Muriel G. Sears obituary, *Tri-City Herald*, March 18, 2018, https://www.legacy.com/obituaries/tricityherald/obituary.aspx?n=muriel-g-sears&pid=188486332 (accessed September 2, 2020).

56 Lisa Brown et al., "Hanford Healthy Energy Workers- Healthcare Needs Assessment and Recommendations," Washington State Department of Commerce Report, June 1, 2021, https://www.commerce.wa.gov/wp-content/uploads/2021/06/Hanford-Healthy Energy-Workers-Healthcare-Needs-Assessment-and-Recommendations.pdf (accessed August 15, 2021).

57 Robin Wojtanik, "DOE Looks to Way to Replace Outgoing Workers," *Tri Cities Business News*, May 2019, https://www.tricitiesbusinessnews.com/2019/05/hanford-workforce/ (accessed August 15, 2021).

58 Annette Cary, "57% of Hanford Nuclear Site Workers Surveyed by WA State Report Toxic Exposures," *Tri-City Herald*, July 7, 2021, https://www.tri-cityherald.com/news/local/hanford/article252613523.html (accessed August 15, 2021).

59 Brown et al., 15, see workers narrative responses on 79–91.

60 Norma Field, "Message from Professor Emeritus Norma Field," *Visual and Material Perspectives on East Asia*, January 6, 2018, https://voices.uchicago.edu/vmpea/2018/01/ (accessed November 25, 2019).

61 See Ralph Vartabedian, "Nation's Most Ambitious Project to Clean Up Nuclear Weapons Waste Has Stalled at Hanford," *Los Angeles Times*, June 4, 2019, https://www.latimes.com/nation/la-na-hanford-nuclear-cleanup-20190604-story.html (accessed June 10, 2019); Ronan Farrow and Rich McHugh, "Welcome to 'the Most Toxic Place in America,'" NBC News.com, November 29, 2016, https://www.nbcnews.com/news/us-news/welcome-most-toxic-place-america-n689141 (accessed January 15, 2017); Doug Pardue, "Deadly Legacy: Savannah River Site Near Aiken One of the Most Contaminated Places on Earth," *Post and Courier*, May 21, 2017, https://www.postandcourier.com/news/deadly-legacy-savannah-river-site-near-aiken-one-of-the-most-contaminated-places-on-earth/article_d325f494-12ff-11e7-9579-6b0721ccae53.html (accessed September 5, 2018); John

Aguilar, "Plutonium in Soil Sample Near Rocky Flats Five Times Higher Than Cleanup Standard," *Denver Post*, August 20, 2019, https://www.denverpost.com/2019/08/20/rocky-flats-plutonium-jefferson-parkway/ (accessed August 25, 2019).

62 Field, "Message from Professor Emeritus Norma Field."

63 Mitsugi Moriguchi spent many years working with the Nagasaki Testimonial Society to document the experiences of Nagasaki residents in the wake of the atomic bomb. An English-language edition of the Nagasaki Testimonial Society's *Shogen: Nagasaki Hiroshima no koe* was published in 2009 as *Nagasaki: Voices of the A-Bomb Survivors* (Nagasaki Testimonial Society, 2009). Mitsugi Moriguchi died in December 2022 at the age of 86.

75 Years after Hiroshima and Nagasaki

Public address on August 6, 2020, for the Oregon Physicians for Social Responsibility Event Commemorating 75 Years after the Bombings of Hiroshima and Nagasaki

YUKIYO KAWANO

My name is Yukiyo Kawano. I was born and raised in Hiroshima, a third-generation survivor of the US atomic bomb.

This year's commemoration is unlike any other. We are in an unprecedented time of human history, as we are facing 150,000 deaths, which disproportionately claim African American, Latino, and Native American people's lives. My heart is wrenching, and I am asking questions: What does the nuclear issue mean to you? And how does it connect to your personal everyday life?

Some of us are starting to recognize that the nuclear issue cannot be solved separately from the issue of racism.

Today, I think of the trauma of Japanese Americans living through the history of their incarceration during the war. They tell me that because of their race—citizens and resident aliens of Japanese ancestry—this grave injustice was done to them. Over the years I learned their journey of fighting against the racial power structure. One of their fights resulted in 1983, the Commission on Wartime Relocation and Internment of Civilians publishing *Personal Justice Denied*, stating that government action had been based on "race prejudice" as opposed to legitimate security reasons.

In June, the Japanese American community unanimously wrote a statement, demanding that Black Lives Matter and stating that the ongoing patterns of attacks and racial discrimination against the Black community must stop.

And today, I think of the trauma of the First Nations. How the Native American boarding school separated their children from the land, and how that brutal act devastated the communities for generations. I think of how the communities continue living on exploited land in the COVID era. How their lives are endangered by the lack of safe running water. In New Mexico, I learned how radioactive land compromises the health of Navajo and Pueblos, and in Oregon, how Hanford contamination is affecting the tribes of Walla Walla, Umatilla, Yakama, and Wanapum.

And they too are putting up Black Lives Matter signs in bold letters.

This year's commemoration is different; after 150,000 deaths, I am asking what does the nuclear issue mean to you? And how does it connect to your personal everyday life?

I am heartbroken and realizing that our issue of racism is the foundation of the suffering of people; as long as we keep the power structure that tolerates racism, we are keeping an impossible distance from solving our nuclear problems.

Many people came to know Hiroshima and Nagasaki as the first victims of the atomic bomb.

Some of us know that the uranium that was used for these bombs was from the Democratic Republic of Congo, and the African miners who worked under unspeakable conditions were generationally exposed to radiation since 1921.

The nuclear detonation and World War II offered America a position to claim its super power, and win the title of the greatest. Now, the country as a whole is dealing with its racist history within, and the oppressed are asking, when has America ever been great?

This year's commemoration is different. It is different for me because I have to ask myself a question as a granddaughter of a Japanese soldier who was deployed to China in 1937, the year of the Nanjing Massacre, and again in the Malay Peninsula in 1942. Grandfather told me, when I was ten, that he won't share his survivor story of August 6 because he said there is no point to tell, that I can never understand him—Japanese violent history forever silenced him.

After his death, I started painting, creating, and performing about Hiroshima, to have finally found a simple answer on the ancient land of Pueblo—the voice of grandfather "in me" recognizes that: "I am in pain three generations later, the act of racism in the war hurts my soul and threatens my body."

It is this pain from which I acknowledge the racism in Japan and its aggression in Asia: I am filled with remorse and sorrow for so much agony and desolation caused by the war crimes of Japan and the government's dismissing the claim of the victims, including those of comfort women.

I regret that for seventy-five years, the commemoration was held without offering the long-awaited apology—I am deeply, deeply sorry.

I will continue my apology and work on this issue, as I know this is where I can honestly start talking about abolishing nuclear weapons and its power structure.

Today, August 6th, I urge you to join me to look directly into the face of the racial injustice—as long as we keep the powerful racial structure in us, the nuclear issue can never be solved.

Please remember Hiroshima and Nagasaki as the new form of race massacre of the nuclear age.

And remember Black Lives Matter to all of us.

Story That Won't End Well

KATHLEEN FLENNIKEN

It begins in a laboratory
under a football field.

While the Axis rolls
over distant continents,

fifty-thousand nomads
journey to the American West

to construct cathedrals in the desert
for Nobel physicists,

performing feats
they're not privileged to understand

to microscopic tolerances
in dust storms the stuff of legend.

Periscopes and code words,
train cars loaded with uranium,

the heroism of a just war—
all prologue to the story

we can't see, smell, or taste,
that seeps underground

and drifts undetected
downstream and downwind,

while the Soviets match us
bomb for bomb,

while we build lives
and more reactors, pledge

allegiance, defend the key,
plant birches in the yard

and a Naugahyde couch
in the family room.

Our story develops
invisibly, incrementally,

until one afternoon
it daylights in town square

and we force ourselves to *read it*
bubbling there—

the ugly, stinking, bitter truth.

And some fall down.
And some go home, unmoved.

Acknowledgments

This book is the result of many years of collaboration, with people who are academic researchers, activists, scientists, students, staff, friends, and family. It was an honor to think together with all of the presenters at our workshops in 2018 ("Telling the Stories of Radiation Exposure"), 2019 ("Ways of Knowing and Radiation Exposure"), and 2020 ("75 Years after Hiroshima and Nagasaki"). These were held at Oregon State University, in lovely Corvallis, Oregon. This book is the second publication to emerge from these workshops—the other was a special issue of the *Journal of the History of Biology* (volume 54, no. 1, 2021). This material is based in part upon work supported by the National Science Foundation under Grant # 1734618. Any opinions, findings, and conclusions or recommendations expressed in this material are those of the author(s) and do not necessarily reflect the views of the National Science Foundation.

We would like to express gratitude to all those who shared their work and experiences in our workshops and publications: Matthew Adamson, David Elijah Bell, Marissa Bell, Austin R. Cooper, Britt Dahlberg, Desmond Narain Doulatram, Kathleen Flenniken, Cindy Folkers, Sarah Fox, Edward Granados, Prerna Gupta, David K. Hecht, Kyle Harvey, Sumiko Hatakeyama, Patricia Hoover, Helen Jaccard, Tatiana Kasperski, Yukiyo Kawano, William M. Knoblauch, Jaroslav Krasny, Valerie Kuletz, Matthew Lavine, Jonathan Luedee, Gisela Mateos, Joshua McGuffie, Joshua McMullan, Robynne Mellor, Keith Meyers, Mary X. Mitchell, Adrian Monty, Laureen Nussbaum, Davide Orsini, Maxime Polleri, Jeffrey C. Sanders, Magdalena Edyta Stawkowski, Sasha Stiles, Edna Suárez-Díaz, Casper Sylvest, Oliver George Tapaha, N. A. J. Taylor, and Emily Yates-Doerr.

In addition to these, some special thanks are in order. We thank Downwinder Patricia Hoover for her essential help and enthusiasm for this project over the years, and artist Yukiyo Kawano for allowing us to share a photograph of her artwork as our cover. We also want to acknowledge writers who have inspired or encouraged our work, especially Trisha Pritikin, Kate Brown, Amy Hay, Bo Jacobs, Mary Jo Nye, Chris Shuey, Maria Rentetzi, and Natasha

Zaretsky. We were lucky to have amazing students who helped to make the workshops successful and worked on our oral histories. These included Anna Dvorak, Kristina Beggen, Adrian Monty, Anthony Vitale, Marcelo Carocci, Celia Eklund Oney, and Mahal Miles. We had great partners on campus: OSU's Special Collections and Archives Research Center, especially Anne Bahde and Chris Petersen; OSU's Spring Creek Project for Ideas, Nature, and the Written Word, especially Carly Lettero and Shelley Stonebrook; the Environmental Arts and Humanities program; and the School of History, Philosophy, and Religion, especially Suzanne Giftai. Of course, we are grateful to OSU Press acquisitions editor Kim Hogeland, for helping us to imagine this as a potential book, and to all the other folks at the press who made it happen: Mary Elizabeth Braun (former acquisitions editor), Micki Reaman, Marty Brown, and Tom Booth. Special thanks to our copyeditor, Teresa Jesionowski, who expertly navigated diverse styles of discourse.

We as editors have worked on this book collaboratively over the years, yet we have our individual acknowledgments to make. Jacob Darwin Hamblin thanks his family—Sara, Sophia, and Harper. They have endured dinner conversations about history longer than anyone should. And he is grateful for the encouragement from his sister Sara, his parents, Sharon and Les, and in-laws, Cathy and Paul. He thanks his co-editor Linda Richards for her unshakable enthusiasm and seemingly infinite patience. Linda Richards wishes to express her debt to all those who shared their stories of radiation exposure, resistance, and resilience with her since 1986, especially her precious late uncle, Veikko Juhola (an atomic veteran), Jay Mullen, Perry H. Charley, Elsie Mae Begay, Roberta Blackgoat, Carrie Dann, Corbin Harney, Hideko Tamura Snider, Lorraine Rekmans, Charmaine Whiteface, Oliver Tapaha, June Ikuko Moore, Patricia Hoover, Jesse Gasper, Lincoln Grahlfs, Keith Kiefer, Fred Schafer, and Michael Driscoll. She thanks Carol Urner of the Women's International League for Peace and Freedom Disarm! committee and those involved with several organizations: the Southwest Research and Information Center (SRIC), the International Campaign to Abolish Nuclear Weapons (ICAN), National Association of Atomic Veterans (NAAV), Consequences of Radiation Exposure (CORE), Oregon Physicians for Social Responsibility (OPSR), and Veterans For Peace. She is deeply grateful to Jacob Darwin Hamblin for collaborating with her in this complex project. She continues to be in awe of his dedication, scholarship, and generosity, teaching her so much more than nuclear environmental history since 2009. She is also thankful for the laughter, guidance, and inspiration given to her by students, colleagues, friends, and her loving family.

Contributors

MATTHEW ADAMSON is a historian living in Budapest, Hungary. His research focuses on the history of nuclear programs, on the worldwide search for nuclear raw materials, and on the diplomacy of nuclear technoscience. He is presently Director of Academic and Student Affairs at McDaniel College's campus in Budapest.

AUSTIN R. COOPER is a historian working on a book about French nuclear weapons development in the Algerian Sahara during the 1960s. He has published research on this topic in *Cold War History, The Nonproliferation Review,* and the *Bulletin of the Atomic Scientists*. He holds a PhD in History and Sociology of Science from the University of Pennsylvania.

DESMOND NARAIN DOULATRAM is a Co-Chair for the Liberal Arts Department at the College of the Marshall Islands, where he also serves as a social science instructor, and is a co-founder of REACH-MI. Desmond also serves as a Public School System National Board Member representing the interests of parents and guardians.

KATHLEEN FLENNIKEN was born and raised in Richland, Washington, and worked as a civil engineer at Hanford in the 1980s. She is the author of three poetry collections including *Plume,* a meditation on the Hanford site. She is a former Washington State poet laureate.

SARAH FOX, author of *Downwind: A People's History of the Nuclear West* (University of Nebraska Press 2014, paperback 2018), is a PhD candidate in history at University of British Columbia and an adjunct instructor in Environmental Policy and Decision Making at University of Puget Sound. She is also a member of the board of the nonprofit organization Consequences of Radiation Exposure, or CORE. Fox's current research examines lived experiences of pollution, knowledge production, and community organizing in the plumes of contamination produced by different settler colonial projects in the Pacific Northwest.

EDWARD GRANADOS received BA, MA, and PhD degrees in chemistry from the State Universities of New York. He did post-doctoral research at the Roswell Park Comprehensive Care Center. The remainder of his career was as a Principal Investigator at Abbott Laboratories where he was involved in the R&D of Immunonassays and Molecular Diagnostic Assays until his retirement in 2015. He served on the Board of Directors of Physicians for Social Responsibility (Colorado) between 2019 and 2021.

PRERNA GUPTA is a PhD candidate at the Institute for Resources, Environment and Sustainability at the University of British Columbia, Vancouver. Her research investigates what cultural, economic and political factors drive people's acceptance or rejection of various energy technologies in India, especially nuclear energy. She has been engaging with nuclear issues for more than six years both academically and through civil action. Currently, she is making a documentary on people's experience of nuclear power in India and is interested in exploring the relevance of the form of documentary for research in social sciences. Her experience with various social movements drives her passion for socially relevant research and creative projects.

JACOB DARWIN HAMBLIN is Professor of History at Oregon State University. His books include *The Wretched Atom: America's Global Gamble with Peaceful Nuclear Technology* (2021), *Arming Mother Nature: the Birth of Catastrophic Environmentalism* (2013), *Poison in the Well: Radioactive Waste in the Oceans at the Dawn of the Nuclear Age* (2008), and *Oceanographers and the Cold War* (2005).

Hanford Downwinder **PATRICIA HOOVER** is a University of Oregon Journalism graduate living in Eugene. A professional radio and TV broadcaster, Hoover was born and raised in Eastern Oregon where, in the 1950s and '60s, she was heavily exposed to the radioactive emissions from the Hanford Nuclear Reservation. She has been an anti-nuclear/environmental activist for over thirty years.

HELEN JACCARD is a peace and nuclear abolition activist, organizer, and researcher. She is the Project Manager, crew, and primary public speaker for the *Golden Rule* anti-nuclear sailboat, which is owned by Veterans For Peace. Helen has written about the environmental costs of war and militarism in

Sardegna, Italy; Puerto Rico; Guatemala; and Hawai'i. She's a member of Women's International League for Peace & Freedom and many other peace, justice, and nuclear abolition organizations.

YUKIYO KAWANO, an artist, activist, and educator, is a third-generation *hibakusha* (atomic bomb survivor) who grew up decades after the bombing of Hiroshima. Her artwork is "a direct response to the tragedy of the past," she states, "but the past I depict wants to remember the present. People say my work is like a dream; it is not meant to deliver certainties. It asks us to remain in the present, so that we can develop a new relationship of mind and body, confront the ongoing deceptive rhetoric that surrounds us, reject violence, and save ourselves from our own extinction." Kawano teaches through Vermont College of Fine Arts studio mentorship program (Artist Teacher) and is an Oregon Physicians for Responsibility Advisory Board Member. Kawano currently lives in Portland, Oregon.

WILLIAM M. KNOBLAUCH is a Visiting Assistant Professor of History at The University of Montana–Western (Dillon, MT).

JAROSLAV KRASNY is a recent graduate of a doctoral program at Hiroshima University and currently works as a Research Associate at the Center for Peace, Hiroshima University (CPHU). His research focuses on the long-term health effects of ionizing radiation caused by a nuclear explosion as a breach of fundamental norms of international humanitarian law. He is also active in advocacy for the victims suffering long-term health effects of particular weapons, including but not limited to nuclear weapons, including their testing, and chemical weapons.

JOSHUA McGUFFIE holds a PhD in History from the University of California, Los Angeles. His research focuses on the Medical Section of the Manhattan Project during World War II as well as on the biological research that took place at US atomic test sites during the Cold War. A pastor in the Evangelical Lutheran Church in America, he serves at the Church of the Ascension in suburban Los Angeles and teaches as an adjunct professor at Occidental College.

JOSHUA McMULLAN did his collaborative PhD at the University of Leicester and National Archives on the UK nuclear industry's public relations strategy in the late twentieth century. He completed in 2021 and started working at Churchill College. Since January 2022, Joshua is now working for the Cabinet Office and the Office for Veterans' Affairs as the Research Lead. Here he has been working extensively with British Nuclear Test Veterans and worked on securing them a commemorative medal and commissioning an oral history project to capture the testimonies of those involved in the tests.

ADRIAN MONTY is a former graduate student of Oregon State University, having completed her project in 2019 titled "The Fissures of Fission: A Look at the Cracks in the Construction of the Atomic Age." As a member of the Affected Communities and Allies Working Group, she is still working within the nuclear activism community. Adrian also enjoys working with children and has recently become a substitute teacher along with working at a children's science museum.

DR. LINDA MARIE RICHARDS writes and teaches about the places where nuclear and environmental history converge with human rights. She received the Oregon State University Phyllis S. Lee University Award for her work on peace and justice in 2018.

JEFFREY C. SANDERS lives and works in the Columbia River Basin about three hours from Hanford, Washington. He is a professor in the Department of History at Washington State University, Pullman, where he teaches and researches environmental history. His most recent book is titled *Razing Kids: Youth, Environment, and the Postwar American West* (Cambridge University Press, 2021). His current project examines the history of strontium-90 and is tentatively titled *Strontium 90: An Unnatural History*.

MAGDALENA EDYTA STAWKOWSKI is a cultural and critical medical anthropologist researching nuclear and militarized spaces in Kazakhstan. Her current research examines Soviet-era nuclear testing on the Semipalatinsk Test Site and how people cope with the aftermath. Specifically, she examines how people navigate everyday life in a damaged environment polluted with residual radioactivity. She is also currently working on a collaborative project that looks at the Anthropocene epoch as the radioactive afterlife of the Cold War. Magdalena is an Assistant Professor at the University of South Carolina and a Researcher at the Danish Institute for International Studies in Copenhagen, Denmark.

DR. SASHA STILES is a resident living within two miles of Rocky Flats. As past president of PSR Colorado, she organized community organizations, epigenetic researchers, and historical experts (including from the Ukraine/Chernobyl area and throughout the United States) to provide a PSR Webinar: "Making the unseen seen at Rocky Flats" is now available for all on the PSR Colorado website. As a member in several local organizations (Peace and Justice Center, the Rocky Flats Downwinders, and Rocky Flats Right to Know), she wishes to acknowledge their ongoing tireless work. Finally, her essay was inspired by her co-author Edward Granados, PhD, and could not have been written without his insight, research, writing, and tireless education to us all.

OLIVER GEORGE TAPAHA (PhD) is an enrolled citizen of the Diné/Navajo tribe from northeastern Arizona. He is currently a Postdoctoral Research Associate in the Department of Education Policy, Organization and Leadership at the University of Illinois, Urbana-Champaign. For the past fifteen years, Oliver has been educating the youth and adults about the negative impacts of uranium contamination on the livelihood of the Diné through social justice projects in schools, conference presentations, and speaking engagements with *The Return of Navajo Boy* documentary film.

Index

N